APPLIED BASIC FOR TECHNOLOGY

THEODORE F. BOGART, JR., P.E.
University of Southern Mississippi

SCIENCE RESEARCH ASSOCIATES, INC.
Chicago, Toronto, Henley-on-Thames, Sydney

A Subsidiary of IBM

Acquisition Editor	Alan Lowe
Project Editor	Jan deProsse
Production Service	Greg Hubit Bookworks
Compositor	Graphic Typesetting Service
Illustrator	Carl Brown
Text Designer	Barbara Ravizza
Cover Designer	Joe di Chiarro

Library of Congress Cataloging in Publication Data

Bogart, Theodore F.
  Applied BASIC for technology.

  Includes index.
  1. BASIC (Computer program language) 2. Technology
—Data processing. I. Title.
QA76.73.B3B62 1984   602′.8′5424   83-14856
ISBN 0-574-21585-9

Copyright © Science Research Associates, Inc. 1984.
All rights reserved.

Printed in the United States of America.

10 9 8 7 6 5 4 3 2 1

CONTENTS

Preface, vii

Chapter 1 COMPUTERS AND COMPUTING, 1
      1 THE IMPORTANCE OF COMPUTERS IN TECHNOLOGY, 1
      2 TYPES OF COMPUTERS, 2
      3 SOME COMPUTER TERMINOLOGY, 3
      4 COMPUTER LANGUAGES, 5
      5 PROGRAMS AND DATA, 7
      6 ALGORITHMS, 8

Chapter 2 INTRODUCTORY BASIC, 11
      1 VERSIONS OF BASIC, 11
      2 NUMBER REPRESENTATION, 12
      3 STATEMENTS, 14
      4 VARIABLES, 14
      5 THE LET STATEMENT, 15
      6 THE PRINT STATEMENT, 18
      7 THE INPUT STATEMENT, 19
      8 THE END STATEMENT, 20
      9 THE KEYBOARD AND HOW TO USE IT, 21
      10 GAINING ACCESS TO BASIC, 23
      11 SIMPLE PROGRAMMING EXAMPLES, 25

Chapter 3 MATHEMATICAL OPERATIONS, 33

1. MATHEMATICAL OPERATORS, 33
2. EXPONENTIATION, 34
3. EXPRESSIONS, 35
4. THE SEQUENCE OF OPERATIONS IN EVALUATION OF EXPRESSIONS, 36
5. THE CALCULATOR MODE, 38
6. THE REM STATEMENT, 39
7. PROGRAMMING EXAMPLES, 40

Chapter 4 INPUT/OUTPUT, 51

1. SYSTEM COMMANDS, 51
2. MORE ABOUT THE PRINT STATEMENT, 52
3. MORE ABOUT THE INPUT STATEMENT, 62
4. THE READ AND DATA STATEMENTS, 64

Chapter 5 BRANCHING AND LOOPING, 79

1. THE GOTO STATEMENT, 79
2. CONDITIONAL BRANCHING: THE IF-THEN STATEMENT, 81
3. LOOPS, 87
4. FLOWCHARTS, 91
5. THE FOR-TO AND NEXT STATEMENTS, 107
6. THE ON-GOTO STATEMENT, 118

Chapter 6 ARRAYS, 127

1. THE NATURE OF AN ARRAY, 127
2. DIMENSIONING ARRAYS, 129
3. SORTING AND ORDERING NUMERIC DATA, 138

Chapter 7 FUNCTIONS AND SUBROUTINES, 153

1. LIBRARY FUNCTIONS, 153
2. USER-DEFINED FUNCTIONS, 166
3. SUBROUTINES, 171
4. A SUBROUTINE FOR GRAPHIC OUTPUT, 180

Chapter 8 SOLVING SIMULTANEOUS EQUATIONS, 205

1. SIMULTANEOUS LINEAR EQUATIONS, 205
2. CRAMER'S RULE, 207
3. MATRIX ALGEBRA, 214
4. MATRIX OPERATIONS IN BASIC, 220

Chapter 9 USING BASIC FOR PROBLEMS IN APPLIED CALCULUS, 229

1. DIFFERENTIATION OF POLYNOMIALS, 229
2. MAXIMUM AND MINIMUM PROBLEMS, 237
3. DIFFERENTIATION OF TRANSCENDENTAL FUNCTIONS, 246
4. DEFINITE INTEGRALS, 250
5. NUMERICAL INTEGRATION, 261

Glossary, 277

Answers to Selected Exercises, 301

Index, 361

PREFACE

BASIC is a computer language that has become increasingly important and is now widely available in schools and colleges. The growing accessibility of this language is due to the large number of time-shared systems now in use and, in particular, to the explosive development of low-cost microcomputers that have put computer programming within the economic means of virtually all students.

Applied BASIC for Technology is designed to (1) teach students how to program in BASIC, and (2) emphasize the computational power of the language in problem-solving applications typical of those found in technology. The book assumes no previous programming experience and therefore carefully introduces fundamental computer concepts that might be taken for granted in more advanced texts. The level is suitable for first-year technology students in two- or four-year programs. Students should have studied some algebra and, for maximum benefit, should have taken or be taking courses in applied physics and introductory courses in their major field. Although the majority of the programming examples and exercises can be solved and understood without extensive knowledge of a particular technology, students' motivation to use BASIC in their field is improved if they learn the language in the context of familiar terminology and equations. This book is intended to provide an alternative to courses in computer science or data processing departments, where students are often not exposed to the relevance of BASIC as a technical problem-solving language.

Each chapter (after the first) develops progressively more advanced techniques in BASIC and uses programming examples from introductory math and physics, courses that are required in all technology curricula. Exercises emphasizing BASIC syntax and structure are dispersed throughout each chapter, and programming exercises from math and physics appear at the end of each chapter. Also, within each chapter there are four sets of programming exercises that draw from four different fields of technology. These exercises are organized by discipline and include (1) architectural/civil/construction technology, (2) electrical/electronics/computer technology, (3) industrial/manufacturing/production technology, and (4) mechanical technology. Each of these exercises contains a brief description of the problem to be solved, including units and terminology used in the area from which it is taken.

The first seven chapters of this book include all the material that should be covered in a one-semester introductory course. Chapter 8, "Solving Simultaneous Equations," is primarily intended for use with computer systems having matrix handling capabilities, although elementary methods are also discussed. Chapter 9 is intended for students with more advanced math backgrounds and shows how BASIC can be used to solve technical problems involving calculus.

Recognizing that minor differences in syntax exist among some versions of BASIC, all programming examples have been constructed to be applicable to all versions of the language. However, variations that are likely to be encountered in some versions are identified at points where they might occur; and, in many cases, these variations are listed by computer manufacturer. Also, we use no program examples whose success relies on a power that one particular computer has relative to others (except Chapter 8, which assumes matrix capabilities).

The author and publisher would like to thank the following reviewers for their valuable feedback:

David M. Hata, *Portland Community College*
Edwin E. Pollock, *Cabrillo College*
John C. Spille, *University of Cincinnati*
Melvin C. Vye, *University of Akron*

CHAPTER 1

COMPUTERS AND COMPUTING

1 THE IMPORTANCE OF COMPUTERS IN TECHNOLOGY

Can anybody doubt that the computer age is upon us? A computer affects our lives one way or another almost every day. If you went to a bank or a supermarket today, made a phone call, earned some money, used a credit card, or played a video game, chances are you were responsible for a computer transaction. When did you last take they do permeate our lives and they are here to stay. Trying to remain aloof from the puter"), in the reporting of grades, during your registration, or in some other way? Whatever your philosophical opinion of the impact of computers on society might be, they do permeate our lives and they are here to stay. Trying to remain aloof from the computer revolution is like shouting the old admonition "Get a horse!" in the dawn of the automobile era, or teaching a college course in how to use a slide rule.

Since all of us in technology are concerned with practical problem-solving through numeric computations, the advent of low-cost, widely available computer power is a godsend. The advantages of having a means to obtain rapid, accurate results from complex computations are obvious. Perhaps more important, the computer relieves us of time-consuming drudgery and tedious computations that might otherwise obscure the true significance of the results. Without the computer it is often difficult to see the results through the computations. By leaving the computational drudgery to the computer, we are better able to focus on the meaning of the results and on the physical relationships under investigation.

As an example, suppose you want to learn how the value of y in the equation $y = 2x^3 - 3x^2 - 12x$ varies as x changes from -4 to $+4$. (This could be an equation showing how a certain force varies with distance, or how the pressure in an engine varies with temperature, or some other physical relationship.) To investigate the variations in y, you might decide to compute y for, say, 101 different values of x, $x = -4, -3.92, -3.84, \ldots, 3.92, 4.00$. After several hours, you would probably decide you had chosen the wrong career, or else you might have forgotten the significance of y in the first place! On the other hand, in about five minutes, a computer programmer can write a simple *program* to compute and list these 101 values. A typical computer can then perform all the computations in less than one second. Therefore, in a very short time we can inspect values of y and quickly conclude that y increases until $x = -2$, after which y decreases until $x = 1$ and then increases again. This insight into how values of y are affected by values of x could be quite important in the investigation of a practical problem. We can see how valuable the computer is in supplying this insight (not to mention restoring faith in your career choice).

2 TYPES OF COMPUTERS

The two major types of computers used by technologists for general programming are the *microcomputer* and the large, *time-shared* "mainframe" computer. The microcomputer is a small, inexpensive "personal" computer designed to be used by only one person at a time. In recent years, there has been a veritable explosion of this type of computer on the market, ranging from the very small (such as the Timex ZX-81 and the TRS-80 pocket computer) to larger and more sophisticated models such as those manufactured by Apple, Atari, Commodore, Heath, and IBM. Figure 1.1 shows a typical microcomputer. Note the self-contained *keyboard* used by the programmer to enter programs into the computer, and the screen, or *video display*, on which results

FIGURE 1.1 A typical microcomputer (Photo courtesy of Radio Shack)

are displayed. The computer itself is contained completely within the unit, and the entire system may be easily moved from one location to another.

In contrast, the time-shared computer system typically is designed around a much larger, much faster, and much more expensive computer that can be used by a number of programmers at (what seems to be) the same time. This type of computer is usually installed in a room or restricted area not directly accessible to programmers: Programmers gain access to the computer by using remote *terminals* that are connected to the computer via telephone lines. Since the computer is so fast, it can communicate with many programmers in succession, with each one for a short time, but in such rapid succession that all programmers seem to be using it simultaneously. Many terminals look very much like a microcomputer, containing both a keyboard and a video display. Figure 1.2 shows a typical remote terminal, connected to a phone line through a conventional telephone receiver, using an electronic device called a *modem*.

3 SOME COMPUTER TERMINOLOGY

As in many other specialized technologies, a great deal of terminology, or technical jargon, is associated with the computer field. To use a computer effectively, it is not necessary to have an in-depth understanding of all the terms used by computer scientists, but it is helpful to be at least superficially acquainted with some of the jargon. The programming manual supplied with every computer is likely to contain some of these terms, and it is quite possible that the manual will be unintelligible to a novice who has never encountered such terminology. Also, prospective microcomputer buyers are not equipped to select the system that best suits their needs if they are confronted with a list of specifications that is more bewildering than informative.

In the remainder of this chapter we will introduce and briefly define some terms frequently encountered in the computing field. Many of these terms will appear again in later chapters. A glossary containing the definitions of these and many other computer science terms is given at the end of the book.

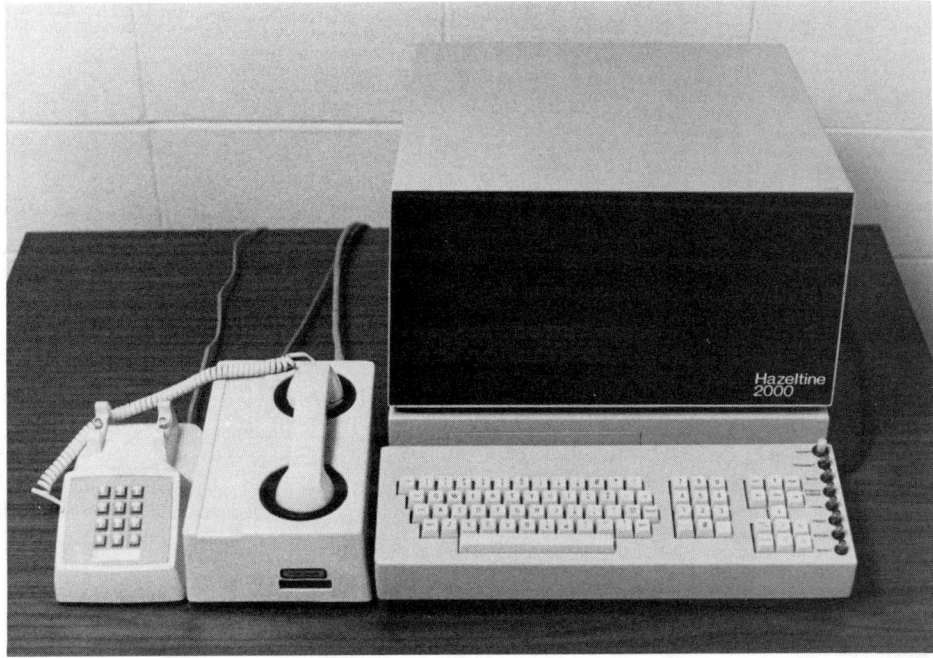

FIGURE 1.2 A remote terminal connected to a time-shared computer system using an acoustic coupler and a modem (Photo courtesy of Hazeltine)

Computer *hardware* is the electronic circuitry, devices, and components from which the computer system is constructed. The term *software* refers to programs that are written and entered into the computer to cause it to perform certain sequences of operations.

The computational "heart" (more appropriately, "brain") of every computer system's hardware is called the *central processing unit* (CPU). In the microcomputer, the CPU is a small, integrated circuit chip called a *microprocessor;* the CPU in a large computer system, however, might consist of a substantial amount of electronic circuitry housed within one cabinet. In either case, the CPU performs all the high-speed computations that allow the computer to solve complex problems very quickly.

The CPU must communicate with a variety of *peripheral* devices—that is, separate units (such as the remote terminals we have already mentioned) that together with the CPU form the complete computer system. "Peripherals," as they are called, are either *input* devices that transmit data *to* the CPU, or *output* devices that receive data *from* the CPU. A keyboard is an example of an input device, and a video display is an example of an output device. Another output device is a *printer,* which can be used to obtain "hard copy" (printouts) of programs and data. Collectively, these components are called *input/output* (I/O) devices. Figure 1.3 is a diagram showing data flow paths between the CPU and various I/O devices.

Another important component of a computer system is its *memory*. Memory can be either of the internal type or connected as a peripheral device. Some internal memory (also called *main* memory) is always required, and many microcomputers use only internal memory. Memory is used to store programs and data (which we will discuss later). Memory *speed* refers to the rate at which memory contents can be transferred to and from the CPU. Internal memory is usually much faster than peripheral memory and is used principally to store programs.

Peripheral memory is often called *mass storage* because it can store very large quantities of data. The two principal types of mass storage used in modern computers are *magnetic tape* and *magnetic disk* storage. In microcomputer systems, these types of mass storage can be optionally provided using *cassette tape* or *floppy disk* units,

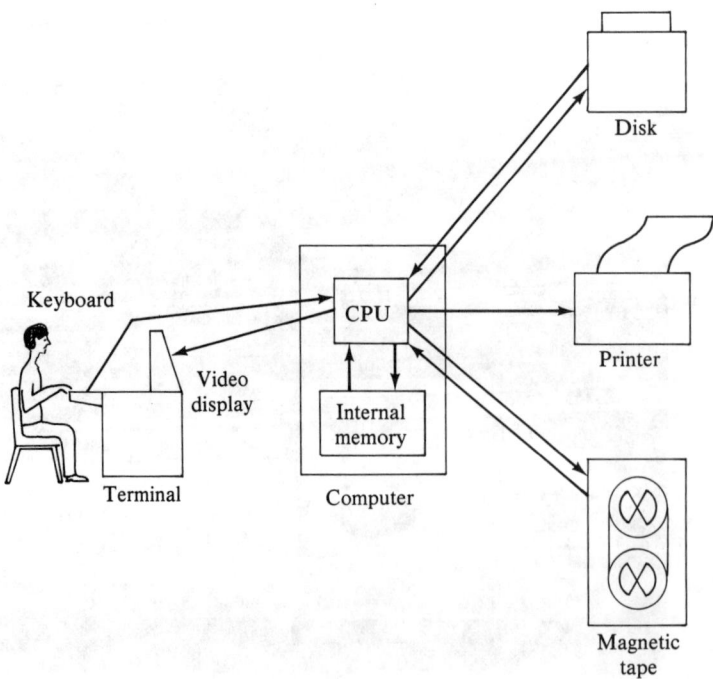

FIGURE 1.3 Data flow paths and I/O devices used in a computer system

which are smaller versions of the tape and disk devices used in a large computer system. Cassette tape storage can be provided using ordinary cassette-tape player/recorders, although special, high-quality tape should be used for best results. The floppy disk (or *diskette*) peripheral units used with microcomputers are enclosed in small cabinets; these are generally much faster than cassette tape. The thin, flexible disk is itself about the size of a 45-rpm record and is purchased separately for insertion into the main unit. Disks and tapes already containing computer programs can be purchased as well. These programs are transferred into the computer's internal memory before being executed by the computer.

All numbers and characters used by a computer are represented inside the computer in the *binary* number system. In this system, all digits are either 0 or 1. Each binary digit is called a *bit;* a group of eight bits is called a *byte*. A *word* consists of a certain number of bytes, depending on the particular computer system. For example, the binary number 0110101110100010 consists of 16 bits, or two 8-bit bytes. This number represents a typical word in a computer system designed to operate with 16-bit words.

Computer words can represent *numeric* data (numbers) or *character* data (letters and special symbols). Numbers can be stored in the computer in either a *fixed-point* or a *floating-point* format, or both. The decimal equivalent of a fixed-point binary number can be found by using a standard arithmetic procedure to convert from binary to the decimal number system. The largest positive number that can be represented by an *n*-bit fixed-point number is $2^n - 1$. Thus, the largest positive 16-bit number is equivalent to $2^{16} - 1 = 65,535$. The floating-point format is similar to scientific notation, in which numbers are represented by a fixed-point number times 10 raised to a (positive or negative) power. For example, $30521 = 3.0521 \times 10^{+4}$. In the binary version of a floating-point number, a certain number of bits in a word are used to specify a power of 2; the remaining bits represent the number that is multiplied by 2 raised to that power.

Memory capacity is often referred to in terms of "K." This indicates how many thousands of words the memory is capable of storing. (Actually, each "K" is 1024 words.) For example, a 16K memory can store approximately 16,000 words. Most microcomputers use 8-bit words, so capacity in words is the same as capacity in bytes. The larger a computer's memory is, the more sophisticated and lengthy are the programs that can be run using the system. A large portion of internal memory must be *dedicated* (allocated) to a special program called a compiler or interpreter (more about this later) whose nature depends on the programming *language* being used.

4 COMPUTER LANGUAGES

A computer program is a set of instructions that cause the computer to perform a desired task. The instructions are written by a programmer and entered into the computer using a special *language*. A so-called *high-level* language uses instructions composed of English words and symbols whose meanings convey the sense of each instruction. For example, PRINT X could be an instruction in a high-level language used to cause the computer to print the value of X. Examples of high-level languages are FORTRAN, PASCAL, PL/1, and BASIC, the language that is the subject of this book. BASIC stands for Beginner's All-purpose Symbolic Instruction Code.

The computer has its own internal set of instructions that are independent of the high-level language used to program it. These instructions are in *machine language,* a *low-level* language. When a program written in a high-level language is to be executed, the computer must first convert the high-level instructions into an equivalent set of machine-language instructions. This conversion is accomplished by a special built-in program called a *compiler* or, in the case of BASIC, an *interpreter.*

The principal difference between a compiler and an interpreter is that a compiler converts the entire high-level program to a machine-language program (called an *object* program) that can be used again, whereas an interpreter must perform this conversion each time the program is run. The interpreter is suited for *interactive* (spontaneous, two-way) communication between a programmer and the computer; we will have more to say about this later. One interactive feature of some BASIC interpreters is the ability to check the user's instructions as they are entered into the computer. In systems with this capability, an *error message* is immediately printed if the system detects anything wrong with the way an instruction is expressed. (In most microcomputer systems, error messages are not printed until program interpretation begins.) Some computer systems have a BASIC compiler, instead of *or* in addition to a BASIC interpreter.

At this point, you might be inclined to abandon ship, in the belief that computer operations are hopelessly complicated. You must realize, however, that in these few paragraphs we have summarized a technology that has been exploding for some 30 years. Fortunately, the ability to *use* (program) a computer depends very little on your comprehension of how it works. For most practical purposes, you can capitalize on the power that a modern computer affords, even if you believe it works as shown in Figure 1.4.

FIGURE 1.4

5 PROGRAMS AND DATA

We have already defined a program as a set of instructions executed by a computer to accomplish a specific task. One of the first great advances in computer technology occurred when designers realized that a computer could be made extremely versatile by storing programs in memory. All modern computers are *stored-program* computers. In normal computer operation, the CPU retrieves one machine-language instruction from memory at a time. After each instruction is executed, the next one in sequence is retrieved and executed. Thus, through a logical succession of steps, the overall goal of a program is accomplished. Of course the programmer's job is to write a high-level program (called the *source* program) that is itself a logical progression of steps leading to the computational goal. Learning how to write such programs in BASIC is the main subject of this book.

We should note that computer programs designed to perform numeric computations must be supplied with *data,* the numeric values on which the computations are to be performed. The data is the raw material that a program must have to produce results. We can think of the program and the data as two separate entities; changing the data supplied to a program does not change the computational procedure it performs, but it does change the final results. As an example, consider a program to find the average of 10 numbers. The program is written so that it can find the average of *any* 10 numbers supplied to it; any particular set of numbers constitutes the data for that particular computation. This is a rather important concept. We rarely write programs that are capable of performing computations on only a single set of data.

Data can be supplied to a program in a number of ways, depending on the nature of the computations being performed and the programming language being used. Each data value is ultimately stored in the computer's internal memory, just as the program itself is stored there, although all data values are not necessarily stored simultaneously. In many languages, including BASIC, the data can be specified somewhere in the program, using special instructions identifying data values as such. In some languages, including BASIC, data values can also be entered from a keyboard by the programmer, after the program has begun execution. Data values can also be "read in" from peripheral mass-storage devices, using special program instructions.

The ability to enter data from a keyboard after the program has begun execution is an important feature of BASIC. Because of this capability, BASIC is said to be an *interactive* language, since the programmer can interact with the computer. Using this feature, the programmer can enter new data based, for example, on the results of computations that have been performed up to a certain point. Video games are an example of interaction between a (micro) computer and a "programmer" (the player). The progress of the action in a game is actually a display of the results of a computer program being executed; the player continually updates the data being supplied to the program. In contrast to this kind of interactive programming is the so-called *batch* programming used for many other computer languages. In batch programming, the programmer submits the entire program and all the data to the computer at one time, waits for the computer to execute the program, and then retrieves or observes the results. No interaction is possible during the time that the program is being run.

As we have already mentioned, a computer program and the data supplied to it are prepared by a programmer and then entered into the computer's main memory when the programmer is ready to "run" (execute) the program. Programs and data can be stored in external (peripheral) memory whenever they are not being used. Programmers use the word *file* as a general term meaning any program or set of data stored in external memory and identified by a *file name.* For example, a program written to find the average of a set of numbers might be called "AVERAGE" and, when that program is stored in external memory, it would be referred to as a program file with the file name AVERAGE. A set of numbers to be averaged might be named "NUMBS" and stored as a file having that name: This is an example of a data file. (Program files and data files are stored exactly the same way; the distinction is made only by the way that the programmer uses them.)

A program file consists of a set of instructions; a data file consists of a set of data values. In either case, the individual instructions or data values that the file is composed of are called *records*. For example, a small data file might contain the following set of three numbers:

$$10,000$$
$$265$$
$$0.004$$

Each of these values constitutes one record in the three-record file. When a programmer is ready to run a program, he or she causes the program file and data file(s), if any, to be moved from external storage into the main memory by issuing certain *system commands* containing the file names of the desired files. Thus, all the records in each specified file are moved into the computer's main memory in preparation for a program run.

6 ALGORITHMS

An *algorithm* is a programming procedure designed to achieve a certain computational goal. Programmers solve problems on a computer by writing instructions that cause the computer to follow the procedures specified by an algorithm. The most convenient way to describe a particular algorithm is to write, in simple English, a list of steps that constitute the procedure to be followed. As a simple example, suppose you want to write an algorithm describing how to find the sum of 50 numbers. Following is a list of steps that constitutes an algorithm for achieving that goal:

(1) Start by assuming that the sum (call it S) is zero.
(2) Get the first number.
(3) Add the number to S.
(4) Get the next number.
(5) Add this number to S.
(6) If all 50 numbers have now been added to S, then stop. Otherwise, go to step 4.

Note that the description of this algorithm has been broken down into a series of very elementary steps. This is the essence of programming: By reducing a computational procedure to a sequence of simple steps, we are able to solve very complex problems.

As a final note on the nature of programming, we should distinguish between *data processing* and *scientific programming*. Programs written for data processing involve relatively simple computations on large quantities of data. An example is the type of program used to compute the account balances for the clients of a large bank. The computations involve only addition and subtraction, but there are a great many of them. In scientific programming, more complex computations are performed on a relatively small amount of data. Consequently, the algorithms tend to be more lengthy and involved than those used in data processing. An example is a program used to find the roots of the equation $Ax^5 + Bx^3 + Cx^2 + Dx + E = 0$. The data consists only of a set of numeric values for the coefficients A, B, C, D, and E, but a considerable number of computations are required to find the solution. In this book, we will be concerned with the types of computations required to solve practical technology problems; the process is therefore more like scientific programming than data processing.

EXERCISES

1.1 What are the two main types of computers used by technologists for general-purpose programming? Describe the major differences in how they are used.

1.2 Name two output devices and one input device used in a typical computer system. Give an example of a device used for both input and output.

1.3 Give two examples of modern mass-storage devices. Can you name two older forms of mass-storage media (involving punched holes)?

1.4 The maximum internal memory capacity of a typical microcomputer is 64K bytes. How many bits does this represent?

1.5 What is the maximum positive (decimal) number that can be represented in fixed-point form by one byte?

1.6 What are the major differences between a compiler and an interpreter?

1.7 Describe three ways that data can be supplied to a program.

1.8 What is interactive programming? Contrast this with batch programming.

1.9 A program is written to find the electrical current I in a circuit as a function of voltages V_1, V_2 and V_3, by solving the equation

$$I = \frac{V_1 + V_2}{300} - \frac{(V_1 - V_3)}{500} + \frac{V_2}{100}$$

What constitutes the data for this program?

1.10 Write an algorithm to find the average of N numbers, where N can be any positive integer (whole number). (Your first step should be to "get the value of N".)

1.11 Explain the difference between scientific programming and data processing.

CHAPTER **2**

INTRODUCTORY BASIC

1 VERSIONS OF BASIC

The *syntax* of a computer language is the set of rules we must follow when writing instructions and statements in that language—including the kinds of symbols that can be used, how punctuation marks are used, and any restrictions imposed on the sequence of words within a statement or on the sequence of statements within a program. As in a human language, the syntax forms the basis for the *grammar* of the language.

Although BASIC is a popular and widely used computer language, there are variations in its syntax, depending on the *version* used. These variations are relatively minor and are mostly due to variations in the software that different computer manufacturers provide for use with their particular machines. Fortunately, there are more similarities than differences among these versions, so if you learn to program in one version you can easily adapt to another. BASIC is also like a human language in another respect: Once you learn the vocabulary and structure, you will have little trouble understanding dialects containing minor variations in vocabulary or pronunciation. In fact, different versions of a computer language are often called dialects. Any syntax that can vary from one version to another is said to be *machine-dependent*. In this book, we will occasionally mention where syntax variations are likely to occur in a particular type of statement. In many cases, we will show just what those variations are for a number of currently popular machines.

We are often told that one version of BASIC is more *powerful* than another. A powerful version has special statements, and/or special variations on standard statements, allowing the programmer to accomplish certain tasks using fewer statements than would otherwise be required. For example, many powerful versions of BASIC have special *matrix*-handling statements that allow the programmer to solve a large number of equations simultaneously. (We will discuss this aspect of BASIC in Chapter 8.) Large time-shared computer systems have this capability, but it is not found with most microcomputer versions of BASIC because of the large memory requirement. Powerful versions of BASIC are often referred to as "extended BASIC" or "BASIC PLUS"; other less powerful versions might be called "tiny BASIC," "minimal BASIC," or "integer BASIC." Most programs written in a less powerful version of BASIC can be used with a computer having a more powerful version, but the reverse is not necessarily true. In this book, we will describe variations available in some powerful versions of BASIC for solving certain kinds of problems, but we will always show how these problems can be solved using less powerful versions of the language.

There is an effort under way to develop a national standard BASIC. The American National Standards Institute (ANSI) has been working toward that goal since January, 1974. The proliferation of machines, especially microcomputers, and the versions of BASIC that accompany them, has been so great since 1974 that the standardization process has been hampered. Thomas E. Kurtz, one of the co-developers of the original BASIC at Dartmouth College, is now chairman of the committee developing the ANSI standard. In his words, "BASIC has been a moving target."* When we describe variations in BASIC syntax, we will occasionally refer to ANSI, meaning the syntax prescribed by that standard in its present form.

Every computer manufacturer publishes a programming manual that describes the capabilities, options, and variations in the particular version of BASIC supplied with the machine. The programmer should consult this manual whenever there is any doubt whether the version in use has a specific capability or whenever a question arises concerning the syntax.

2 NUMBER REPRESENTATION

There are several ways to write numbers in BASIC. We can, for example, write a number the way we normally would when doing pencil-and-paper arithmetic: 25, 118.3, 0.06, and so forth. Numbers represented this way are called *fixed-point constants*. If no plus or minus sign is written in front of the number, the computer assumes that the number is positive. Thus, +140 is equivalent to 140; we can write it either way. A negative sign must be written in front of a negative number (−200, for example).

Another way to express numbers in BASIC is to use the *floating-point* representation, a method that is equivalent to *scientific notation*. Recall that a number written in scientific notation has one digit to the left of the decimal point and all other digits to the right, and it is multiplied by an appropriate power of 10. For example,

$$10743 = 1.0743 \times 10^4$$
$$87.615 = 8.7615 \times 10^1$$
$$0.0000053 = 5.3 \times 10^{-6}$$

In BASIC, the power of 10 (exponent) is specified by writing it after the letter E. For example, we write 3.3E2 for 3.3×10^2, or 3300. Either a positive or negative sign (or else no sign at all) can follow the letter E to indicate the sign of the exponent. If no sign follows E, the computer assumes a positive exponent. When the computer displays a floating-point number on the screen of a user's terminal, it will always show one digit to the left of the decimal point. However, the programmer can type floating-point numbers into the computer as data, or as part of the program, without that

---

*Kurtz, Thomas E., "On the Way to Standard BASIC," *Byte* magazine, June 1982.

restriction. Following are some examples of fixed-point numbers that are correctly expressed using BASIC floating-point notation:

$$
\begin{aligned}
400.72 &= 4.0072\text{E}2 &&= 4.0072\text{E}+2 &&= 0.40072\text{E}3 \\
-1890 &= -1.890\text{E}3 &&= -189\text{E}1 &&= -0.00189\text{E}6 \\
0.0651 &= 6.51\text{E}-2 &&= 0.651\text{E}-1 &&= 651\text{E}-4 \\
-0.00044 &= -4.4\text{E}-4 &&= -4400\text{E}-7 \\
2.3805 &= 2.3805\text{E}0 &&= 0.23805\text{E}1 &&= 0.23805\text{E}+1
\end{aligned}
$$

The number that follows E in scientific notation *must* be an integer (whole number); it cannot contain a decimal fraction or be a decimal fraction. Thus, 4E2.5 and 7.1E−.3 are *not* allowed.

In many practical technology problems, we encounter numbers equal to 1 multiplied by a power of 10. For example, 1 km = 10^3 meters, 0.1 mg = 10^{-4} grams, and so forth. We must be careful to include 1 when writing such numbers in BASIC's scientific notation; that is, we must write 1E3 and 1E−4 for 10^3 and 10^{-4}, *not* E3 or E−4, since BASIC does not assume that E3 is the same as 10^3 or that E−4 is the same as 10^{-4}.

The total number of digits that can be used to represent a number depends on the version of BASIC in use. Many versions permit the programmer to use 16 or more digits. The range of number values permitted (the smallest and largest numbers that can be represented) depends on the version and also depends on whether the number is expressed in floating-point notation. Much greater ranges of numbers are permitted when floating-point notation is used. It should be noted that the computer rounds off numbers when it displays data or results of computations, but the full number of digits is retained internally for maximum accuracy. How numbers are rounded and the form in which they are displayed (fixed-point or floating-point) depends on the version of the language. Your programming manual will contain a list of rules that are followed for rounding and displaying numbers, but these will be of little consequence to you in the practical problems you will be solving in this book.

EXERCISES

2.1 Write the following fixed-point constants in BASIC floating-point format (scientific notation) with one digit to the left of the decimal point.

(a) 483.99
(b) −10766.25
(c) 0.00042
(d) −200.383
(e) −0.000001
(f) 143.9 × 10^5
(g) 10^5
(h) 0.07518 × 10^{-4}

2.2 Write the following floating-point numbers as fixed-point constants:

(a) 3.3E3
(b) 2.1E−2
(c) 1E4
(d) −6.0485E+6
(e) −4.2073E−3
(f) −7.007E−1

2.3 Determine which of the following BASIC constants are written correctly. State the error(s) in those that are written incorrectly.

(a) 1.117E−2
(b) −8.04E−0.2
(c) 25.61E+4
(d) −3.000E−3
(e) 125E5.5
(f) π

3 STATEMENTS

A BASIC program consists of a number of *statements* written in a certain sequence by the programmer to cause the computer to accomplish a desired task. A single statement might instruct the computer to evaluate a mathematical expression, print a result, change the value of a variable, or perform any of a number of other functions that we will be studying. In other words, each statement causes the computer to perform a specific, limited task. When all the statements have been interpreted and executed in the sequence specified by the programmer, the larger goal of the program has been accomplished. The programmer's job is to select the right statements from those available in the language and write them in the logical sequence that best achieves the overall goal of the program.

To specify the sequence of statements that you want the computer to follow, you must use *statement numbers*. In BASIC every statement must be preceded by a statement number that tells its location in the sequence. The statement numbers occur in the order that you want the statements to be executed; they must all be integers (whole numbers) but not necessarily successive integers. You could, for example, assign statement numbers 1, 2, 3, . . . , etc., or 10, 20, 30, . . . , etc., or any other ascending sequence, such as 5, 20, 40, 100, . . . , etc. Most programmers prefer to use the sequence 10, 20, 30, . . . , since this permits them to insert new statements between existing ones, if necessary.

When you enter a program into the computer by typing statements on a keyboard, the computer *prompts* you to provide statement numbers. It displays the prompt symbol > (] in the Apple computer), whereupon you type in the statement number, leave a space, and then type the statement. Once the statement has been typed, you must depress the "carriage return" or ENTER key on the keyboard to enter the statement into the computer. It is important to realize that nothing typed on the keyboard is actually transmitted to the computer until it is entered by depressing the RETURN or ENTER key. After a statement has been entered, the computer generates another prompt symbol and you can proceed to type another statement number. If you want to delete a statement, simply type in the statement number following a prompt, then depress ENTER or RETURN. To change a statement, type in the statement number, type the new statement, and then enter it. This deletes the old statement and replaces it with the new one. You can insert new statements and delete or alter old ones any time, regardless of their location in the program or the current statement number.

Variations

In a BASIC program, there is usually one statement per line—that is, one statement per statement number. In many versions of BASIC it is permissible to type more than one statement per line. These so-called *multiple statements* must be separated from each other by a special symbol, called a delimiter. A widely used delimiter symbol is the colon (:). The advantage of using multiple statements is that they can save memory space, which is often at a premium in microcomputer systems.

4 VARIABLES

As in algebra, we use *variables* in BASIC to represent quantities whose values might change (or be changed as the result of a computation). In algebra, we might let x stand for the unknown in an equation and then proceed to find the value of x by solving the equation. Similarly, we use letters in BASIC to stand for quantities whose values might change during the execution of a program. Suppose, for example, you want to find the sum S of three numbers—say, 5, 8, and 10. One way to do this is to start by setting S equal to 5; then change its value to $5 + 8 = 13$ and finally to $13 + 10 = 23$. The last value assigned to S would be the sum we are seeking, but the value of S changed three times during the computation. Of course constants (numbers) never change their values, but variables can be assigned (set equal to) the value of a constant.

In BASIC we specify a variable by giving it a *variable name*. As we have already indicated, a variable name can be simply a single letter. It can also consist of two characters, but *the first character must be a letter.* In some versions of BASIC, the second character (if any) must be a single-digit integer number. Thus A2, Z9, and F5 would be valid variable names. Many versions of BASIC (for example, Apple, Commodore, and TRS-80) permit the second character to be either a letter or an integer digit. In these versions, the variable names AA, TF, R3, and MZ would all be valid. Many versions also permit a variable name to consist of more than two characters but treat only the first two as significant. Thus, SUM and SUBSTANCE would be regarded as the same variable. The ANSI standard allows variable names to have up to 31 characters.

Following are examples of names that are *not* valid in *any* version of BASIC:

```
3F         X-7
A2.5       3E+4A
```

If you always represent a variable name by a single letter or a single letter followed by a single-digit integer (as we will do in this book), then you can be certain that you are using valid names, no matter what version of BASIC you use.

EXERCISES

2.4 Determine which of the following are valid variable names in BASIC. State the error(s) in those that are invalid.

(a) I
(b) A7
(c) 9Z
(d) B5
(e) X;Y
(f) 25
(g) T
(h) R-1

5 THE LET STATEMENT

Every statement in BASIC begins with a *keyword*. A keyword is a special word that uniquely identifies the kind of statement associated with it. The programmer can usually choose the symbols or numbers that follow the keyword, within whatever limits the syntax imposes on the form of a statement, but the keyword must appear first so that the computer will know what to expect.

The first BASIC statement we will study has the keyword LET. This statement is used to *assign* a value to a variable. In algebra we write: Let $x = 3$, to assign the value of 3 to the variable x. Similarly, in BASIC we write

```
LET X=3
```

The variable name (X in this example) must always follow the keyword, since whatever follows LET is the quantity whose value is changed. We *cannot* write LET 3 = X, since this statement would attempt to change the value of the constant 3. Once X has been assigned the value 3, it will retain that value in all future computations, unless another LET statement modifies it. In other words, the computer will treat the symbol X as it would the constant 3, when performing any computations involving X (unless X is assigned a new value before these computations are performed).

A LET statement can also be used to assign the value of one variable to another variable. Consider, for example, the following sequence of statements:

```
10 LET X=2
20 LET Y=X
```

The LET in statement 10 assigns X the value of 2, and the LET in statement 20 assigns the value of X to Y. Thus Y also becomes 2. The value of X is not affected by the statement LET Y = X. Note again that *it is the quantity to the left of the equal sign whose value is changed*.

The general form of a LET statement is

$$\text{LET } v = \textit{const} \text{ or } \textit{exp}$$

where *v* is a valid variable name, *const* is any numeric constant, and *exp* is an expression. (We will define and discuss expressions in Chapter 3. A variable name is one form of an expression.)

In this book, we will use lowercase (uncapitalized) letters in program statements to indicate quantities, symbols, or names supplied by the programmer. Capital letters are used for words or symbols that must appear in exactly the form shown. (We use this convention to show the general *form* of a statement; when a statement is actually typed into the computer, all letters appear as capitals.)

A LET statement can also be used to assign a new value to a variable in terms of its old value. In other words, it is permissible for the same variable name to appear on *both* sides of the equal sign in a LET statement. Consider the following sequence:

```
10 LET X=50
20 LET X=X+1
```

Statement 10 assigns X the value 50, and statement 20 assigns X its old value (50) plus 1, or 51. The variable X will then retain its new value of 51 unless a subsequent LET statement modifies it again. Note that statement 20 has no meaning in ordinary algebra, since X cannot "equal" $X + 1$. It is permissible in BASIC because we use the LET statement to *assign* a value to a variable. LET statements are often referred to as *assignment* statements. (To understand the meaning of the equal sign in a LET statement, think of it as equivalent to the words "is replaced by." Thus, for example, statement 20 above can be read as "X is replaced by $X + 1$.")

All variables must be assigned values before the computer can evaluate any expression, formula, or equation that contains their variable names. In most versions of BASIC, the computer assumes a value of zero for any variables that have not been assigned values.

Study the following sequence of statements for practice in interpreting LET statements. What are the values of *A* and *B* after all statements have been executed?

```
10 LET A=-5
20 LET B=A+1
30 LET A=A+B
40 LET B=B+2
```

In statement 10, *A* is set equal to -5. In statement 20, *B* is set equal to $A + 1 =$

$-5 + 1 = -4$. In statement 30, A is set equal to $A + B = -5 - 4 = -9$. Finally, in statement 40, B is set equal to $B + 2 = -4 + 2 = -2$. So after all statements have been executed, $A = -9$ and $B = -2$.

Variations

In some versions of BASIC (but *not* in most microcomputer versions), it is permissible to use one LET statement to assign the same value to more than one variable. For example, in these versions it is permissible to write

or
```
LET X=Y=2
LET A1=A2=A3=10
```

The first statement assigns the value 2 to both X and Y; the second statement assigns 10 to A1, A2, and A3. These are called multiple assignment statements. In versions that permit multiple statements per line (see section 3, "Variations"), we could also write

and
```
LET X=2: LET Y=2
LET A1=10: LET A2=10: LET A3=10
```

In many versions of BASIC (but *not* in the ANSI standard), it is permissible to omit the keyword LET in an assignment statement. In these versions, an assignment statement begins with a variable name followed by an equal sign. The computer interprets this combination as equivalent to the keyword LET. For example, we could write either

```
X=16   or   LET X=16
```

and the computer would in each case assign the value 16 to X.

EXERCISES

2.5 Determine which of the following are valid LET statements. State the error(s) in those that are invalid.

(a) LET B5=29   (e) LET P=π
(b) LET R=4E-2   (f) LET X+1=X
(c) LET 13=Q   (g) LET 2R=2R
(d) LET T=T+T

2.6 Determine the values of A, B, and C after each of the following sequences is executed.

(a) 10 LET A=0
    20 LET B=A
    30 LET C=B+1

(b) 50 LET C=1E+2
    60 LET A=C+2E+2
    70 LET B=A+C

(c) 25 LET A=-1
    30 LET B=-1
    40 LET C=A+B
    45 LET B=A+C
    56 LET A=B+C

6 THE PRINT STATEMENT

The PRINT statement is used in a BASIC program whenever the programmer wants results or messages to be displayed on the screen of the terminal (or microcomputer). In some systems, the statement will also activate a printer that produces a permanent copy of what is displayed on the screen. The keyword PRINT can be followed by a variable name; in this case the computer displays the current *value* of that variable. Consider, for example, the following sequence:

```
10 LET W=25
20 PRINT W
```

In this example, the computer simply displays the number 25 when it executes statement 20. As another example, consider the following:

```
10 LET M=12
20 PRINT M
30 LET M=M+5.7E-2
40 PRINT M
```

In this example, the first PRINT statement causes the current value of M (12) to be displayed. Since the value of M is modified by statement 30, the second PRINT statement causes the new value of M (12.057) to be displayed. When the computer has executed these four statements, the screen will display:

```
   12
   12.057
```

Note that every time a PRINT statement is executed, the result is shown on a new line. One or more PRINT statements can appear anywhere in a BASIC program.

A *string* is a set of characters (letters, numbers, or symbols) that is treated as text material rather than as numeric data or variables. When we want to print a string, the keyword PRINT is followed by the string characters *enclosed within quotation marks*. Following is an example:

```
100 PRINT "THE ANSWER IS:"
```

After executing statement 100, the computer displays

```
THE ANSWER IS:
```

The following sequence illustrates the use of a string in a PRINT statement:

```
10 LET V1=5
20 LET V2=10
30 LET V=V1+V2
40 PRINT "THE SUM OF V1 AND V2 IS:"
50 PRINT V
```

In statement 30, V is assigned the value $V1 + V2 = 15$. Hence the two PRINT statements cause the following display to appear on the screen:

```
THE SUM OF V1 AND V2 IS:
15
```

EXERCISES

2.7 Show the results that would be displayed if the following sequences were executed in BASIC:

(a)
```
10 LET A=5
20 LET A=A+1
30 PRINT A
```

(b)
```
100 PRINT "I="
110 LET I=500
120 LET J=100
130 LET I=I+J
140 PRINT I
150 PRINT "DONE"
```

(c)
```
1 LET X=25
2 PRINT "X=25"
3 PRINT X
```

2.8 The following sequence of statements is supposed to print the sum of X and Y. Find all errors, and correct the sequence.

```
10 LET X=500
20 LET Y=50E1
30 PRINT THE SUM OF X AND Y IS:
40 PRINT "S"
```

7 THE INPUT STATEMENT

In Chapter 1, we said the BASIC is an *interactive* computer language—that is, a language that permits the programmer to enter data while a program is running or alter the program after it has started running. The statement that permits us to interrupt a program this way is called the INPUT statement. Its format is

INPUT *v*

where *v* is a variable name. When the computer executes an INPUT statement, it displays a question mark on the screen. This question mark is the INPUT prompt symbol, and it means the computer is waiting for the user to enter a number. When the user types in a numeric value in response to the prompt, that value is assigned to the variable (*v*) whose name is written in the INPUT statement.

Consider for example the following sequence:

```
10 INPUT K
20 PRINT K
```

After execution of these statements, the display on the screen might look like this:

```
?75
75
```

The question mark is generated when the INPUT statement at 10 is executed. The programmer then types in a value (75 in this example) and that value is assigned to the variable K. The PRINT statement at 20 then causes the value of K to be displayed. You can see that the INPUT statement gives us another way, besides using the LET statement, to assign a numeric value to a variable.

When using an INPUT statement in a program, it is good practice to precede it with a PRINT statement specifying what the user is supposed to enter. Otherwise, the user sees only the ? prompt symbol, and might not remember what quantity is expected. This is especially important if there is more than one INPUT statement in a program. Following is an example:

```
10 PRINT "ENTER MASS IN KILOGRAMS"
20 INPUT M
```

EXERCISES

2.9 Write a sequence of BASIC statements to find the sum of three numbers. Write the sequence so that the user can type in the three numbers from a terminal, when prompted to do so. The result should be displayed at the terminal, along with the statement THE SUM IS:.

2.10 Modify your solution to Exercise 2.9 so that it prints the value of each number entered by the user immediately after the number is entered.

2.11 Find all errors in the following sequence:

```
10 PRINT ENTER ONE NUMBER
20 INPUT 40
30 LET X=X+40
40 PRINT X
50 INPUT X+Y
60 PRINT Y
```

8 THE END STATEMENT

The END statement should be the last statement in a BASIC program. This statement consists only of the keyword END. It is called a *nonexecutable* statement because the computer is not required to perform any computations or accomplish any other task related to the program's objective. The END statement simply tells the computer that there are no more executable statements in the program. Many versions of BASIC (including the Apple, Commodore, TRS-80 and Xerox versions) do not require the use of an END statement; the computer simply halts when it has executed the highest-numbered statement.

Some versions of BASIC allow only one END statement in a program, although others permit as many as desired. (You will learn later that it is sometimes desirable to be able to halt execution before reaching the last statement.) In some versions, execution of an END statement causes the computer to print a message telling where it encountered the END.

In the preceding part of this chapter, we have referred to lists of BASIC statements as "sequences" rather than as "programs." In this book, we will always use an END statement to terminate a program and will henceforth use the word *program* when a sequence is terminated that way.

Following is our first example of a complete program. The program prompts the user to enter two numbers, $X1$ and $X2$, and then computes and displays the value of $(X1-2) + (X2+4)$.

```
10 PRINT "ENTER X1"
20 INPUT X1
30 PRINT "ENTER X2"
40 INPUT X2
50 LET X1=X1-2
60 LET X2=X2+4
70 LET S=X1+X2
80 PRINT "THE VALUE OF (X1-2)+(X2+4) IS:"
90 PRINT S
100 END
```

If this program is entered and executed and if the values entered for $X1$ and $X2$ are 8 and 10, the display on the screen will look like this:

```
ENTER X1
?8
ENTER X2
?10
THE VALUE OF (X1-2)+(X2+4) IS:
20
```

9 THE KEYBOARD AND HOW TO USE IT

Figure 2.1 shows a typical keyboard used for entering BASIC programs into a computer. The arrangement of the keys in Figure 2.1 is very similar to that of a typewriter keyboard. Indeed, the ability to type on a conventional typewriter is a skill that can be transferred to a computer keyboard; however, even a non-typist can soon gain an instinctive feel for the location of the keys, with some practice.

FIGURE 2.1 A typical keyboard (Photo courtesy of Tandy Corporation)

One difference between this keyboard and a typewriter is the absence of lowercase letters on the computer keyboard. All alphabetic characters used in BASIC are uppercase, so lowercase keys are not needed. The keys used for entering the numeric digits 0 through 9 are in the same locations as they are on a typewriter—along the upper row. The set of alphabetic and numeric characters together is called the *alphanumeric* character set. Note that the computer keyboard also has a number of keys with special symbols, including punctuation marks and math symbols. As on the typewriter, many of these symbols share keys with numeric characters, so they are selected by using the SHIFT key. For example, depressing the 4 key while holding down the SHIFT key will enter the symbol $. Other special symbols appear together on the same key, so symbols on the lower half of the key are entered directly, and those on the upper half are selected by using the SHIFT key.

The space bar at the bottom of the keyboard is used to enter *blanks*. The BASIC language ignores blanks typed between statement numbers, keywords, variable names and constants, so blanks can be used to insert spaces in a statement for clarity, appearance, and ease of reading. For example, the appearance of the statement

```
10 LET X=1
```

is preferable to

```
10LETX=1
```

although both would be treated as the same by the computer. A blank typed within a string is treated as a blank *character* (space) and is displayed as such by a PRINT statement. For example, the statement

```
10 PRINT "TIME    TEMPERATURE"
```

(which might be used to print a table heading) would result in the display

```
TIME    TEMPERATURE
```

Note the key labeled ENTER in Figure 2.1. This key is used to enter a statement or command (such as RUN) into the computer after typing it on the keyboard. On some computer keyboards, this key is labeled RETURN (or RET), a carryover from typewriter nomenclature.

We note in Figure 2.1 that there are other keys with special symbols or words on them. Other computer keyboards will have different symbols or words, since the functions of these keys depend on the particular computer system. For example, there is usually a key that allows the user to "backspace" or erase a character from a line before that line is entered. In Figure 2.1, the ← key is used for this purpose but, on some keyboards, the word ERASE is used. The keyboard in Figure 2.1 also has a key labeled BREAK. Once a program has commenced a run, the user can interrupt it by depressing this key. This feature is particularly useful when a program seems to be taking an excessive amount of time to run. Programmers sometimes inadvertently write an "infinite loop" into a program (this will be discussed in Chapter 4); in this case, the BREAK key is the only way to interrupt the run. In some systems, an ESCAPE (or ESC) key is used for this purpose. In many systems, when this external means of control is used to terminate a program run (as opposed to using an END statement in the program), the program is erased. Unless the program is "saved" in memory before the run begins (we will cover this in Section 2.10), the program may have to be retyped after an external BREAK.

Finally, most keyboards have a *control* key labeled CNTL, CONT, or some other special characters; this key is used in conjunction with other keys to perform special functions. For example, if the control key and the C key are depressed simultaneously, this might cause an external program break (like the BREAK key does). Consult your programming manual for information on these special control functions.

10 GAINING ACCESS TO BASIC

The two major kinds of computers used for BASIC programming are microcomputers and the large time-shared systems. We will discuss separately how to gain access to the BASIC programming capability of each type of computer. You will only need to study the material related to your computer, since there is considerable duplication of information in these two sections.

Many beginning programmers are unnecessarily timid when using a computer keyboard for the first time. It is important to realize that *programming cannot damage a computer.* You will make mistakes, but none will be fatal. Computers are cleverly designed to help you detect every conceivable kind of programming error but, even if you invent a new one, no damage will ensue. Don't let the computer intimidate you!

(a) Using Large (Time-Shared) Computer Systems

A large computer installation whose computer is time-shared by a number of programmers (using remote terminals, as described in Chapter 1) generally requires users to *log on* before access to any computer language is granted. The log-on procedure varies according to the installation but usually requires the user to enter an authorized *account number.* The account number identifies the programmer as a legitimate user and is also used for recordkeeping: The total computer time is reported and/or charged to the user's account. When the user types in the account number at the terminal, the screen (which usually displays what is typed) can suppress the display of the account number, to maintain confidentiality. Remember to depress the RETURN or ENTER key after typing an account number.

Once the system has recognized and acknowledged the user's account number, it will display a message to that effect. Since a large computer can generally be programmed in several different languages, the user must next inform the computer which language is to be used. In many systems, this can be accomplished by simply typing BASIC. The computer will then usually display a message acknowledging the user's access to BASIC along with the standard BASIC prompt symbol (>). Remember again that every instruction typed in from the keyboard must be followed by pressing the RETURN or ENTER key.

Each time the user enters a BASIC statement, the computer will check the statement for syntax errors and then display an error message, if any are detected. The programmer must retype the statement number and reenter the (corrected) statement, if it is found to contain errors. Once the entire program has been entered, the programmer will usually want to save it in the computer's memory for future use. The procedure for saving a program depends on the system and involves what are called *system commands*. System commands are different from BASIC statements; they do not affect the structure of a program or what it accomplishes. (System commands will be discussed in detail in Chapter 4.) In a large computer system, the command SAVE—followed by whatever name the programmer has chosen for the program and possibly other system-dependent words—will usually cause a program to be saved in memory. If a BASIC program is not saved, it will be lost after the user logs off the system or exits the BASIC programming capability. However, the program can be run (by typing RUN) and altered and run again as many times as desired without being erased, provided the BASIC programming capability is still in operation.

Typically, the programmer will test run the program to make sure it works properly, correct any errors that appear, and then save the final version. After a run has commenced, the computer will detect any errors that occur as a result of the run (these are called *run-time errors,* as opposed to syntax errors), then interrupt the run, and print an appropriate error message. An example of a run-time error that might not be detected before an actual run is an attempt in the program to divide by zero.

If a program has been properly saved, the programmer can then log off the system and return later to retrieve the program from memory. This retrieval again requires a system-dependent command (often simply the word LOAD) followed by the name of the program.

Before you attempt to use a time-shared system for BASIC programming, you should consult your system programming manual to learn the details of the log-on and log-off procedures and the commands used for saving and loading programs.

(b) Using Microcomputers

The procedure for gaining access to a microcomputer's BASIC programming capability varies somewhat among machines of different manufacture. Depending on the size and complexity of the computer, the procedure can be as simple as turning on the power and depressing a single key. Unlike large, time-shared systems, no account number is required to identify an authorized user. (Consult the user's manual for the procedural details applicable to your particular machine.)

As the programmer types a BASIC statement from the keyboard, the screen displays what has been typed. However, that statement is *not* entered into the computer until the RETURN or ENTER key has been depressed.

Once a complete program has been entered, the programmer can run it by typing the command RUN after the prompt symbol (>), instead of typing a statement number. The program will move off the screen but will not be lost. It can be run again as often as desired. If the program contains errors that can only be detected after a run has begun (*run-time errors*), the computer will interrupt the run and display the appropriate error message. This message might actually be an abbreviation or code whose meaning can be found in the programming manual. If an error is detected, it can be corrected by simply retyping the statement number, followed by the statement in its correct form. An example of a run-time error in the program is an attempt to divide by zero. If errors are detected, the programmer proceeds to *debug* the program (find and correct the sources of error) and runs it again.

After the program has run satisfactorily, the programmer will probably save it for future use. The internal memory of a microcomputer is erased whenever the power is turned off, so programs must be saved on external memory such as cassette tape or floppy disk. Most microcomputers are designed to operate with this type of equipment, but special *system commands* are required to gain access to it. System commands are instructions that differ from BASIC statements in that they do not affect the structure of a program or what it accomplishes. (This will be discussed further in Chapter 4.) Additional system commands are used to retrieve programs that have been previously saved on tape or disk. In many systems, programs are saved under a name chosen by the programmer and retrieved by the same name. The words SAVE and LOAD, or some variations of these, are usually used in the system commands for saving and retrieving programs.

Since the system commands are machine-dependent, we will not discuss them further here. Consult your programming manual for details about saving and loading programs in your computer system.

11 SOME SIMPLE PROGRAMMING EXAMPLES

The symbol for multiplication in BASIC is the asterisk (*). This symbol can be used to show multiplication of a constant by a constant, a constant by a variable, or a variable by another variable. Examples are 5*7, 40.2*X, A3*6E−5, and I*J2. Note that we *cannot* omit the multiplication symbol when expressing a product, as we often do in ordinary algebra. For example, in algebra we write 5Y to mean 5 times Y, but in BASIC we must show the product as 5*Y.

We will study multiplication and other mathematical operations in greater detail in Chapter 3. We introduce multiplication here so we can use it to show applications of BASIC in some practical technology problems. The example problems are all quite simple and could be solved quickly and easily using a pocket calculator rather than a computer. We chose these problems to illustrate basic programming principles (using the statements we have learned so far), but you won't appreciate the true power of the computer until you have a greater variety of statements at your disposal. Study the following programs carefully. To practice using your system, enter and run at least one of them on your computer.

Example 2.1 To convert velocity V given in miles per hour to velocity in feet per second, we compute

$$\left(V\frac{\text{miles}}{\text{hour}}\right)\left(5280\frac{\text{feet}}{\text{mile}}\right)\left(\frac{1\text{ hour}}{3600\text{ sec}}\right) = 1.46667V\text{ ft/sec}$$

To convert velocity V given in miles per hour to velocity in meters per second, we compute

$$\left(V\frac{\text{miles}}{\text{hour}}\right)\left(1609.34\frac{\text{m}}{\text{mile}}\right)\left(\frac{1\text{ hour}}{3600\text{ sec}}\right) = 0.447V\text{ m/sec}$$

The BASIC program shown in Figure 2.2 converts a velocity entered by the user in miles per hour to equivalent velocities in feet per second and meters per second. Figure 2.3 on page 26 shows the results of two runs of the program in Figure 2.2. In the first run, a velocity of 30 miles per hour was entered; this velocity is seen to equal 44.001 ft/sec or 13.41 m/sec. In the second run, the velocity of light (6.706 × 10^8 miles/hr) was entered and this velocity is seen to equal 9.83549 × 10^8 ft/sec or 2.99758 × 10^8 m/sec.

Example 2.2 Newton's second law states that the force required to accelerate a mass m is given by $F = ma$, where a is the acceleration. In the metric (SI) system of units, F is in newtons (N), mass in kilograms (kg), and acceleration in m/sec^2. The program

```
10 PRINT "ENTER VELOCITY IN MILES/HR."
20 INPUT V
30 LET V1=1.46667*V
40 LET V2=.447*V
50 PRINT "THE VELOCITY IN FT/SEC IS:"
60 PRINT V1
70 PRINT "THE VELOCITY IN METERS/SEC IS:"
80 PRINT V2
90 END
```

FIGURE 2.2 Program for Example 2.1

```
ENTER VELOCITY IN MILES/HR.
? 30
THE VELOCITY IN FT/SEC IS:
44.0001
THE VELOCITY IN METERS/SEC IS:
13.4100

ENTER VELOCITY IN MILES/HR.
? 670600000
THE VELOCITY IN FT/SEC IS:
9.83549E+08
THE VELOCITY IN METERS/SEC IS:
2.99758E+08
```

FIGURE 2.3

shown in Figure 2.4 allows the user to enter mass in kilograms and acceleration in m/sec^2 and then computes the force in both newtons and pounds. (Force in pounds = 0.22408 times force in newtons.)

Figure 2.5 shows the results of one run of the program in Figure 2.4. In this run, the value of mass entered is the average mass of a man (73 kg) and the acceleration is that due to gravity on the earth (9.81 m/sec^2). We see that the average man weighs 716.13 N or 160.47 lb.

Example 2.3 The density D of a body is its mass per unit volume

$$D = M/V$$

where M is the mass of the body in kilograms,
        V is its volume in cubic meters, and
        D is its density in kilograms per cubic meter.

The volume of a cube is $V = l^3$ cubic meters, where l is the length of one side. Therefore, the mass of a cube is $M = Dl^3$, where D is its density. The program shown in Figure 2.6 computes the mass of a cube in kilograms when the user enters the length of one side in *centimeters* and the density in kilograms per cubic meter.

Figure 2.7 shows the results of one run of the program in Figure 2.6. The mass of a cubic foot of air (30.48 cm per side) having a density of 6.5 kg/m^3 is seen to be 0.18406 kg.

```
10 PRINT "ENTER MASS IN KG"
20 INPUT M
30 PRINT "ENTER ACCELERATION IN METERS PER SEC-
SQUARED"
40 INPUT A
50 LET F=M*A
60 PRINT "THE FORCE IN NEWTONS IS:"
70 PRINT F
80 LET F=F*.22408
90 PRINT "THE FORCE IN POUNDS IS:"
100 PRINT F
110 END
```

FIGURE 2.4 Program for Example 2.2

INTRODUCTORY BASIC

```
ENTER MASS IN KG
? 73
ENTER ACCELERATION IN METERS PER SEC-SQUARED
? 9.81000
THE FORCE IN NEWTONS IS:
716.130
THE FORCE IN POUNDS IS:
160.470
```

FIGURE 2.5 The results of a run of the program in Figure 2.4

```
10 PRINT "ENTER LENGTH OF ONE SIDE IN CM"
20 INPUT L
30 PRINT "ENTER DENSITY IN KG PER CUBIC METER"
40 INPUT D
50 LET L1=L*1E-2
60 LET L2=L1*L1
70 LET L3=L1*L2
80 LET M=L3*D
90 PRINT "THE MASS IN KG IS:"
100 PRINT M
110 END
```

FIGURE 2.6 Program for Example 2.3

```
ENTER LENGTH OF ONE SIDE IN CM
? 30.4800
ENTER DENSITY IN KG PER CUBIC METER
? 6.50000
THE MASS IN KG IS:
.184060
```

FIGURE 2.7 The results of a run of the program in Figure 2.6

Example 2.4 The equation of a straight line is $y = mx + b$, where m is the slope and b is the y intercept. See Figure 2.8. The program shown in Figure 2.9 on page 28 computes the y coordinate of a point on a straight line, given the x coordinate, the slope of the line, and the y intercept.

Figure 2.10 on page 28 shows the results of two runs of the program in Figure 2.9. In the first run, we used $x = 4.08$, $m = -2.5$ and $b = 17.66$. We see that the y coordinate is 7.46. In the second run, we used $x = 0$, $m = 0.95$, and $b = 6.4$. We see that $y = 6.4$, as we would expect (since the y intercept is by definition the point on the line where $x = 0$).

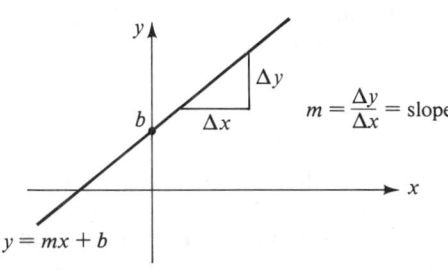

FIGURE 2.8 The equation of a straight line is $y = mx + b$ (Example 2.4).

```
10 PRINT "ENTER X-COORDINATE"
20 INPUT X
30 PRINT "ENTER SLOPE"
40 INPUT M
50 PRINT "ENTER Y-INTERCEPT"
60 INPUT B
70 LET Y=M*X+B
80 PRINT "Y="
90 PRINT Y
100 END
```

FIGURE 2.9 Program for Example 2.4

```
ENTER X-COORDINATE
? 4.08000
ENTER SLOPE
? -2.50000
ENTER Y-INTERCEPT
? 17.6600
Y=
7.46000

ENTER X-COORDINATE
? 0
ENTER SLOPE
? .950000
ENTER Y-INTERCEPT
? 6.40000
Y=
6.40000
```

FIGURE 2.10 The results of two runs of the program in Figure 2.9

GENERAL EXERCISES

2.12 To practice loading and running programs on your computer system, select the appropriate program from Examples 2.1 through 2.4 and run it to find the following:

(a) the speed of sound (741.5 MPH) in feet per second and meters per second;
(b) the weight in newtons and pounds corresponding to the mass of an average man on the surface of the moon (where the acceleration due to gravity is 1.67 m/sec^2);
(c) The y-coordinate of a point whose x coordinate is 15, if the line passes through the origin (0,0) and has a slope of 4; and
(d) the mass of a block of ice with dimensions of 2 m × 2 m × 2 m, if its density is 920 kg/m^3.

2.13 To familiarize yourself with the error messages generated by your computer system, type and enter each of the following statements into the computer. If an error message is not displayed when a statement is entered, then type and enter RUN after entering each statement. Record the error message displayed in either case, give its meaning, and explain why the statement is in error.

(a) 10 LET 3X=100
(b) 10 LET 4=Y
(c) 10 LET X+Y=12
(d) 10 PRINT THE VALUE IS
(e) 10 INPUT 35
(f) 10 LET Z=52E1.5
(g) 10 INPUT A1+B1
(h) 10 WRITE G
(i) 10 PRINT X="
(j) 10 LET R=4 S=10
(k) 10 INPUT 40*F5

2.14 Write a BASIC program to find the area of a triangle. The user enters the base and height in centimeters and the program prints the area in square centimeters. Run your program to find the area of a triangle with a base of 25.061 cm and a height of 157.2044 cm.

2.15 Write a BASIC program to find the area of a circle. The user enters the radius in *inches* and the program prints the area in square *centimeters*. Run your program to find the area of a circle with a radius of (a) 10 inches (b) 0.3937008 inches and (c) 0.164 feet.

2.16 Write a BASIC program to convert temperature in degrees Celsius (°C) to temperature in degrees Fahrenheit (°F). (°F = 32 + $\frac{9}{5}$°C). Run your program to determine the temperature of the freezing point of water (0°C) and the boiling point of water (100°C) in degrees Fahrenheit.

2.17 Write a BASIC program to determine the total distance a body travels, given its speed and time of travel. The user enters speed in *miles per hour* and time of travel in *minutes,* and the program prints distance traveled in *meters*. Run your program to determine how many meters an automobile travels in 10 minutes when its speed is 45 MPH.

EXERCISES FOR ARCHITECTURAL/CIVIL/CONSTRUCTION TECHNOLOGY

2.18 Write a BASIC program to determine and print the total construction cost of a building, when the user enters cost per square foot and total area in square feet. Run your program to determine the total construction cost of

(a) a 15,550-square-foot building at $6.32 per square foot, and
(b) a 64,077-square-foot building at $4.95 per square foot.

2.19 The moment of the force on a beam about a given axis is $M = Fd$, where F is the magnitude of the force and d is the perpendicular distance between the axis and the line of action of the force. (See Figure 2.11.) Write a BASIC program to determine and print the moment in *foot-pounds* after the user enters the force in *newtons* and the distance d in *meters* (1 N = 0.2248 lb and 1 m = 3.281 ft). Run your program to determine the moment of (a) an 8064.3 N force acting 1.95 m from an axis, and (b) a 15.2 kN force acting 4083 mm from an axis.

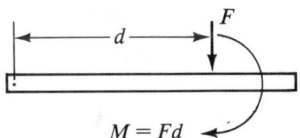

FIGURE 2.11 Exercise 2.19

2.20 According to Hooke's law, the stress σ on a material is $\sigma = E\epsilon$ where E is the modulus of elasticity and ϵ is the strain. (Strain ϵ is a dimensionless quantity, defined as the change in length per unit length of the material under stress.) Write a BASIC program to determine and print the stress in kilonewtons per square meter when the user enters the strain and the value of E in pounds per square inch ($1\,N = 0.2248$ lb and $1m = 39.37$ in). Run your program to determine the stress on (a) a steel beam ($E = 30 \times 10^6$ lbs/in^2) when $\epsilon = 0.0016$ and on (b) a concrete cylinder ($E = 4 \times 10^6$ lbs/in^2) when $\epsilon = 0.0008$.

EXERCISES FOR ELECTRICAL/ELECTRONICS/COMPUTER TECHNOLOGY

2.21 The voltage drop V across a resistance R that has current I in it is, by Ohm's law, $V = IR$, where V is in volts, I in amperes, and R in ohms. (See Figure 2.12.) Write a BASIC program to determine and print the voltage drop after the user enters resistance in ohms and current in amperes. Run your program to find the voltage when (a) $I = 0.26045$ amps and $R = 330.94$ ohms; and (b) $I = 15.2$ mA and $R = 68$ kΩ.

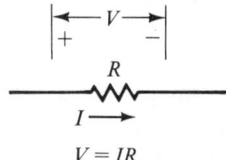

FIGURE 2.12 Exercise 2.21

2.22 The power P dissipated in a resistance R with current I is $P = I^2R$, where P is in watts, I in amperes, and R in ohms. Write a BASIC program to determine and print the power in watts, after the user enters resistance in ohms and current in *milliamperes*. Run your program to determine the power for both inputs (a) and (b) in Exercise 2.21.

2.23 The efficiency of an electric motor is $\eta = P_o/P_i$, where P_o is output power and P_i is input power. Write a BASIC program to determine and print the output power in *horsepower*, after the user enters the input power in *watts* and the efficiency η (1 watt = 0.0013405 HP). Run your program to find the horsepower of a motor with (a) efficiency 0.94 and input power 793.617 watts and with (b) efficiency 0.85 and input power 54.22 kW.

EXERCISES FOR INDUSTRIAL/MANUFACTURING/PRODUCTION TECHNOLOGY

2.24 The number of parts produced by a machine tool is $N = Rt$, where N is the total number of parts produced in time t, and R is the rate of part production. Write a BASIC program to determine the total number of parts produced, after the user enters the rate of production in parts per hour and the total time t. Run your program to determine the total number of parts produced (a) in 24 hours, if $R = 627.5$ parts/hr; and (b) in one 5-day work week, if $R = 32$ parts/hr.

2.25 A manufacturing facility employs 153 production workers and 32 supervisory personnel. Write a BASIC program to determine and print the total *monthly* payroll, after the user enters the average *weekly* wage for production workers and the average *annual* wage for supervisors. (Assume 40 hours per week, 4.3 weeks per month, and 12 months per year.) Run your program to determine the monthly payroll when the average wages are (a) $325.00 per week for production workers and $26,500 per year for supervisors; and (b) $9.20 per hour for production workers and $2,150.00 per month for supervisors.

2.26 The maximum outside diameter (OD) of a steel tube is $d + (0.01T)d$, where d is the nominal OD and T is the percent tolerance. Write a BASIC program to determine and print the maximum OD of a tube in *millimeters* after the user enters the nominal OD in *mils* and the percent tolerance (1 mil = 0.001 inches and 1 inch = 25.4 mm). Run your program to determine the maximum OD of a tube with (a) a nominal OD of 0.125 inches and a tolerance of 0.01%; and (b) a nominal OD of 0.075 inches and a tolerance of 0.15%.

EXERCISES FOR MECHANICAL TECHNOLOGY

2.27 The angular velocity ω_2 of a gear having N_2 teeth when it is driven by a gear having N_1 teeth is $\omega_2 = \omega_1(N_1/N_2)$, where ω is in radians per second and (N_1/N_2) is the gear ratio. (See Figure 2.13.) Write a BASIC program to find the angular velocity of a *driven* gear after the user enters the angular velocity of the *driving* gear and the gear ratio. Run your program for the case where the driving gear has (a) an angular velocity of 2.54 rad/sec if the gear ratio is 1.52; and (b) an angular velocity of 13.704×10^3 rad/sec, if the gear ratio is 0.000025.

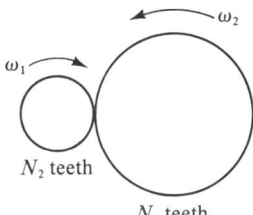

FIGURE 2.13 Exercise 2.27

2.28 The efficiency of a machine is $\eta = P_o/P_i$, where P_o is the power developed by the machine and P_i is the power supplied to it. Write a BASIC program to determine and print the power developed in *horsepower*, after the user enters the power supplied in foot-pounds per second and the efficiency (1 ft-lb/sec = 1.818×10^{-3} HP). Run your program to determine the horsepower developed by (a) a machine with an efficiency of 0.76 when the power supplied is 723.69 ft-lbs/sec, and (b) a machine with an efficiency of 0.54 when the power supplied is 12.05 HP.

2.29 The force developed by a stretched spring having one fixed end is $F = Kx$, where K is the spring constant and x is its displacement. Write a BASIC program to determine and print the force developed by the spring in *pounds*, after the user enters the spring constant in newtons per meter and the displacement in *inches* (1 newton = 0.225 lb; 1 inch = 2.54×10^{-2} meters). Run your program for (a) K = 0.063 N/m, x = .25 inches; and (b) K = 3.779 N/m, $x = 10^{-4}$ ft.

CHAPTER 3

MATHEMATICAL OPERATIONS

1 MATHEMATICAL OPERATORS

In Chapter 2, we used the standard BASIC symbols for addition (+), subtraction (−), and multiplication (∗) in several introductory programs. These are examples of mathematical *operators*. They are called operators because they can be used with constants, variables, or combinations of constants and variables, to produce mathematical operations between those quantities. For example, the multiplication operator (∗) can be applied to the constant 5 and the variable Y to produce 5∗Y, the product obtained when 5 is multiplied by whatever value Y may have. Y∗5 is of course the same as 5∗Y. Another mathematical operator, division, is represented in BASIC by a slash (/). Thus, A/2 means the quotient obtained when A is divided by 2.

Parentheses are used in BASIC in the same way that they are used in algebra, to show the order in which mathematical operations are performed. For example, we write

```
(A+B)/2
```

when we want to divide the sum of A and B by 2. If we did not use parentheses, but instead wrote

```
A+B/2
```

then BASIC would interpret this to mean "add *A* to one half of *B*." In the example (A + B)/2, the addition is performed *first* (A + B), and the division is performed second. In the example A + B/2, the division is performed first (B/2) and the addition second. Thus you can see that parentheses affect the order of the mathematical operations. As another example, suppose we wrote

```
A+B*2-C
```

If $A = 1, B = 2$, and $C = 3$, BASIC would evaluate this expression as $1 + (2)(2) - 3 = 1 + 4 - 3 = 2$. However, suppose we inserted parentheses as follows:

```
(A+B)*2-C
```

In this case, BASIC would compute $(1 + 2)(2) - 3 = 3(2) - 3 = 3$. Finally, if parentheses were inserted in the original expression like this

```
A+B*(2-C)
```

BASIC would compute $1 + 2(2 - 3) = 1 + 2(-1) = -1$.

It is important to remember that parentheses *cannot* be used to imply multiplication, as they do in algebra. For example, if we want to multiply 5 times the sum of A and B, we *must* write

```
5*(A+B)
```

If we wrote 5(A + B), BASIC would generate an error message.

EXERCISES

3.1 Find the value that BASIC would compute for each of the following, when $A = 2$, $B = 4, X = 1$, and $Y = 5$.

(a) `Y/(B+X)` (e) `(X-B)/(Y*B)`
(b) `A*(B-A)` (f) `(B*B)*(X*2)`
(c) `(X+Y-1)/10` (g) `5/(2-Y)`
(d) `A+(3/A)` (h) `(A*A-B)/Y`

3.2 Write the BASIC expression that is equivalent to each of the following algebraic expressions:

(a) $(5 - X)(Y + 2)$ (e) $V/(BC)$
(b) $(A + B)^2$ (f) $M(N - 1)$
(c) $23Z$ (g) $(G - 12)R$
(d) $6XY$

---

2 EXPONENTIATION

Exponentiation means raising to a power. The following expressions each show an example of exponentiation in ordinary algebra: X^2, $(A - B)^{1/2}$, $W^{4.5}$, $(Z - 10)^{-3}$, 10^X, X^y. In some versions of BASIC, the exponentiation operator is shown by the symbol **. In these versions, we would express X^2 as X**2, $(A - B)^{1/2}$ as (A − B)**0.5, 10^X as 10**X, and so forth. Symbols used in several different computers to show exponentiation are shown in Figure 3.1.

Apple	Atari	Commodore	IBM/PC	TRS-80	Xerox
∧	∧	↑	↑	↑	**

FIGURE 3.1 Symbols used to represent exponentiation in several computer models.

In this book, we will use the double asterisk ** symbol for the exponentiation operator.

As shown in the preceding examples, the power to which a number is raised can be a constant or a variable, positive or negative, and can either contain a fraction or be an integer. However, *a negative number can be raised only to an integer power.* Thus, $(-5)**3$ is permitted, but $(-5)**3.5$ is not. When a number and/or its power are represented by variables, we must be sure the program does not assign variable values that result in raising a negative number to a noninteger power. If, for example, we wrote

```
(X-7)**(Y/2)
```

and if the program assigned the value 4 to X and the value 3 to Y, the expression would then be equivalent to the illegal computation $(-3)^{1.5}$.

3 EXPRESSIONS

We have used the word "expression" somewhat loosely in our previous discussions. This is justified because the definition of an expression makes it applicable to many combinations of different kinds of quantities. We will define an expression as any valid combination of constants, variables, and mathematical operators. In its simplest form, an expression can be a single number or a single variable. Thus 25 and X are expressions. Examples of more complex expressions are: $X+3*Y$, $25**(Z-2)$, and $M/(3*T+R)$.

In Chapter 2, you saw examples in which the LET statement was used to assign the value of an expression to a variable. We wrote, for example, LET X = Y + 2. In this case, the value of the expression $Y+2$ is assigned to the variable X. In general, the quantity to the right of the equal sign can be any valid expression, but the quantity to the left of the equal sign must always be a variable.

Another valid use of an expression is in exponentiation. Either the quantity raised to a power or the power itself, or both, can be an expression. We could write, for example, $(2*A-X)**(Y-4)$. If $A = 2$, $X = 1$, and $Y = 3$, this quantity would be evaluated as $(4-1)**(3-4) = 3^{-1} = 1/3$. In later work we will give more examples showing where expressions can be used in BASIC statements.

EXERCISE

3.3 Find the value assigned to X by each of the following programs:

(a)
```
10 LET Z=64
20 LET A=Z**.5
30 LET X=Z+A
40 END
```

```
(b) 10 LET X=12
    20 LET X=(3*X)**(-.5)
    30 LET X=12/X
    40 END

(c) 100 LET R1=4E6
    110 LET R2=2*R1-R1
    120 LET X=(R1+R2)**(R1/R2)
    130 END

(d) 50 LET M=3
    60 LET M1=M*(M**2)
    70 LET X=M1**(1/M)
```

4 THE SEQUENCE OF OPERATIONS IN EVALUATION OF EXPRESSIONS

We have shown that the result obtained from the evaluation of an expression depends heavily on the order in which the operations within the expression are performed. We have also shown that we can use parentheses to prescribe the order in which we want the operations to be performed. In the absence of parentheses, the order in which operations are performed depends on the so-called *hierarchy* of the operators. The hierarchy specifies the priority that each operator has in the overall sequence of operations within an expression; thus, operations with a higher priority are performed before those with a lower priority. The following list shows the hierarchy of BASIC operators in order of priority (the first has highest priority):

**	exponentiation
*, /	multiplication and division
+, −	addition and subtraction

Note that multiplication and division have the same priority, and so do addition and subtraction. In expressions containing two operators with the same priority, BASIC assigns the higher priority to the leftmost operator—that is, operators of equal priority are processed from *left to right*.

These concepts are best explained by example. We will show how BASIC evaluates each of the following expressions when $A = 1$, $B = 2$, $X = 3$, and $Y = 4$.

(1) **Y/B + A** Since division has a higher priority than addition, the operation $Y/B = 4/2 = 2$ is performed first. Therefore, $Y/B + A = 2 + 1 = 3$. Contrast this result with what would have resulted if the addition had been performed first: $B + A = 3$, $Y/3 = 4/3$.

(2) **X/B*Y** Since division and multiplication have the same priority and since the division operator appears to the left of the multiplication operator, the division $X/B = 3/2$ is performed first. Thus, $X/B*Y = (3/2)4 = 6$. Contrast this result with what would have resulted if the multiplication had been performed first: $B*Y = 8$, $X/8 = 3/8$.

(3) **A − B + X** Since subtraction and addition have the same priority and since the subtraction operator appears to the left of the addition operator, the subtraction $A − B = 1 − 2 = −1$ is performed first. Thus $A − B + X = −1 + 3 = 2$. Contrast this result with what would have resulted if the addition $B + X = 5$ had been performed first: $A − 5 = −4$.

(4) **X**B*2** Since exponentiation has a higher priority than multiplication, the operation $X**B = 3^2 = 9$ is performed first. Therefore, $X**B * 2 = 9*2$

= 18. Contrast this result with what would have resulted if the multiplication had been performed first: B*2 = 4, X**4 = 81.

As we have mentioned, parentheses can be used to override the hierarchy of the operators. In the last of the preceding examples, we could write X**(B*2), if we wanted the multiplication to be performed before the exponentiation—i.e., if we wanted to evaluate the algebraic expression X^{2B}. Note that we *can* place parentheses in the expressions in Examples 1 through 4 so that the order of operations is the same as it would be without parentheses. For example, BASIC evaluates Y/B + A the same way that it evaluates (Y/B) + A. Similarly, X/B*Y is the same as (X/B)*Y; A − B + X is the same as (A − B) + X; and X**B*2 is the same as (X**B)*2. The use of parentheses that are not required often improves the *readability* of an expression, thus making it easier to discern the order in which operations should take place. Parentheses can also be used to ensure a certain order whenever the programmer has any doubt about how the hierarchy affects an expression.

More than one set of parentheses can appear in an expression. In these cases, BASIC evaluates the contents of the *innermost* set of parentheses first. Suppose, for example, that X = 1 and Y = 2 in ((X + 2)*Y)/2. The innermost parentheses contain X + 2, so this is evaluated first: 1 + 2 = 3. The expression is then equivalent to (3*Y)/2 = 6/2 = 3. If the innermost set of parentheses had been omitted, the expression would have been evaluated as (X + 2*Y)/2 = (1 + 4)/2 = 5/2. The following examples show how multiple sets of parentheses can be used in BASIC to write some algebraic expressions:

	Algebra	BASIC
(1)	$\dfrac{x(y - 3)}{4}$	(X*(Y-3))/4
(2)	$\dfrac{a}{(3 + b)^2}$	A/((3+B)**2)
(3)	$(\sqrt{x - y^2})z$	((X-Y**2)**.5)*Z
(4)	$\dfrac{p}{3}\left[\dfrac{q}{2}(r + 4)\right]$	(P/3)((Q/2)*(R+4))

When you are writing a BASIC expression containing several sets of parentheses, always make sure that the total number of left parentheses equals the total number of right parentheses. This check does not guarantee that the expression is written correctly but, if the check fails, the expression must surely be wrong. In the BASIC expression of Example 4 above, there are a total of four left parentheses and four right parentheses.

EXERCISES

3.4 Find the value of each of the following expressions, assuming that A = 2, B = 5, G = −1, and T = 7.

(a) B/A+G
(b) A**2+B
(c) G*T/B
(d) A+T−B
(e) T/B**A
(f) A+B/2*T
(g) A**B**G
(h) T*3/B*A
(i) T/7*A/B
(j) 5/G−A**A*B

3.5 Find the value of each of the following expressions when $W = 10, X = 20, Y = 40$, and $Z = 50$:

(a) `Z-W**(W-8)`
(b) `(Y/(X*2))*W`
(c) `(Z+(3*Y))/W`
(d) `(Y/W*2)**(X/W)`
(e) `(X-Y*(-3))/7+W`
(f) `(Z-(W/X-1)/X**(W-12))*2`
(g) `W/(X+Y)/W`
(h) `Z*(W**3)*2`
(i) `(200/(W+Y)/X**2)**(1/2)`
(j) `Z-(Y-(W+(X-5)))`

3.6 Write each of the following algebraic expressions in BASIC:

(a) $a + (b - c)d/e$

(b) $\sqrt{x^2 + y^2}$

(c) $\dfrac{(4 \times 10^6)m}{p + r}$

(d) $xy^2(z + 4)^{-1}$

(e) $(f + g)/b$

(f) $\dfrac{u - v^2 + w}{(z + 1)^2}$

(g) $\{a/(r - s)\}x$

(h) $(v + w/y)^2$

(i) $0.0001\left(e - \dfrac{f}{2g}\right)$

(j) $\left(\dfrac{t + p}{r + s}\right)\dfrac{x}{y^2}$

5 THE CALCULATOR MODE

In most versions of BASIC, it is permissible to enter a statement without typing a statement number. When this is so, the statement is executed the instant that the RETURN or ENTER key is depressed. For example, if we type

```
PRINT 3**2
```

then after this statement is entered, the computer will display 9. This mode of operation is called the *calculator* mode (also known as direct mode or immediate execution mode). This mode is useful for performing the kinds of computations that a pocket calculator can do and for checking BASIC expression evaluation whenever there is any doubt about how an expression will be interpreted.

The calculator mode is particularly useful for program debugging. After a program has been run, you can examine the final values of any variables used in the program by simply entering PRINT statements containing the variable names. Suppose, for example, that you are running the following program:

```
10 LET X=10
20 LET Y=(X**2)/(X-5)
30 LET Z=Y/X
40 END
```

After the program has been run, you could type PRINT Y and PRINT Z, in the calculator mode, and the computer would display the values 20 and 2. You could also type

```
LET A=X+Z
PRINT A
```

and the computer would display 22.

In Chapter 4 we will study branching statements, including the GOTO statement. We will learn that this statement can be used to cause the computer to skip to a specified statement number and execute statements beginning with that number. The format is GOTO *sn*, where *sn* is a statement number. We mention the GOTO statement at this point because it can also be used in the calculator mode as a valuable debugging tool. By typing GOTO *sn* after a program has run, we can make the computer reexecute any portion of the program that we want to check.

6 THE REM STATEMENT

BASIC permits a programmer to insert a *remark* (comment) in a program by using the REM statement. The keyword REM can be followed by any characters, symbols, words, or statements that the programmer wants to add. Remarks are used at the beginning of a program to describe the purpose of the program, and they can be inserted anywhere in the body of a program to explain procedures or computations performed in the program.

It is important to realize that remarks do not affect the operation of a program in any way. In fact, remarks are not even displayed when a program is run. They appear only on a program *listing* (a list of all the statements in the program). Of course, the PRINT statement can be used to display a message when the program is run so, if the programmer wants to display a remark *during* a run, the PRINT rather than the REM statement should be used.

Programmers should use remarks liberally, especially if there are complex computations, expressions, or procedures in a program. Remarks help other users understand how a program works, and they are invaluable when the program must be modified by someone other than the original programmer. Also, since programs are frequently set aside for a long period of time, the programmer could easily forget the procedures used in the original construction of the program. In these cases, particularly if a program is long and involved, even the original programmer would have trouble modifying or adding to it without the benefit of remarks.

A remark can be inserted on the same line as another statement. Many versions of BASIC have a special symbol (statement delimiter) used to separate an executable statement from a remark that follows it on the same line. Also, since most versions of BASIC permit more than one statement on a line, the statement delimiter (usually a colon) can be typed between an executable statement and a REM statement occurring on the same line. If no special remark symbol is used, then the REM keyword must be used to begin the remark statement. In ANSI standard BASIC, an exclamation mark (!) can be used in place of the REM keyword.

Following is a program that shows correct use of the REM statement:

```
10 REM     THIS PROGRAM CONVERTS DISTANCE IN MILES
20 REM     TO DISTANCE IN FEET AND TO DISTANCE
30 REM     IN INCHES. USER ENTERS DISTANCE IN MILES.
40 PRINT "ENTER DISTANCE IN MILES"
50 INPUT M
60 LET F=M*5280      :REM CONVERT MILES TO FEET
70 LET I=F*12        :REM CONVERT FEET TO INCHES
80 PRINT "THE DISTANCE IN FEET IS:"
90 PRINT F
100 PRINT "THE DISTANCE IN INCHES IS:"
110 PRINT I
120 END
```

7 PROGRAMMING EXAMPLES

Technologists use computers to solve problems involving equations. The solutions are usually in the form of numbers showing the value(s) of an "unknown" variable, when the values of the other variables that affect it are known. When you set up this sort of a problem for solution by computer, it is important to realize that you, the programmer, must perform any algebra necessary to obtain an *explicit* expression for the unknown: *The computer cannot solve algebraic equations*. For example, if you could somehow enter the equation $2x^2 - 128 = 0$ into a computer, you could not expect the machine to find the value of x that satisfies the equation. Computers do not "solve" equations in that sense. Instead, you must first obtain an expression for x and then write a program that evaluates that expression. In the preceding example, you would perform the algebraic calculation necessary to arrive at the expression $x = \pm \sqrt{128/2}$ and then write a program that performs that calculation.

In this simple example, the equation can be solved easily by using a calculator ($x = \pm \sqrt{64} = \pm 8$) and there is no practical reason for using a computer. You can capitalize on the power a computer gives you when you write a program that solves any equation of the general *form* given in the example—namely, the form

$$Ax^2 - B = 0$$

where A and B are both positive constants. Solving this equation algebraically for x, we find

$$x = \pm \sqrt{B/A}.$$

You can then write a program to calculate the value of x when any positive values for the constants A and B are supplied by the user. Thus, you could write:

```
10 INPUT A
20 INPUT B
30 LET X1=(B/A)**.5
40 LET X2=-X1
50 PRINT "THE SOLUTIONS ARE:"
60 PRINT X1
70 PRINT X2
80 END
```

We can see that the availability of a computer in no way reduces the level of mathematical skill that a programmer must have to use it effectively for solving algebraic problems. Indeed, it is usually more difficult for beginning students to solve equations with nonnumeric coefficients (and thus obtain the general form necessary for computer solution) than it is for them to obtain a numeric solution for a specific equation. As another example, most technology students can readily solve the equation

$$3y^2 - 12 = 2y^2 + 4$$

by writing

$$y^2 - 2y^2 = 12 + 4$$
$$y^2 = 16$$
$$y = \pm 4$$

In practice, the general form of this equation might be

$$Qy^2 - A^2 f = xy^2 + M$$

In this case, the programmer must be able to solve for y in the form

$$y = \pm \sqrt{\frac{M + A^2 f}{Q - x}}$$

so that he or she could write the BASIC expression necessary to calculate y when the user supplies values for M, A, f, Q, and x.

The following example shows how to program a computer to solve the equation $Ax^2 + Bx + C = 0$.

Example 3.1 The roots of the quadratic equation $Ax^2 + Bx + C$ are given by the quadratic formula

$$x = \frac{-B \pm \sqrt{B^2 - 4AC}}{2A}$$

Figure 3.2 shows a BASIC program to find and print the roots, when the user enters values for A, B, and C. Note that the program cautions the user to enter values that satisfy the relation $B^2 > 4AC$; otherwise the value of $B^2 - 4AC$ would be negative, and it would thus be impossible to calculate its square root. Note that in line 40 the INPUT statement is used for entering three variable values (A, B, and C). When the ? prompt is displayed, the user enters values separated by commas. Note also that the PRINT statement in line 90 is used to print two variable values ($X1$ and $X2$) on the same line. The variable names are separated by commas. These uses of the INPUT and PRINT statements, and other variations in their formats, will be covered in Chapter 4.

Figure 3.3 on page 42 shows the results of three runs of the program in Figure 3.2. In the first run, the roots of the equation $3x^2 + 57x + 252$ are found to be -7 and -12. In the second run, the roots of $4x^2 + 72x + 324$ are both found to be -9 (in this case $B^2 - 4AC = 0$). The results of the third run show what happens when the condition $B^2 > 4AC$ is violated. In this run, we let $A = 2$, $B = 3$, and $C = 4$; the error message printed by the Xerox Sigma IX (the computer used for this example) is seen to be: NEG BASE TO NON-INTEGER POWER.

```
10 REM   THIS PROGRAM FINDS THE ROOTS OF AX**2+BX+C
20 REM   BY SOLVING THE QUADRATIC FORMULA.
30 PRINT "ENTER A,B,C. B**2 MUST BE > 4AC."
40 INPUT A,B,C
50 LET R=((B*B-4*A*C)**.5)/(2*A)
60 LET X1=-B/(2*A)+R
70 LET X2=-B/(2*A)-R
80 PRINT "THE ROOTS ARE:"
90 PRINT X1,X2
100 END
```

FIGURE 3.2 Program for Example 3.1

```
ENTER A,B,C.    B**2 MUST BE > 4AC.
?3              57              252
THE ROOTS ARE:
-7.00000        -12.0000

ENTER A,B,C.    B**2 MUST BE > 4AC.
?4              72              324
THE ROOTS ARE:
-9              -9

ENTER A,B,C.    B**2 MUST BE > 4AC.
?2              3               4
NEG BASE TO NON-INTEGER POWER
```

FIGURE 3.3 The results of three runs of the program in Figure 3.2

Example 3.2 The displacement s of a body having an initial velocity of v_0, after it has traveled for time t with an acceleration a, is $s = v_0 t + \frac{1}{2} at^2$. Figure 3.4 shows a program to find the displacement after the user enters the time of travel, the initial velocity, and the acceleration. Note that the units of the result will be consistent with the units in which the time, velocity and acceleration are entered. For example, if time is in seconds, velocity in meters per second, and acceleration in meters/sec^2, the displacement will then have the unit of meters.

Figure 3.5 shows the results of two runs of the program in Figure 3.4. In the first run, $v_0 = 0$, $a = 32.2$ ft/sec^2, and $t = 2$ seconds. The displacement is therefore equal to the distance traveled in 2 seconds by a free-falling body under the influence of gravity, after having been released from rest. We see that this distance is 64.4 feet. In the second run, $v_0 = 25$ m/sec, $a = -5$ m/sec^2, and $t = 12$ seconds. In this case the body is decelerating (since the acceleration is negative) and we see that its displacement is -60 m. This result means that the body decelerated until it came to rest, then reversed direction, and after 10 seconds was located 60 meters "behind" its original position.

Example 3.3 Given the coordinates (x_1, y_1) and (x_2, y_2) of two points on a straight line, the distance between the points is $d = \sqrt{(x_1 - x_2)^2 + (y_1 - y_2)^2}$ and the slope of the line is $m = (y_2 - y_1)/(x_2 - x_1)$. Figure 3.6 shows a BASIC program to find the distance between two points on a line and the slope of the line, when the coordinates of the points are entered by the user.

```
10  REM     THIS PROGRAM FINDS THE DISTANCE S THAT A BODY
20  REM     TRAVELS IN TIME T GIVEN THE INITIAL VELOCITY
VO
30  REM     AND ACCELERATION A: S=(VO)T+.5A(T**2).
40  PRINT "ENTER INITIAL VELOCITY"
50  INPUT V
60  PRINT "ENTER ACCELERATION"
70  INPUT A
80  PRINT "ENTER TIME OF TRAVEL"
90  INPUT T
100 LET S=V*T+.5*A*T**2
110 PRINT "THE DISTANCE TRAVELED IS:"
120 PRINT S
130 END
```

FIGURE 3.4 Program for Example 3.2

```
ENTER INITIAL VELOCITY
?0
ENTER ACCELERATION
?32.2000
ENTER TIME OF TRAVEL
?2
THE DISTANCE TRAVELED IS:
64.4000

ENTER INITIAL VELOCITY
?25
ENTER ACCELERATION
?-5
ENTER TIME OF TRAVEL
?12
THE DISTANCE TRAVELED IS:
-60
```

FIGURE 3.5 The results of two runs of the program in Figure 3.4

Figure 3.7 on page 44 shows the results of two runs of the program in Figure 3.6. In the first run, $(x_1,y_1) = (3,8)$ and $(x_2,y_2) = (5,12)$. We see that the distance between the points is 4.47214, and the slope of the line joining them is 2. In the second run, $(x_1,y_1) = (-4,3)$ and $(x_2,y_2) = (0,-1)$. In this case, the distance between the points is 5.65685 and the slope is -1. A negative slope means that the line has a graph that goes in a downward direction from left to right.

Example 3.4 The mass of a body depends on its velocity according to the equation

$$m = m_0 / \sqrt{1 - (v/v_c)^2}$$

where m_0 is the rest mass (the mass at zero velocity), v is the velocity of the body, and v_c is the speed of light. (Note that the mass becomes infinitely large as the velocity

```
10  REM    THIS PROGRAM FINDS THE DISTANCE D BETWEEN TWO
20  REM    POINTS (X1,Y1) AND (X2,Y2) ON A STRAIGHT LINE
30  REM    AND THE SLOPE M OF THE LINE.
40  PRINT "ENTER X-COORDINATE OF FIRST POINT"
50  INPUT X1
60  PRINT "ENTER Y-COORDINATE OF FIRST POINT"
70  INPUT Y1
80  PRINT "ENTER X-COORDINATE OF SECOND POINT"
90  INPUT X2
100 PRINT "ENTER Y-COORDINATE OF SECOND POINT"
110 INPUT Y2
120 LET D=((X1-X2)**2+(Y1-Y2)**2)**.5
130 LET M=(Y2-Y1)/(X2-X1)
140 PRINT "THE DISTANCE IS:"
150 PRINT D
160 PRINT "THE SLOPE IS:"
170 PRINT M
180 END
```

FIGURE 3.6 Program for Example 3.3

```
ENTER X-COORDINATE OF FIRST POINT
?3
ENTER Y-COORDINATE OF FIRST POINT
?8
ENTER X-COORDINATE OF SECOND POINT
?5
ENTER Y-COORDINATE OF SECOND POINT
?12
THE DISTANCE IS:
4.47214
THE SLOPE IS:
2

ENTER X-COORDINATE OF FIRST POINT
?-4
ENTER Y-COORDINATE OF FIRST POINT
?3
ENTER X-COORDINATE OF SECOND POINT
?0
ENTER Y-COORDINATE OF SECOND POINT
?-1
THE DISTANCE IS:
5.65685
THE SLOPE IS:
-1
```

FIGURE 3.7 The results of two runs of the program in Figure 3.6

approaches the speed of light.) Figure 3.8 shows a BASIC program to compute the mass of a body in kilograms, after the user enters the rest mass in kilograms and the velocity of the body in meters per second.

Figure 3.9 shows the results of two runs of the program in Figure 3.8. In the first run, the rest mass is that of an average man (73 kg) and the velocity is 1 million miles per hour (4.47×10^5 m/sec). We see that the man's mass increases to 73.0001 kg. In the second run, we compute the mass of a proton (the rest mass is 1.67×10^{-27} kg) when traveling at 99 percent of the speed of light (2.97×10^8 m/sec). We see that the mass is $1.18383E-26$ kg, an increase of 41 percent.

```
10 REM    THIS PROGRAM FINDS THE MASS OF A BODY GIVEN
20 REM    ITS VELOCITY AND ITS REST MASS.
30 PRINT "ENTER REST MASS IN KG"
40 INPUT M
50 PRINT "ENTER VELOCITY IN M/SEC"
60 INPUT V
70 LET C=3E8
80 LET M1=M/((1-(V/C)**2)**.5)
90 PRINT "THE MASS IN KG IS:"
100 PRINT M1
110 END
```

FIGURE 3.8 Program for Example 3.4

MATHEMATICAL OPERATIONS

```
ENTER REST MASS IN KG
?73
ENTER VELOCITY IN M/SEC
?447000
THE MASS IN KG IS:
73.0001

ENTER REST MASS IN KG
?1.67000E-27
ENTER VELOCITY IN M/SEC
?297000000
THE MASS IN KG IS:
1.18383E-26
```

FIGURE 3.9 The results of two runs of the program in Figure 3.8

EXERCISES

3.7 To test your ability to write expressions correctly, write each of the following as a BASIC expression and evaluate it using the calculator mode of your computer. The results you should obtain after entering your expressions are shown on the right in parentheses.

(a) $\dfrac{\sqrt{4^2 + 3^2}}{5}$ (1)

(b) $\dfrac{(3)(2 + 4)^2}{(2^2 + 5)}$ (12)

(c) $\sqrt[4]{4^3 + \sqrt{200}}$ (2.97318)

(d) $\dfrac{(3.29)(7.41)}{(1.09)(4.66)}$ (4.79956)

(e) $\dfrac{(39.36 - 12)^{1/5}}{(\sqrt{7})(8 - 4.05)}$ (0.185471)

(f) $\dfrac{10^6 - (5 \times 10^5 - 20 \times 10^4)}{10^7(1.55 - 3.92)}$ ($-2.95359\text{E}-2$)

(g) $\dfrac{(2.025)^3(2^7 - 11)}{13.88 - (4.709)^2}$ (-117.128)

3.8 If the velocity of a body with mass m changes from v_1 to v_2, the total change in its kinetic energy is $\Delta KE = \left(\tfrac{1}{2}\right)m(v_1^2 - v_2^2)$, (neglecting any change in mass caused by the change in velocity). Write a BASIC program to find and print the change in kinetic energy, after the user enters the mass and velocities v_1 and v_2. Use your program to find (a) the change in the kinetic energy of an average man (73 kg) going from rest to the speed of a typical jet aircraft (720 m/sec). Also find (b) the change in kinetic energy of a comet with a mass of 2×10^4 kg when it strikes the earth with a velocity of 500 m/sec.

3.9 The square root of the sum of the squares of all integer numbers from N_1 to N_2 is

$$\sqrt{\frac{(N_2)(N_2 + 1)(2N_2 + 1) - (N_1 - 1)(N_1)(2N_1 - 1)}{6}}$$

For example, if $N_1 = 2$ and $N_2 = 4$, then $\sqrt{2^2 + 3^2 + 4^2} = \sqrt{29} = 5.385$. Write a BASIC program to find the square root of the sum of the squares from N_1 through N_2 after the user enters N_1 and N_2. Include a warning that both N_1 and N_2 must be positive integers and that N_2 must be greater than N_1. Run your program for (a) $N_1 = 1$, $N_2 = 10$ and (b) $N_1 = 5$, $N_2 = 50$.

3.10 If a rock is thrown downward from a height x with a velocity v_0, then the velocity with which it strikes the earth is

$$v = \sqrt{v_0^2 + 2gx}$$

where g is the gravitational acceleration (9.81 m/sec^2). Write a BASIC program to determine and print the impact velocity v in meters per second after the user enters the height x and velocity v_0. Run your program to determine (a) the impact velocity when $v_0 = 5$m/sec and the height is 33 meters. Also determine (b) the impact velocity when the rock is dropped (not thrown) from the top of the Washington Monument (169.2 meters high).

3.11 When $a < 1$, a good approximation for the quantity e^{-a}/b, where e is the base of the natural logarithm (approximately 2.72), is:

$$\frac{e^{-a}}{b} \simeq \frac{1}{b} - \frac{a}{b} + \frac{a^2}{2b} - \frac{a^3}{6b}$$

Write a BASIC program to find and print an approximate value for e^{-a}/b after the user enters values for a and b. Run your program to find approximations of the following:

(a) $\dfrac{e^{-0.5}}{2}$ (b) $e^{0.1}$

EXERCISES FOR ARCHITECTURAL/CIVIL/CONSTRUCTION TECHNOLOGY

3.12 The stress on a rod having a diameter of d is

$$\sigma = \frac{P}{(\pi d^2/4)}$$

where P is the axial load and d is the diameter of the rod. The units of σ are consistent with those of P and d: If P is in pounds and d in inches, then σ will be in pounds per square inch. Write a BASIC program to determine and print the stress after the user enters the load P and diameter d. Run your program for each of the following cases: (a) $P = 4900$ lbs, $d = 0.505$ inches; and (b) $P = 45$ kN, $d = 25$ mm.

3.13 In uniform, cohesionless soil, the settlement of a foundation having a width of less than 20 feet is estimated by

MATHEMATICAL OPERATIONS

$$H = \frac{4pB^2}{K(B + 1)^2}$$

where H is the settlement in feet,
           B is the width of the foundation in feet,
           p is the pressure imposed by the foundation in kips per square foot, and
           K is the modulus of vertical subgrade reaction for a one-square-foot plate bearing on the ground surface, in kips per cubic foot.

Write a BASIC program to compute and print this estimate of the settlement, after the user enters values for p, K, and B. Run your program to estimate (a) the settlement in dense sand ($K = 400$ kips/ft^3) of a 6-foot wide foundation having a soil bearing pressure of 8000 lbs/ft^2. Run your program again to estimate (b) the settlement in loose sand ($K = 100$ kips/ft^3) of an 8-foot wide foundation having a soil bearing pressure of 6 kips/ft^2.

3.14 The sag correction in surveyor's tape is given by

$$C = \frac{W^2 L}{24 P^2}$$

where C is the correction,
           W is the weight of the tape between supports,
           L is the length of the interval between supports, and
           P is the tension in the tape.

The units of C are consistent with those of W, L, and P. Write a BASIC program to determine the correction factor after the user enters values of W, L, and P. Run your program to find (a) the correction in a 100-foot steel tape weighing 2 lbs and supported at each end, when the tension is 12 lbs. Run your program again to determine (b) the correction in a 30-meter steel tape weighing 0.336 kg, when it is supported at the 0- and 15-meter points under a tension of 5 N.

EXERCISES FOR ELECTRICAL/ELECTRONICS/COMPUTER TECHNOLOGY

3.15 The force between two charges is given by

$$F = \frac{kQ_1 Q_2}{r^2}$$

where F is the force in newtons,
           $k = 9 \times 10^9$,
           Q_1 and Q_2 are the magnitudes of the charges in coulombs, and
           r is the distance between the charges in meters.

Write a BASIC program to determine and print the force between two charges, after the user enters values for Q_1, Q_2 and r. Run your program to find (a) the force when $Q_1 = 25$ μcoulombs, $Q_2 = 100$ μcoulombs, and $d = 0.03$ m. Run your program again to determine (b) the force when $Q_1 = Q_2 = 1$ coulomb and $d = 15$ mm.

3.16 The power dissipated in the resistor R_2 of the series circuit shown in Figure 3.10 is

$$P = \frac{\left[\left(\frac{R_2}{R_1 + R_2}\right)E\right]^2}{R_2}$$

where R_1 and R_2 are in ohms,
        E is the applied voltage in volts, and
        P is the power in watts.

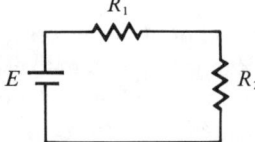

FIGURE 3.10   Exercise 3.16

Write a BASIC program to determine and print the power P after the user enters values for E, R_1, and R_2. Run your program to find the power in each of the following cases: (a) $E = 12V$, $R_1 = 1$ kΩ, $R_2 = 470$ Ω; and (b) $E = 100V$, $R_1 = 330$ Ω, $R_2 = 220$ kΩ.

3.17 The total equivalent resistance of three parallel-connected resistors is

$$R_T = \frac{R_1 R_2 R_3}{R_1 R_2 + R_1 R_3 + R_2 R_3}$$

Write a BASIC program to find and print R_T after the user enters values for R_1, R_2, and R_3. Run your program for each of the following cases: (a) $R_1 = R_2 = R_3 = 33$ kΩ and (b) $R_1 = 1$ Ω, $R_2 = 100$ Ω, $R_3 = 10$ k Ω.

EXERCISES FOR INDUSTRIAL/PRODUCTION/MANUFACTURING TECHNOLOGY

3.18 The upper control limit on a control chart for the fraction of defective units in a production process is

$$\text{UCL} = p + 3\sqrt{\frac{p(1-p)}{n}}$$

where n is the sample size and
        p is the proportion of defective units in the population.

Write a BASIC program to determine and print the upper control limit after the user enters values for p and n. Run your program for the following cases: (a) $p = 0.02$, $n = 100$; and (b) $p = 0.04$, $n = 500$.

3.19 The economic lot size in inventory control theory is given by

$$Q = \sqrt{\frac{2CB}{E(1 - C/R)}}$$

where C is the consumption rate per day,
       B is the ordering cost per lot,
       E is the carrying cost per unit per day, and
       R is the delivery rate per day.

Write a BASIC program to determine and print the economic lot size when the ordering cost is $10 per lot, after the user enters values for C, E and R. Run your program for each of the following cases:

(a) $C = 600$ units/day, $E = \$0.001$ per unit per day, and $R = 1000$ units per day
(b) $C = 250$ units/day, $E = 0.2¢$ per day, and $R = 400$ units/day.

3.20 The total volume of metal removed from a steel shaft having a diameter of d_1, when it is machined to diameter d_2 is

$$V = \frac{l\pi(d_1 - d_2)^2}{4}$$

where l is the length of the shaft. The units of V are consistent with those of l, d_1, and d_2. Write a BASIC program to determine and print the volume of metal removed, after the user enters values for l, d_1, and d_2. Run your program for each of the following cases: (a) $l = 36$ inches, $d_1 = 0.1$ inch, $d_2 = 0.075$ inch; and (b) $l = 0.8$ meter, $d_1 = 50$ mm, $d_2 = 38$ mm.

EXERCISES FOR MECHANICAL TECHNOLOGY

3.21 The polar moment of inertia for a hollow circular shaft is given by

$$J = \frac{\pi(d_o^4 - d_i^4)}{32}$$

where d_o is the outer diameter, and d_i is the inner diameter. The units of J are consistent with those of d_o and d_i. Write a BASIC program to determine and print J after the user enters values for d_o and d_i. Run your program to find J (a) when $d_o = 0.1$ m, $d_i = 0.09$ m; and (b) when $d_o = 0.595$ inches, $d_i = 0.42$ inches.

3.22 The torque required to cause slippage of a hollow shaft in contact with a fixed surface is

$$M = f_k P \frac{2(R_o^3 - R_i^3)}{3(R_o^2 - R_i^2)}$$

where M is the torque,
       f_k is the kinetic coefficient of friction,
       P is the axial force,
       R_o is the outer radius of the shaft, and
       R_i is the inner radius of the shaft.

The units of M are consistent with those of P, R_o, and R_i. See Figure 3.11.

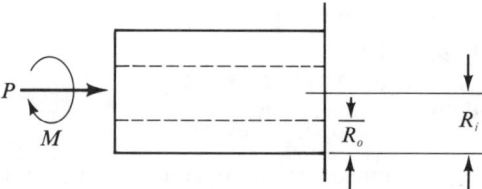

FIGURE 3.11 Exercise 3.22

Write a BASIC program to determine and print the torque required for slippage of a metal shaft on a wood surface ($f_k = 0.25$) after the user enters values for P, R_o, and R_i. Run your program for (a) $P = 1600$ lbs, $R_o = 2.75$ inches, $R_i = 1.5$ inches; and (b) $P = 3.25$ kN, $R_o = 48$ mm, $R_i = 16$ mm.

3.23 The (damped natural) frequency of the vibration of a spring-mass-damper system is given by

$$\omega_d = \frac{1}{2}\sqrt{\frac{4K}{M} - \left(\frac{B}{M}\right)^2}$$

where ω_d is the vibration frequency in radians per second,
K is the spring-constant in newtons per meter,
B is the damping in newtons per meter per second, and
M is the mass in kilograms.

Write a BASIC program to determine and print the vibration frequency after the user enters values for K, B, and M. Run your program to find ω_d for each of the following cases: (a) $K = 12$ N/m, $M = 1$ kg, $B = 6$ N/m/sec; and (b) $K = 105$ N/m, $M = 0.5$ kg, $B = 4$ N/m/sec.

CHAPTER 4

INPUT/OUTPUT

1 SYSTEM COMMANDS

We have already discussed two fundamental input/output statements in BASIC: the INPUT statement, used for supplying input data interactively; and the PRINT statement, used for output. We will examine these statements in detail and study some variations of them later in this chapter. We now propose to introduce a different kind of I/O operation, one that is accomplished by use of *system commands*.

Although system commands are available for operations other than input/output operations, the most important ones are used for storing (saving) programs and loading (retrieving) previously stored programs. It is convenient to think of the computer as providing us with a temporary *workspace* where we can develop BASIC programs. While in this workspace, we can change statements at will, correct syntax errors, perform test runs, and so forth, until we are satisfied with the program's performance. If the power is then turned off, or in the case of a large, time-shared system, if we "log off," then the workspace and the program it contains will be erased. To save the program for future use, we must use a system command to store it in some form of external memory, usually either tape or disk. This process is an output operation, since we send information from the workspace to an output storage device. Similarly, system commands are used to retrieve previously stored programs—that is, to input programs from a storage device to the workspace. A program retrieved this way can be altered like any other program in the workspace, but the original program remains in storage and is unaffected by any such alterations.

System commands are different from BASIC statements, since they are not written as a part of a program. They have no effect on the structure of a program or the computations it performs. Instead, system commands are typed in directly from the keyboard and are executed as soon as the RETURN key is depressed. Of course, no statement numbers are required.

System commands are machine-dependent—that is, their format depends on the computer system being used. In many systems, the words SAVE and LOAD, or slight variations of these words, are used as commands for storing and retrieving programs. For example, in some microcomputer systems the command CLOAD means "load a program from cassette tape." Programs are often assigned names so they can be stored along with other programs having different names and then later retrieved by name. In a typical time-shared system, the command SAVE ON *name* (where *name* is a name assigned to the program by the programmer) will cause the computer to store the program (on disk) under that name. The command LOAD *name* will cause the program having that name to be transferred to the workspace. If the programmer modifies the program while it is in the workspace and then wishes to replace the original version with the new one (under the same name), a command such as SAVE OVER *name* can be used. This command erases the old program having that name and replaces it with the new one. Note that different systems impose different restrictions on the number and kinds of characters that can be used to form a program name. Consult your programming manual to determine the restrictions that apply to your particular system.

Another important system command is LIST. This command is used to obtain a line-by-line display of the current program (the one in the workspace). Most systems also permit the listing of a statement at a specified statement number or all statements in a specified range of statement numbers.

Table 4.1 summarizes the most important system commands for some currently popular computers. This table is intended as a convenient reference, but it does not include every command, nor every option and variation of every command available for the computers shown. Consult the programming manual for your computer for additional details. If an entry appears inside brackets { }, it is optional.

2 MORE ABOUT THE PRINT STATEMENT

We have already shown how the BASIC print statement can be used to display the value of a variable or to display a string. Recall that you need only write the statement PRINT X to obtain a display of the value of X, or to PRINT "string," where "string" is a string that you want to be displayed. In many programming problems, you will want to display more than one variable value or string, or combinations of values and strings, on a single line. You might, for example, want to create a table of values containing several entries on each line, with several strings used as headings across the top. In these cases, you must somehow control the *spacing* between the displayed items.

A semicolon inserted between consecutive items in a PRINT statement signifies that these items are to be printed on the same line with no spaces between them. Following is an example:

```
10 LET X=4
20 PRINT "X=";X;"=X"
30 END
```

When this sequence is run, the PRINT statement causes the following display:

```
X= 4 =X
```

Note that a leading space is automatically inserted in front of a positive number. (In some versions, including Apple and TRS-80, a trailing blank is also automatically

Table 4.1 System Commands for Various Popular Computers

| \multicolumn{2}{c}{APPLE (APPLESOFT II)} ||
Command	Description
CLEAR	Sets all numeric variables to zero and all string variables to null
CONT	Causes execution of a program to resume after being halted by a STOP, END, or Control C
Control C	Interrupts program execution
DEL n_1, n_2	Deletes lines n_1 through n_2 from a program
FRE(0)	Displays the number of bytes of memory still available to the user
LIST	Causes the entire program to be displayed
LIST n_1, n_2	Causes line numbers n_1 through n_2 to be displayed
LOAD	Loads a program from cassette tape
NEW	Deletes the current program and all variables
RUN $\{n_1\}$	Begins execution of a program at line n_1—If no n_1 is specified, execution begins at the lowest line number.
SAVE	Stores a program on cassette tape
SPEED $= n$	Sets the rate at which characters are sent to the video display or other I/O device (n is between 0 and 255)
TRACE	Causes the line number of each statement to be displayed as it is executed

| \multicolumn{2}{c}{ATARI 400/800} ||
Command	Description
BYE	Causes an exit from BASIC and allows use of the keyboard without disturbing any programs in memory
CLOAD	Loads a program from cassette tape
CONT	Causes program execution to resume after a STOP, END, or break
CSAVE	Saves a program on cassette tape
ENTER	Causes a cassette tape to play back a program originally recorded, using LIST
LIST	Displays the entire program
LIST n_1, n_2	Displays line numbers n_1 through n_2
LIST "P"	Prints the program on a line printer
LIST "P", n_1, n_2	Prints line numbers n_1 through n_2 on a line printer
LIST "filename"	Stores a program on cassette tape under the designation "filename" (quotation marks must be used)

(continued)

Table 4.1 System Commands for Various Popular Computers (continued)

Command	Description
LOAD *"filename"*	Loads a program designated *"filename"* (quotation marks must be used)
RUN {*"filename"*}	Causes the current program to be executed if no filename is specified—Otherwise, retrieves and runs specified file (quotation marks must be used).
SAVE *"filename"*	Saves a program designated *"filename,"* which must be enclosed in quotation marks

COMMODORE (CBM and PET)

Command	Description
CLR	Sets all numeric variables to zero and all string variables to null
CONT	Resumes program execution after a STOP or END statement has been executed, or after the stop key has been depressed
DLOAD *"filename"*	Loads a file named *"filename"* from disk
DSAVE *"filename"*	Saves a file named *"filename"* on disk
LIST	Displays the entire program
LIST n_1–n_2	Displays line numbers n_1 through n_2
NEW	Deletes the current program and clears all variables
RENAME *"oldname"* TO *"newname"*	Changes the name of a disk file
RUN {*n*}	Runs a program beginning at line *n* or the lowest line number, if *n* is not specified
SAVE {*"filename"*}	Saves a program on cassette tape or disk—If *"filename"* is used, it must be enclosed by quotation marks.
VERIFY {*"filename"* {*,device*}}	Compares the contents of the program in memory to a file on disk or tape and reports differences—*Device* defaults to cassette number 1; *filename* defaults to null

HEATH HDOS

Command	Description
AUTO {*initial sn, incr*}	Causes automatic numbering of statements, beginning with initial *sn* and using *incr* as increment—If neither of these are specified, it is assumed that both equal 10.

Table 4.1 System Commands for Various Popular Computers (continued)

Command	Description
CLEAR	Sets all program variables to zero
CONT	Causes program execution to resume after having ceased due to the execution of a STOP or END statement or a CONTROL/C command
CONTROL/*ch*	Simultaneous depression of the control key and a *ch* character key, resulting in one of the following actions.
CONTROL/A	Allows use of EDIT commands on the line currently being typed
CONTROL/C	Interrupts execution of a program
CONTROL/O	Suppresses all output until an INPUT statement is executed
CONTROL/Q	Causes program execution to resume after a CONTROL/S command
CONTROL/S	Causes program execution to pause until a CONTROL/Q is entered
CONTROL/U	Erases current line and executes a carriage return
FRE(0)	Causes display of the total number of bytes still available in memory
KILL *"filename"*	Deletes file named *"filename"* from disk (quotation marks must be used)
LIST	Displays the program currently in memory, statement by statement, beginning with the first statement number
LIST *sn1–sn2*	Displays all statements from statement number *sn1* to statement number *sn2*, inclusive
LIST *sn*	Lists statement with statement number *sn*
LIST *sn–*	Lists all statements from statement number *sn* through end of program
LIST–*sn*	Lists all statements from the first through statement number *sn*
LOAD *"filename"*	Loads a file named *"filename"* from disk (quotation marks must be used)
LOAD *"filename"*, R	Same as LOAD *"filename"*, except a program loaded from disk is also run
NAME *"oldname"* AS *"newname"*	Assigns the name *"newname"* to a file on disk already having the name *"oldname"*—An *"oldname"* file must exist, and there cannot already be a *"newname"* file.
NEW	Deletes program currently in memory and sets all variables to zero
RENUM	Renumbers all statement numbers in a program. The first new statement number is 10, and there is an increment of 10 between each new statement number.

(continued)

Table 4.1 System Commands for Various Popular Computers (continued)

Command	Description
RENUM sn , , incr	Renumbers all statement numbers in a program using sn as the first new statement number with increment incr between each new statement number
RENUM sn1, sn2, incr	Same as RENUM sn , , incr except that renumbering begins with the old statement number sn2
RESUME	Causes program execution to resume after an error recovery procedure has been performed—Execution resumes at the statement that caused the error.
RESUME NEXT	Same as RESUME except that execution resumes at the statement immediately following the error-causing statement
RESUME sn	Same as RESUME except that program execution resumes at statement number sn
RUN {sn}	Causes a program to be executed, beginning with statement number sn—If sn is omitted, execution then begins at the lowest-numbered statement number.
SAVE "filename"	Saves a file on disk under the name "filename" (quotation marks must be used)
TRON	Causes the statement numbers of a program to be displayed as each corresponding statement is executed
TROFF	Disables TRON—that is, eliminates display of line numbers caused by TRON
WIDTH int	Sets the width of the display on the screen of a terminal—int is the (integer) number of characters per line that are displayed; it must be less than 255.

IBM PERSONAL COMPUTER

Command	Description
AUTO	Generates line numbers automatically
CLEAR	Clears program variables
CONT	Resumes program execution after an interrupt
DELETE	Deletes a range of program lines
LIST	Displays a program or portions thereof
LLIST	Prints a program on a line printer
LOAD	Loads a program file
NEW	Deletes the current program
RUN	Loads and runs a program

Table 4.1 System Commands for Various Popular Computers (continued)

Command	Description
SAVE	Saves the current program
SYSTEM	Exits BASIC and returns control to the operating system

TRS-80	
Command	Description
AUTO	Generates line numbers automatically in increments of 10, beginning with the number 10
AUTO n_1, n_2	Generates line numbers automatically, beginning with n_1 and having an increment of n_2
CLEAR	Sets all numeric variables to zero and all string variables to null
CLEAR n	Makes n bytes available for string storage
CLOAD "*filename*"	Loads a program from cassette tape (quotation marks must be used)
CLOAD? "*filename*"	Compares a program stored on cassette with a program in the computer, and displays BAD if there are any discrepancies
CONT	Resumes program execution after a STOP instruction or after depressing the break key
CSAVE "*filename*"	Stores the current program on cassette tape under *filename* (must be enclosed by quotation marks)
DELETE n_1-n_2	Deletes line numbers n_1 through n_2
EDIT n	Puts computer in edit mode for editing line n
LIST n_1-n_2	Displays line numbers n_1 through n_2
LIST $n-$	Displays line number n and all higher line numbers
RUN $\{n\}$	Runs current program beginning at line n—If no n has been specified, the run begins at the lowest line number.
SYSTEM	Causes exit from BASIC and returns control to the operating system for loading programs

XEROX SIGMA IX	
Command	Description
Break key	Stops current operation and generates > prompt symbol—A second depression of the break key returns control to the system with the ! prompt.

(continued)

Table 4.1 System Commands for Various Popular Computers (continued)

Command	Description
CATALOG	Lists the names of all program files in the user's account
CLEAR	Sets numeric variables to zero and string variables to null
CLEAR ARRAYS	Sets only array variables to zero
CLEAR STRINGS	Sets string variables to null
DELETE n	Deletes line n
DELETE n_1-n_2	Deletes lines n_1 through n_2
DELETE *filename*	Deletes the program named *filename*
EXECUTE n_1 $\{-n_2\}$	Begins program execution at line n_1—If n_2 is specified, program execution halts at line number preceding n_2.
EXTRACT n	Deletes the entire program except line n
EXTRACT n_1-n_2	Deletes the entire program except line numbers n_1 through n_2
LIST n	Displays line n
LIST n_1-n_2	Displays line numbers n_1 through n_2
LOAD *filename*	Loads a program called *filename*
RENUMBER $\{n_1\{,n_2\{,n_3\}\}\}$	Renumbers lines—n_1 is the lowest new line number; n_2 is the lowest old line number; n_3 is the increment. n_1, n_2, and n_3 are 100, 1, and 10, respectively, by default.
RUN	Begins execution of a program at the lowest line number
STATUS	Returns EDITING, COMPILING, or RUNNING
WIDTH n	Changes the print width from 72 characters to n characters, where n is 0 to 255.

inserted after every number printed.) Semicolons can be omitted between the items in a data statement with the same results that occur above (no spacing), provided their omission does not result in an ambiguous or illegal combination of characters*. In the previous example, if statement 20 were changed to

```
20 PRINT "X="X"=X"
```

the display would then appear exactly as shown before. However, if line 20 were changed to

```
20 PRINT XX
```

the combination XX would then be interpreted as the variable name XX, to which no value has previously been assigned. In some systems, XX is an illegal variable name so this could result in an error message. (See Section 2.4).

---

*VAX 11 BASIC *always* requires a semicolon to suppress spacing.

As another illustration of spacing in a PRINT display, consider the following program:

```
10 LET X=5
20 LET Y=-X
30 LET Z=X**2
40 PRINT "X="X";MINUS X IS ";Y";X SQUARED IS "Z","
50 END
```

The display that results from running this program is

```
X=5;MINUS X IS -5;X SQUARED IS 25,
```

Note that some of the semicolons in statement 40 are enclosed within quotation marks and are therefore treated as string characters.

One important feature of BASIC, found in few other computer languages, is that an expression can appear in a PRINT statement. Consider, for example, the program

```
10 LET A=2
20 PRINT "RESULT="A**3-2*A
30 END
```

This program results in the display

```
RESULT= 4
```

Blank lines can be inserted between other lines displayed by PRINT statements by simply writing the keyword PRINT without anything following it. Here is an example:

```
10 LET W=12
20 PRINT "3W=" 3*W
30 PRINT
40 PRINT "4W= "4*W
50 END
```

This program creates the display:

```
3W= 36

4W=  48
```

Note the blank line in the display inserted as a result of statement 30. Note also the extra blank inserted in front of the number 48: This was caused by the space inside the quotation marks in statement 40.

Another way to control the spacing between items displayed on a line is by using commas in the PRINT statement. BASIC divides the display space into horizontal "zones." The width of each zone (the number of characters that can be displayed in a zone) and the total number of available zones depends on the version of BASIC. In a typical example, each line of display might be divided into 16-character zones. When commas are placed between the variable names, expressions, or strings in a PRINT statement, each item separated by a comma from the previous item is displayed at the beginning of a new zone. Consider, for example, the following program:

```
10 LET X=-12.5
20 LET Z=2*X
30 PRINT X,Z
40 END
```

Assuming 16-character zones, this program will create the following display:

```
-12.5           -25
```

In this display, since the number -12.5 represents five characters (including the minus sign and decimal point), a total of 11 ($16 - 5 = 11$) spaces are inserted before the number -25 is displayed.

Two consecutive commas can be used in a PRINT statement to skip a zone entirely. Semicolons can also be mixed with commas in a PRINT statement, as illustrated by the following example:

```
10 LET A=13
20 PRINT "A=";A,,"A**2=";A**2
30 END
```

This program generates the following display:

```
A=13                            A**2=169
```

Another way to control the spacing of displays is by using the TAB function. We can think of any display as being generated by a moving *cursor* that travels from left to right and occupies successively numbered positions as it displays each character. Using the TAB function, we can position the cursor at any point along a line. This operation is similar to setting the tab on a conventional typewriter. The desired cursor position is specified by writing it in parentheses after the keyword TAB: for example, TAB(5). The first (leftmost) position is usually numbered zero, so TAB(5) places the

```
10 PRINT "ENTER MASS IN KG."
20 INPUT M
30 PRINT "ENTER FIRST ACCELERATION"
40 INPUT A1
50 PRINT "ENTER SECOND ACCELERATION"
60 INPUT A2
70 PRINT
80 PRINT "MASS,KG" TAB(10) "ACCEL,M/SEC**2" TAB(27) "FORCE,N"
90 PRINT M; TAB(10); A1; TAB(27); M*A1
100 PRINT M; TAB(10); A2; TAB(27); M*A2
110 END
>

ENTER MASS IN KG.
?10
ENTER FIRST ACCELERATION
?1.5
ENTER SECOND ACCELERATION
?2.5

   MASS,KG    ACCEL,M/SEC**2    FORCE,N
     10         1.50000            15
     10         2.50000            25
>
```

FIGURE 4.1 Program for Example 4.1 and the results of a program run

cursor at the sixth column from the left. The cursor position given inside the parentheses can also be an integer-valued expression as, for example, TAB(3 * I − 1). One or more TAB specifications can be written on the same line as the PRINT statement; each occurrence will cause the display of the next item printed to begin at the cursor position given by the previous TAB. However, TAB cannot be used to move the cursor to the left of its current position.

Example 4.1 To illustrate the use of the TAB function, suppose that you want to create a table containing the values of force in Newtons on a mass M (in kilograms) whose value is entered by the user for two different user-entered values of acceleration A, in meters per second per second. Table headings should identify the columns containing the mass values, accelerations, computed forces, and the units of each quantity. The required calculations ($F = MA$) should be performed in PRINT statements.

Figure 4.1 shows the program listing and the results of a program run in which a mass of 10 kg was entered, along with accelerations of 1.5 m/sec^2 and 2.5 m/sec^2. Note that the same TAB values were used to space the table headings and the data entries in the table. The data entries are shifted to the right one space from the headings, because of the leading blank automatically inserted in front of positive numbers.

EXERCISES

4.1 Show the display that would result if each of the following programs were run:

(a)
```
10 LET A1=16
20 LET A2=35
30 PRINT "A1="A1;" A2="A2
40 END
```

(b)
```
10 LET X=9
20 PRINT "X**2=";X**2"X**.5="X**.5
30 END
```

(c)
```
10 LET R=2
20 PRINT R;R**2;R**3
30 PRINT
40 PRINT R**4;R**5
50 END
```

(d)
```
10 LET F4=400
20 PRINT "FORCE="F4**.5" NEWTONS"
30 PRINT
40 PRINT
50 PRINT "MASS="F4/10"KG"
60 END
```

4.2 Write BASIC programs to accomplish each of the following tasks:

(a) Assign N the value 100 and display

```
N IS 100
N/10 IS 10
```

(b) Allow the user to input a value for T and display

```
THE INPUT IS t DEGREES (C)
```

where t is the value entered by the user for T.

(c) Allow the user to input values for X and Y and display

```
THE INPUT VALUES ARE

x=X    y=Y
```

where x and y are the values entered by the user for X and Y.

(d) Allow the user to enter values for A, B, and C and then solve the equation $AX^2 + BX + C = 0$ for its roots R_1 and R_2. Display the results in the form

```
FIRST ROOT IS: r₁
SECOND ROOT IS: r₂
```

where r_1 and r_2 are the computed values of R_1 and R_2.

4.3 Show the display that would result if each of the following programs were run (assume 16-character zones):

(a)
```
10 LET X=64
20 PRINT X,X**.5,X/32
30 PRINT
40 PRINT X,,X/32
50 END
```

(b)
```
10 LET A=1E4
20 PRINT "ROOT A=",A**.5
30 PRINT "SIZE B=",A/(A-5E3)
40 END
```

(c)
```
10 LET T3=-6
20 LET V4=30
30 PRINT T3;TAB(8);V4" VOLTS"
40 PRINT TAB(8);V4-T3
50 END
```

(d)
```
10 LET X1=1
20 LET X2=2
30 LET X3=3
40 PRINT "ONE" TAB(6) "TWO" TAB(12) "THREE"
50 PRINT
60 PRINT X1;TAB(6);X2;TAB(12);X3
70 END
```

4.4 Write a BASIC program to create a table of the square roots and squares of three numbers entered by the user. Table headings should be NUMBER, ROOT, and SQUARE.

3 MORE ABOUT THE INPUT STATEMENT

We have shown how the INPUT statement is used to assign a value to a variable. A single INPUT statement can also be used to assign values to two or more variables by simply writing each variable name, separated by commas, after the INPUT keyword. When the system prompts for input, the user enters as many values as there are variable names in the INPUT statement and types a comma between each new value entered. As usual, the user actually enters the values into the computer by depressing the RETURN or ENTER key. As an example, the following display would result upon execution of an INPUT statement used to assign values to four different variables:

```
10 INPUT A1,B,X,Y

 ?5,-3,161,2E5
```

This example shows that $A1$ is assigned a value of 5, B is assigned -3, X is assigned 161, and Y is assigned 2×10^5.

A *string variable* is a variable that can be assigned strings rather than numeric values. A string variable is distinguished from a numeric variable by placing a dollar sign ($) at the end of its variable name. For example, the string variable A$ could be assigned any legitimate string, such as "YES", "NO", or "MAYBE AND MAYBE NOT". An INPUT statement can be used to assign a string to a string variable just as it is used to assign a value to a numeric variable. Consider the following sequence of statements and the display that results when they are executed:

```
10 INPUT N$
20 PRINT N$" OH "N$
30 END

 ?"BOY"
 BOY OH BOY
```

When the INPUT statement in line 10 is executed, the string variable N$ is assigned the string "BOY". (In many versions of BASIC, it is not necessary to enclose a string in quotation marks when it is entered in response to an input statement.) One INPUT statement can be used to make assignments to more than one string variable or to combinations of string and numeric variables, as in the following example:

```
INPUT B$,W,C$,Z2
```

As with numeric variable names, the number and kinds of characters that can be used to form a string variable name depends on the version of BASIC being used. Consult your programming manual for details. In this book we will always use a single letter followed by a $ sign for a string variable name; this is legitimate in any version.

In many versions of BASIC, a *prompting message* can be included as part of an INPUT statement. This message is displayed when the INPUT statement is executed and is used to remind the user to supply a value for a variable. The message must be enclosed in quotation marks, must immediately follow the INPUT keyword, and must be separated from the variable name by a semicolon. Following is an example:

```
10 INPUT "ENTER VELOCITY";V
```

When this statement is executed, the display looks like this:

```
ENTER VELOCITY?
```

This use of the INPUT statement eliminates the need for a separate PRINT statement containing a prompting message.

EXERCISES

4.5 Write one BASIC statement that can be used to assign values to the variables A, B, A$ and B$.

4.6 Show the display that would result if the following program were executed, assuming that the user types 25 when prompted by statement 10 and "MEN" when prompted by statement 20.

```
10 INPUT "NUMBER OF PEOPLE";N
20 INPUT "SEX";S$
30 PRINT "THERE ARE"N" "S$" IN THIS CLASS."
40 END
```

4 THE READ AND DATA STATEMENTS

In Chapter 1, we mentioned that data can be supplied to a program by including it as part of the program and identifying it as such. The DATA statement allows us to identify data in a program, and the READ statement is used to assign these data values to variables.

A DATA statement simply consists of the keyword DATA followed on the same line by data values, each value separated from the next by a comma. There can be as many data values as the line will hold, and any number of DATA statements can appear in a program. Like all BASIC statements, each DATA statement must have a statement number. They can appear anywhere in the program, but most programmers prefer to place all DATA statements together at the end of the program to improve program legibility.

For each data value in a DATA statement, a corresponding READ statement must appear somewhere in the program to assign that value to a variable. Variable names and data used in READ-DATA combinations can be in either numeric or string form. When READ statements are executed, data values are assigned in the left-to-right order in which they appear in a DATA statement. If there is more than one DATA statement, data values are taken first from the statement with the smallest statement number. When all these values have been used by READ statements, the next READ will use the first data value in the next DATA statement.

As an example, suppose that the following READ-DATA combination occurs in a BASIC program:

```
10 READ X
20 READ Y
30 READ Z
  . . .
  . . .
  . . .
100 DATA 1,2
101 DATA 3
```

In this program, X will be assigned the value 1, Y will be assigned the value 2, and Z will be assigned the value 3. The same assignments would have been made if the DATA statement had been written as

```
100 DATA 1,2,3
```

or again as

```
100 DATA 1
101 DATA 2,3
```

More than one variable name can appear in a READ statement, and both numeric and string variables can be assigned values by the same READ. Variables are assigned values in the same left-to-right order as their names appear in the READ statement, using the left-to-right order of values in the DATA statement(s), as previously described.

To illustrate, the following READ-DATA combinations all result in the same value assignments:

```
(1) 10 READ A,B2,C$
    20 READ X
     . . .
     . . .
     . . .
    100 DATA 73,102.5,"GIRL"
    110 DATA-3E4

(2) 10 READ A,B2
    20 READ C$
    30 READ X
     . . .
     . . .
     . . .
    100 DATA 73
    110 DATA 102.5,"GIRL"
    120 DATA-3E4

(3) 10 READ A,B2,C$,X
     . . .
     . . .
     . . .
    100 DATA 73
    110 DATA 102.5,"GIRL",-3E4
```

In every case, A is assigned the value 73, $B2$ is assigned 102.5, $C\$$ is assigned "GIRL", and X is assigned -3×10^4.

In some programming applications, it is desirable to be able to use the data values appearing in one or more DATA statements again. In other words, we might want one or more READ statements to use the *same* data. This reuse of data contained in DATA statements can be accomplished with the RESTORE statement, which effectively restores data availability from the first value in the first DATA statement. Following is an example:

```
10 READ A,B
20 RESTORE
30 READ X,Y,Z
 . . .
 . . .
 . . .
100 DATA 50,100,150
```

In this example, A is assigned the value 50 and B is assigned the value 100. Then, by virtue of the RESTORE statement, the READ statement at line 30 assigns X the value 50, Y the value 100, and Z the value 150. Thus A and B are assigned the same set of values as X and Y.

Some final comments on input/output operations and computing in general are in order now. By this time, we trust that you realize a computer is not a "giant brain" capable of thinking or making independent decisions, as it is often popularly portrayed. You have seen that it simply performs computations on data supplied to it, by following an orderly set of steps called a program. A computer is incapable of making mistakes. Inconsistent output is due to either faulty input or programming errors, and both of these can only be attributed to human beings. For example, if you write a program to compute the area of a triangle $\left(A = \frac{1}{2}BH\right)$ and then supply a negative value for B, the computer will compute a negative area, and no one should condemn the machine for producing a ridiculous answer. A common slogan among computer scientists is "GIGO", which stands for "garbage in/garbage out."

EXERCISES

4.7 Show the display that would result if each of the following programs were executed:

(a)
```
10 READ A,B$
20 LET W=A**2-A
30 LET Y=W/10
40 PRINT B$;W;Y
50 DATA 10, "ANSWERS:"
60 END
```

(b)
```
10 READ X,M2,T
20 READ B
30 LET P=X-M2+B*T
40 DATA 25,4
50 DATA 2,7
60 PRINT P
70 END
```

(c)
```
10 READ A,B,C
20 RESTORE
30 READ D
40 PRINT A-B; B-C; A-D
50 DATA 50,20,5
60 END
```

4.8 Write a BASIC program to display a table of values of N, N^2, and N^3 for $N = 3$, $N = 4$, and $N = 5$. Use a READ-DATA combination to supply the values of N to the program.

4.9 Write a BASIC program to compute and print values of X_1, X_2, and X_3, based on the following sets of equations:

(a) $X_1 = 4 - X_1$; $X_2 = X_2 - 5$; $X_3 = 7X_3$ (b) $X_1 = X_1 + 1$; $X_2 = X_2$; $X_3 = X_3^2$

(A total of six values printed.) Use the data values $X_1 = 7$, $X_2 = -4$, and $X_3 = 2$ for each set of equations. Use DATA, READ, and RESTORE statements to supply data to the program.

Example 4.2 If a body is at rest, the sum of the horizontal forces upon it must equal zero, and the sum of the vertical forces upon it must also equal zero. This is called an *equilibrium* condition. If all forces are known except one, we can solve for the unknown force by applying the equilibrium condition. Consider the "free body" diagram shown in Figure 4.2

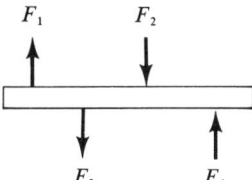

FIGURE 4.2 Forces in a free-body diagram (Example 4.2)

There are no horizontal forces. If we assume upward forces are positive and downward forces are negative, then we must have

$$F_1 + F_4 - F_2 - F_3 = 0 \tag{1}$$

This equation can be solved for any unknown force. In this example, suppose that F_1, F_2, and F_3 are known and F_4 is unknown.

Figure 4.3 shows a BASIC program used to print the value of F_4, after the user enters the known values of F_1, F_2, and F_3. Note that one INPUT statement is used to enter all three force values, and the computation of F_4 is performed in the PRINT statement.

```
10 PRINT "ENTER F1,F2, AND F3."
20 INPUT F1,F2,F3
30 PRINT "F4=" F2+F3-F1
40 END
>
```

FIGURE 4.3 Program for Example 4.2

Figure 4.4 shows the results of two runs of this program. In the first run, the values $F1 = 8.5$ newtons (N), $F2 = 17N$, and $F3 = 9.1N$ were entered, and we see that the computed value of $F4$ is 17.6N. In the second run, the values $F1 = 0.34N$, $F2 = 0.01N$, and $F3 = 0.26N$ were entered, and the computed value of $F4$ is $-0.06N$. The minus sign means that $F4$ actually has a downward direction, rather than the upward direction assumed in Figure 4.2.

Example 4.3 Figure 4.5 shows a lever supporting a rock that exerts a force of F_1 newtons at a point d_1 meters from the fulcrum. A force F_2 is applied on the other end of the lever, at a distance of d_2 meters from the fulcrum. For the force F_1 to balance the rock, we must have

$$F_1 d_1 = F_2 d_2 \tag{2}$$

```
ENTER F1,F2, AND F3.
?8.5,17,9.1
F4= 17.6000

>

ENTER F1,F2, AND F3.
?.34,.02,.26
F4=-6.00000E-02

>
```

FIGURE 4.4 The results of two runs of the program in Figure 4.3

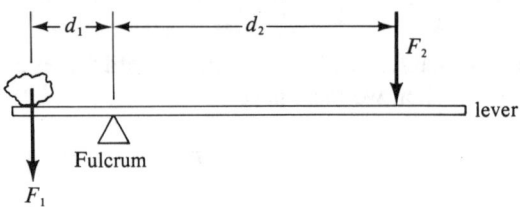

FIGURE 4.5 The force F_2 balances the force F_1 of a rock on a lever (Example 4.3)

We wish to write a BASIC program that prompts the user to enter values for the force of the rock (its weight in newtons) and the force F_2 that is available to lift it and then computes the distance d_2 at which F_2 must be applied. The distance d_1 is fixed at 0.4 meters. From Equation 2,

$$d_2 = \frac{F_1 d_1}{F_2} = \frac{0.4 F_1}{F_2}$$

Figure 4.6 shows the program. The INPUT statements contain the appropriate prompt messages, and the computation of d_2 is performed in the PRINT statement.

```
10 INPUT "ENTER ROCK WEIGHT"; F1
20 INPUT "ENTER FORCE"; F2
30 PRINT "DISTANCE FROM FULCRUM IS" .4*F1/F2 "METERS"
40 END
```

FIGURE 4.6 Program for Example 4.3

Figure 4.7 shows the results of two runs of this program. The results of the first run show that a rock weighing 3.6N can be lifted by a force of 0.8N applied 1.8 meters from the fulcrum. The second run shows that a rock having a weight of an average automobile (15,555N ≈ 3500 lbs) can be lifted by a force of 1N (about 0.225 lbs) provided that the force is applied 6222 meters (about 3.87 miles) from the fulcrum!

```
ENTER ROCK WEIGHT? 3.6
ENTER FORCE? .8
DISTANCE FROM FULCRUM IS 1.8 METERS

ENTER ROCK WEIGHT? 15555
ENTER FORCE? 1
DISTANCE FROM FULCRUM IS 6222 METERS
```

FIGURE 4.7 The results of two runs of the program in Figure 4.6

Example 4.4 Following is a list of seven angles between 0 and 90 degrees and their trigonometric sines:

Angle (degrees)	Sine
0	0.0
15	0.258819
30	0.5
45	0.7071068
60	0.8660255
75	0.9659259
90	1.0

We wish to write a BASIC program to create a table of angles and their sines and cosines, where the cosines are to be computed using the trigonometric identity

$$\sin^2 \theta + \cos^2 \theta = 1 \tag{3}$$

or

$$\cos \theta = \sqrt{1 - \sin^2 \theta}$$

Figure 4.8 shows the program and the table that is displayed when the program is run. Note the use of READ-DATA statements for assigning the known sine values to the variables S1, S2, ..., S7. Also note the use of the TAB function in the PRINT statement. (This program is rather lengthy and repetitious. Later we will study a more efficient method for creating data tables.)

GENERAL EXERCISES

4.10 Write a BASIC program to solve the equations

$$Y_1 = 4X_1^3 - 3X_2^2 + 2X_3 - 1$$
$$Y_2 = X_1^3 + 2X_2^2 - 8X_3 + 5$$

for Y_1 and Y_2 using the data $X_1 = 3.75$, $X_2 = 0.894$, and $X_3 = 1.77$. Use a single INPUT statement to furnish data to the program. Perform all computations in PRINT statement(s). The results should be displayed in the format

$$Y_1 = y_1 \text{ and } Y_2 = y_2$$

where y_1 and y_2 are the computed solution values.

```
10 READ S1,S2,S3,S4
20 READ S5,S6,S7
30 PRINT "ANGLE(DEG)" TAB(16) "SINE" TAB(28) "COSINE"
40 PRINT TAB(4) "0" TAB(15); S1; TAB(26);
(1-S1**2)**.5
50 PRINT TAB(4) "15" TAB(15); S2; TAB(26);
(1-S2**2)**.5
60 PRINT TAB(4) "30" TAB(15); S3; TAB(26);
(1-S3**2)**.5
70 PRINT TAB(4) "45" TAB(15); S4; TAB(26);
(1-S4**2)**.5
80 PRINT TAB(4) "60" TAB(15); S5; TAB(26);
(1-S5**2)**.5
90 PRINT TAB(4) "75" TAB(15); S6; TAB(26);
(1-S6**2)**.5
100 PRINT TAB(4) "90" TAB(15); S7; TAB(26);
(1-S7**2)**.5
110 DATA 0,.258819,.5,.7071068,.8660255,.9659259,1
120 END
>

ANGLE(DEG)     SINE          COSINE
    0           0              1
   15          .258819        .965926
   30          .500000        .866025
   45          .707107        .707107
   60          .866025        .500000
   75          .965926        .258819
   90           1              0
```

FIGURE 4.8 Program for Example 4.4 and the results of one program run

4.11 Following is a list of horizontal and vertical forces measured on a certain body undergoing stress testing. Each set of measurements was obtained for a different test condition.

Horizontal force (N)	Vertical force (N)
153.4	171.9
208.0	188.6
537.9	600.2
1082.2	824.5
1925.5	1098.0

In each case, the resultant force R is

$$R = \sqrt{F_H^2 + F_V^2}$$

where F_H is the horizontal force and F_V is the vertical force. Write a BASIC program to generate a table showing each horizontal force, the corresponding vertical force, and the resultant force. Use READ and DATA statements to supply the data to the program. Include appropriate table headings, and use the TAB function to space the table entries.

4.12 Modify the program in Example 4.4 so that each angle is displayed in *radians* instead of degrees (2π radians = 360 degrees) and so that a new column is displayed showing the tangent of each angle ($\tan\theta = \sin\theta/\cos\theta$). The entry corresponding to tan 90° should be INFINITY. Your conversions of the angle data (given in the DATA statement of the example) to angles in radians should be done in PRINT statements.

4.13 Following is a table of experimental data that were recorded to verify the value of the gravitational constant g. Each set of measurements represents the distance traveled in time *t* by a body dropped from rest. The acceleration is calculated from

$$a = 2s/t^2$$

where *s* is the distance traveled and *t* is the time of travel.

Time (sec)	Distance (ft)
1.04	16.8
2.11	72.5
3.84	245.8
5.26	421.9
9.05	1402.0

Write a BASIC program to generate a table showing the following quantities: time, distance, calculated acceleration, and percent difference from theoretical acceleration. Assume that the theoretical acceleration is g = 32.2 ft/sec² (% difference = (theoretical − calculated)/theoretical × 100%). Use READ and DATA statements to supply data to the program.

EXERCISES FOR ARCHITECTURAL/CIVIL/CONSTRUCTION TECHNOLOGY

4.14 In a percolation test of soil to determine its suitability for a septic field, a hole is dug in the soil and then filled with water. See Figure 4.9. The rate at which the water level falls in the hole is then determined by measuring its fall in 30-minute intervals. To determine the required area of a septic field, we must calculate the time required for the water level to drop 1 inch. If the water level changes from d_1 to d_2 feet in 30 minutes, then the time required for the level to drop 1 inch is

$$t = \frac{30}{12(d_1 - d_2)} \text{ minutes}$$

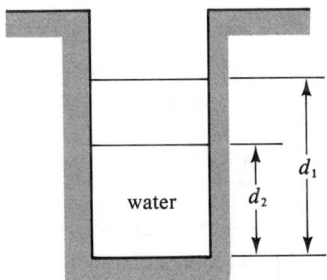

FIGURE 4.9 Exercise 4.14

Write a BASIC program to calculate the time (in minutes) required for the level to drop 1 inch after a user enters the depths d_1 and d_2 in feet. Use one INPUT statement to supply data to the program. The answer should be displayed in the format

TIME REQUIRED IS n MINUTES

where n is the computed time. Perform computations in the PRINT statement. Run your program for the following sets of data:

(a) d_1 = 3.8 ft, d_2 = 3.1 ft
(b) d_1 = 4 ft 3 in, d_2 = 3 ft 6 in

4.15 The heat that must be supplied to a room to overcome infiltration losses caused by air changes (the exchange of warmer inside air with cooler outside air) is

$$H = \frac{C(t - t_o)n}{55.2}$$

where H = BTU/hour of heat required,
C = the volume of the room in cubic feet,
t = the temperature of the room (°F),
t_o = the outside air temperature (°F), and
n = the number of air changes per hour.

The following table contains the results of a survey that was conducted to determine infiltration losses in five different rooms at different times of the day. The table shows the room sizes in cubic feet and the measured temperature differentials (values of $t - t_o$). In each case, the air was changed 1.5 times per hour.

C (ft^3)	$t - t_o$ (°F)
240	42.5
112	39.1
466	57.7
357	49.8
265	29.9

Write a BASIC program to generate a table showing values of C, $t - t_o$, and calculated values of H. Include appropriate table headings. Use READ and DATA statements to supply data to the program, and perform all computations in PRINT statements.

4.16 The following table contains experimental data that was obtained to verify the Young's modulus (or elastic modulus) E of a steel beam. By definition,

$$E = \sigma/\epsilon$$

where E is the Young's modulus in newtons per square meter,
σ is the stress in newtons per square meter, and
ϵ is the strain (dimensionless).

σ (N/m²)	ϵ
1.8×10^8	1.4×10^{-3}
3.6×10^8	3.0×10^{-3}
5.4×10^8	4.0×10^{-3}
7.2×10^8	6.0×10^{-3}
9.0×10^8	6.8×10^{-3}

Write a BASIC program to generate a table displaying the following quantities: σ, ϵ, calculated E, and the percent difference between the calculated E and the nominal E. Assume that the nominal E is 1.3×10^{12} N/m².

$$\text{Percent difference} = \frac{E_N - E_C}{E_N} \times 100\%$$

where E_C = the calculated E and E_N = the nominal E. Use READ and DATA statements to supply data to the program.

EXERCISES FOR ELECTRICAL/ELECTRONICS/COMPUTER TECHNOLOGY

4.17 According to Kirchhoff's voltage law, the sum of the voltage drops around the closed loop in Figure 4.10 is equal to the voltage rise E.

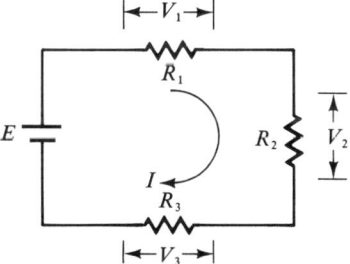

FIGURE 4.10 Exercise 4.17

In equation form,

$$\begin{aligned} E &= V_1 + V_2 + V_3 \\ &= IR_1 + IR_2 + IR_3 \end{aligned} \qquad (4)$$

If E, V_1, V_2, and R_3 are known, the current I in the circuit can be found from Equation 4 by

$$I = \frac{E - V_1 - V_2}{R_3}$$

Write a BASIC program to find the current I when a user enters values for E, V_1, V_2, and R_3. Use a single INPUT statement to supply this data to the program. The result should be displayed in the format

```
THE CURRENT IS i AMPS.
```

where i is the computed current. Perform computations in the PRINT statement. Use the following sets of input data to run your program:

(a) $E = 25V$, $V_1 = 6V$, $V_2 = 9V$, $R_3 = 4.7$ kΩ
(b) $E = 6V$, $V_1 = 1V$, $V_2 = 2.8V$, $R_3 = 1.5$ MΩ

4.18 The following table lists measurements of the voltage across five different resistors. The power dissipated in each resistor is given by

$$P = \frac{V^2}{R} \text{ watts}$$

where V is the measured voltage in volts and R is the resistance in ohms.

Resistance (Ω)	Voltage (V)
100	20.2
1.5K	15.6
22K	46.8
330K	75.0
0.47M	100

Write a BASIC program to generate a table showing the resistance, voltage, and computed power dissipation for each resistor. Include appropriate table headings. Use READ and DATA statements to supply data to the program and the TAB function to space entries in the table.

4.19 The following table shows experimental voltage and current measurements that were made to verify Ohm's law ($R = E/I$) for a 1000 Ω resistor.

Voltage (V)	Current (mA)
5	4.87
10	11.09
15	14.42
20	22.11

Write a BASIC program to generate a table showing the following quantities: voltage, current, computed resistance, and percent difference from nominal value (1000 Ω).

$$\text{Percent difference} = \frac{R_N - R_C}{R_N} \times 100\%$$

where R_C = the calculated resistance and R_N = the nominal resistance. Use READ and DATA statements to supply data to the program.

EXERCISES FOR INDUSTRIAL/PRODUCTION/MANUFACTURING TECHNOLOGY

4.20 The quantity of steel stock consumed by a certain machine operation is given by

$$Q = 3.7(N_1 t_1 + N_2 t_2) + \frac{L}{2}(t_1 + t_2)$$

where Q is the total quantity consumed, in cubic feet;
   N_1 and N_2 are the numbers of type 1 and type 2 machines in operation;
   t_1 and t_2 are the total times that each machine type is in operation, in hours;
and
   L is the average loss rate, in cubic feet per hour.

Write a BASIC program to compute Q when the user enters values for N_1, N_2, t_1, and t_2. Assume that $L = 0.25$ ft^3/hr. Use a single INPUT statement to supply data values to the program, and compute Q in the PRINT statement. The results should be displayed in the format

```
QUANTITY=q CU.FT./MIN
```

where q is the computed value of Q. Run your program for the following data values:

$$N_1 = 4, \ N_2 = 6, \ t_1 = 7.1, \ t_2 = 3.5$$

4.21 The following table contains values of $\Sigma(X_i - \bar{X})^2$ for five different sets of experimental observations, where Σ means "the sum of," X_i stands for the experimentally observed values, and $\bar{X}$ is the mean (average) of the values. The table also shows the total number of observations (n) in each sum. The standard deviation is defined to be

$$SD = \sqrt{\frac{\Sigma(X_i - \bar{X})^2}{n - 1}}$$

n	$\Sigma(X_i - \bar{X})^2$
30	185.5
47	201.9
25	62.6
139	277.0
52	118.4

Write a BASIC program to create a table showing each n, $\Sigma(X_i - \bar{X})^2$, and the corresponding computed value of standard deviation. Include appropriate table headings. Use READ and DATA statements to supply data to the program, and perform all computations in PRINT statements.

4.22 The following table contains experimental data that were gathered to verify the efficiency of a power plant. The table shows values of total energy consumption E_{in} and total energy output E_o measured by five different industrial technicians. The efficiency is defined to be

$$\eta = \frac{E_o}{E_{in}}$$

E_o(kw-hrs)	E_{in}(kw-hrs)
1.8×10^6	2.1×10^6
1.6×10^6	1.9×10^6
2.0×10^6	2.3×10^6
1.7×10^6	1.8×10^6
1.6×10^6	2.0×10^6

Write a BASIC program to create a table showing the following quantities: E_o, E_{in}, the computed value of efficiency, and the percent difference from the nominal efficiency. Assume that the nominal efficiency is 0.86.

$$\text{Percent difference} = \frac{\eta_N - \eta_C}{\eta_N} \times 100\%$$

where η_N is the nominal efficiency and η_C is the calculated efficiency. Use READ and DATA statements to supply data to the program.

EXERCISES FOR MECHANICAL TECHNOLOGY

4.23 The (weight) flow rate of a fluid having velocity V in a duct of cross-sectional area A is

$$Q = \frac{AV}{s}$$

where s is the specific volume of the fluid.

Write a BASIC program to determine the flow rate in pounds per second, when the user enters the cross-sectional area A in square feet, velocity V in feet per second, and specific volume in cubic feet per pound. Use one INPUT statement for supplying data to the program. Perform the computation in a PRINT statement. The result should be displayed in the format

```
FLOW RATE=q LB/SEC
```

where q is the computed value of Q. Run your program to find the flow rate of air through a 1.7 ft² duct when the velocity is 30.02 ft/sec. The specific volume of the air is 13.09 ft³/lb.

4.24 The following table contains experimentally measured values of the angular acceleration α and the torque T on five different wheels. Torque and angular acceleration are related by the equation

INPUT/OUTPUT

$$T = J\alpha$$

where J is the inertia of the wheel.

T (N-m)	α (rad/sec^2)
75.32	11.09
143.55	24.17
32.80	57.61
308.65	129.52
166.96	33.24

Write a BASIC program to create a table showing the torque, angular acceleration, and computed inertia for each wheel. Include appropriate table headings. Use READ and DATA statements to supply data to the program, and perform all calculations in PRINT statements.

4.25 The following table contains experimentally measured values of the pressure p and volume v in a constant-temperature chamber that were gathered to verify the gas law

$$pv = nRT = \text{constant}$$

p (N/m^2)	v (m^3)
1.05×10^6	5.2×10^{-3}
2.11×10^6	2.0×10^{-3}
4.77×10^6	1.7×10^{-3}
8.63×10^6	0.8×10^{-3}

Write a BASIC program to create a table showing the following quantities for each measurement: pressure, volume, computed value of the product pv, and the percent difference between the computed product and the theoretical constant 5.5×10^3.

$$\text{Percent difference} = \frac{(\text{theoretical} - \text{computed})}{\text{theoretical}} \times 100\%$$

Use READ and DATA statements to supply data to the program.

CHAPTER 5

BRANCHING AND LOOPING

1 THE GOTO STATEMENT

You have learned that a computer executes statements in a BASIC program by following the order of the statement numbers. In many programming situations, however, it is useful to be able to change the order of execution without changing the statement numbers. This alteration of the normal sequence in which statements are executed is called *branching*. For example, you might want the computer to repeat the execution of a certain set of statements. After the last statement in the set is executed, it would be necessary for the computer to go back to the first statement in the set, rather than continue in its normal sequence. In another situation, you might want the computer to skip over a certain set of statements—that is, jump to a statement further along in the program.

You can use the statement

GOTO sn

(where *sn* is a statement number) to tell the computer the number of the statement you want to be executed next. Suppose, for example, that the computer encounters the following sequence:

```
50 GOTO 70
60 PRINT "NOT HERE"
70 PRINT "HERE"
```

79

After statement 50 is executed, the computer jumps to statement 70 and prints HERE; it skips statement 60 entirely.

Following is another example of the use of a GOTO statement:

```
10 LET X=1
20 PRINT "X EQUALS:"
30 PRINT X
40 LET Y=X
50 GOTO 10
```

After statements 10 through 40 have been executed, the computer is told in statement 50 to go back to statement 10. You can easily see that this whole process would repeat itself indefinitely, resulting in a continuous repetition of the display

```
X EQUALS:
1
X EQUALS:
1
X EQUALS:
    . . .
    . . .
    . . .
```

This is an example of an *infinite loop* that can only be stopped by depressing a special key on the keyboard (BREAK, ESCAPE, or some other key, depending on the system). Except for special cases when user intervention is expected, infinite loops should obviously be avoided. When writing complex programs, programmers sometimes have difficulty detecting the presence of an infinite loop until the program is actually run. As far as any program or programming exercise in this book is concerned, you can assume that an infinite loop is present if the program runs for as long as several minutes.

EXERCISES

5.1 Show the display that would result if the following program were run:

```
10 LET X=10
20 LET Y=X**2
30 GOTO 70
40 PRINT Y
50 LET Y=Y+1
60 GOTO 100
70 PRINT X,Y
80 LET Y=Y+1
90 GOTO 40
100 PRINT Y
110 END
```

5.2 Determine whether an infinite loop is present in the following program:

```
10 LET A=25
20 LET B=50
30 GOTO 80
40 LET B=B+1
50 LET A=A+1
60 PRINT A,B
70 GOTO 90
80 LET A=A+B
90 PRINT A
100 GOTO 40
110 END
```

2 CONDITIONAL BRANCHING: THE IF-THEN STATEMENT

The GOTO statement we have just discussed is called an *unconditional branch*, because the computer *always* goes to the statement number specified by the GOTO, regardless of the outcome of any computation performed up to that point. A *conditional branch*, on the other hand, is like a GOTO statement that will be executed only if a specified condition is satisfied. The format of a conditional branch statement is

IF *condition* THEN *sn*

where the *condition* can be any of several we will describe presently, and *sn* is the statement number to which the computer branches if the condition is satisfied. If the condition is *not* satisifed, the next statement following the conditional branch statement is executed.

The condition imposed in a conditional branch statement is arithmetic in nature (when imposed on a numeric variable, which is most often the case). The condition is applied to the result of a previous computation or to a data value supplied to the program. As an example, the conditional branch

```
60 IF X>Y THEN 100
```

means that the computer will branch to statement 100 if the value of X at that point in the program (at statement 60) is greater than ($>$) the value of Y at that point. To illustrate, consider the following sequence:

```
10 INPUT X
20 LET Y=X**2
30 IF Y>X THEN 100
40 PRINT X
```

If the user enters a value of X greater than 1, then $Y = X^2$ will be greater than X and the condition imposed in statement 30 will be satisfied. (For example, if $X = 2$, then $Y = X^2 = 4$ and, since $4 > 2$, the condition is satisfied.) In this case, the computer will branch to statement 100. On the other hand, if the user enters a value of X less than 1, $Y = X^2$ will then be less than X. (For example, if $X = 1/2$, then $Y = X^2 = 1/4$ is less than X.) In this case, the condition in statement 30 is *not* satisfied, so the next statement to be executed will be the one following statement 30: PRINT X.

Table 5.1 shows the symbols used to express the various arithmetic conditions that can be imposed in an IF-THEN statement. These are called *relational operators*.

Note that several of the symbols in Table 5.1 must be entered from the keyboard by depressing two keys in succession. For example, $<=$ is entered by typing $<$ followed by $=$. In some versions of BASIC, symbols different from those shown in Table 5.1 might be used for some of the conditions. For example, the combination $><$ is sometimes used for "not equal to" and $=>$ for "greater than or equal to."

The following example shows how a conditional branch is used in a program that computes the square roots of numbers entered by a user. If the number entered is negative (less than 0), a conditional branch statement will cause an error message to be generated.

```
10 INPUT N
20 IF N<0 THEN 50
30 PRINT "ROOT=" N**.5
40 GOTO 10
50 PRINT "ERROR: NEGATIVE NUMBER"
60 GOTO 10
70 END
```

In this example, statement 20 causes a branch to statement 50 if a negative N is entered. Statement 50 prints the error message. If N is greater than or equal to 0, statement 30 is executed and the square root of N is printed. Note that this program contains an infinite loop: Regardless of the sign of N, the computer branches back to statement 10 and prompts the user to enter a new number. As discussed previously, the user must depress a special key to escape this loop.

The *condition* part of a conditional branch statement can also contain expressions. Thus, branching can depend on the outcome of computations expressed within the statement itself, as in the following example:

```
IF X-A<=Y**2/10 THEN 50
```

This statement means that a branch to statement 50 will take place only if the computed value of $X - A$ is less than or equal to the computed value of $Y^2/10$. Of course, the current values of the variables X, A, and Y are used when the computations are performed. In the next example, we will show how this kind of conditional branch statement can be used to determine whether the roots of a quadratic equation are (1) real and unequal, (2) real and equal, or (3) imaginary. The user enters values for the coefficients A, B, and C (in the equation $Ax^2 + Bx + C = 0$), and the program compares the value of B^2 to $4AC$. We know that $B^2 > 4AC$ means the equation has real and unequal roots; $B^2 = 4AC$ means it has real, equal roots; and $B^2 < 4AC$ means that it has imaginary roots.

Table 5.1

Symbol	Condition
<	less than
<=	less than or equal to
>	greater than
>=	greater than or equal to
=	equal to
<>	not equal to

```
10 PRINT "ENTER A,B,C"
20 INPUT A,B,C
30 IF B*B>4*A*C THEN 70
40 IF B*B<4*A*C THEN 90
50 PRINT "THE ROOTS ARE REAL AND EQUAL"
60 GOTO 100
70 PRINT "THE ROOTS ARE REAL AND UNEQUAL"
80 GOTO 100
90 PRINT "THE ROOTS ARE IMAGINARY"
100 END
```

Statement 30 determines if B^2 is greater than $4AC$ and, if so, causes a branch to statement 70, where the appropriate message is printed. If B^2 is *not* greater than $4AC$, statement 40 is executed next. Here we must determine if B^2 is less than $4AC$. If so, a branch to statement 90 takes place, and the correct message is printed. If B^2 is *not* less than $4AC$, then at this point we know that it must be true that $B^2 = 4AC$, since neither of the conditions in statements 30 or 40 were satisfied. Consequently, the next statement (50) prints the appropriate message. Note the need for unconditional GOTO statements 60 and 80. These are used to cause branches to the end of the program, once a message has been printed. Without these statements, the computer would continue to execute the succeeding statements in the program and more than one message would be printed.

We must be careful to choose the correct condition from Table 5.1 to control the branching in a program. For example, if we had used $>=$ rather than $>$ in statement 30 of the preceding example, we would never detect the presence of equal roots and the program would not run properly.

There are usually several equivalent ways to express a condition or set of conditions. For example, both of the following sequences cause a branch to statement 200 if X is in the range $-1 \leq X < 3$.

```
10 IF X>= 3 THEN 40          10 IF X>=-1 THEN 30
20 IF X< -1 THEN 40          20 GOTO 40
30 GOTO 200                  30 IF X<3 THEN 200
40 . . .                     40 . . .
```

EXERCISES

5.3 Write the statement number of the next statement to be executed after each of the following sequences:

(a)
```
10 LET X=1E4
20 LET Y=X**.5
30 IF Y>50 THEN 100
40 GOTO 85
```

(b)
```
10 LET A=-2.79
20 IF A>0 THEN 70
30 IF A<=-2 THEN 90
40 GOTO 110
```

(c)
```
10 LET M=-120
20 LET N=1200
30 IF M-N<>N-M THEN 60
40 IF M-N<N-M THEN 80
```

(d)
```
10 LET B=4
20 LET C=5
30 IF (C**2-B**2)**.5>C-B THEN 50
40 IF (C**2-B**2)**.5>C THEN 60
50 IF B**2>C**2 THEN 200
60 GOTO 150
```

(e)
```
10 LET X1=1E6
20 LET X2=2.5E3
30 IF 10*X2<X1 THEN 60
40 IF 1E3*X2>X1 THEN 70
50 GOTO 160
60 GOTO 40
70 GOTO 220
```

5.4 Write BASIC program segments using IF-THEN statements to accomplish each of the following, after the user enters a value for N:

(a) Branch to 170 if N is in the interval $0 < N \leq 20$; otherwise branch to 200.

(b) Branch to 200 if $N > 5$ or $N \leq -5$; otherwise branch to 260.

(c) Branch to 160 if N^2 is greater than $100N$ but less than $1000N$; to 190 if N^2 is greater than or equal to $1000N$ but less than $5000N$; and to 230 if N^2 is greater than or equal to $5000N$.

(d) Branch to 80 if N does not equal 12.5; then to 120 if N is less than 12.5 or to 200 if N is greater than 12.5. Branch to 250 if N equals 12.5.

Variations

There are numerous variations in the form of the IF-THEN statement, depending on the version of BASIC being used, although all versions permit the statement to be used as we have already described. Many versions allow the keyword THEN to be replaced by GOTO, so, for example, the statements

```
IF X<1 THEN 50
```

and

```
IF X<1 GOTO 50
```

would be equivalent. Other versions permit the programmer to write a statement after THEN. As an example, the statement

```
IF Y>=2 THEN LET X=5
```

assigns X a value of 5 if $Y \geq 2$; otherwise, the next statement in sequence is executed.

Another common variation on the IF-THEN statement is IF-THEN-ELSE. This statement has the following format:

IF *condition* THEN sn_1 ELSE sn_2

If the condition is satisfied, statement number sn_1 is executed next; otherwise, statement number sn_2 is executed next. In a previous example, we showed how conditional branch instructions can be used to cause a branch to 200 if X is in the range $-1 \leq X < 3$. This conditional branch could also be performed using IF-THEN-ELSE, as follows:

```
10 IF X>= -1 THEN 20 ELSE 30
```

```
20 IF X<3 THEN 200
30 . . .
```

Versions of BASIC that permit a statement to follow THEN in the IF-THEN statement usually also allow a statement to follow ELSE. As an example, the statement

```
IF N<>2 THEN PRINT "N NOT EQUAL TO 2" ELSE PRINT "N
EQUALS 2"
```

prints the string N NOT EQUAL TO 2 if the condition (N $<>$ 2) is satisfied; otherwise, it prints N EQUALS 2.

Many versions of BASIC permit *compound logical expressions* to be used in the conditional branch statement. A compound logical expression is formed by using the keywords AND, OR, and NOT. An example of a logical expression is:

```
X<2 AND X>1
```

X satisfies this expression if the value is in the range $1 < X < 2$. Another example is

```
X<1 OR Y>X+1
```

This expression is satisfied if X is less than 1 or if Y is greater than $X + 1$. To illustrate the use of logical expressions in conditional branch instructions, we will write a program that reads ten numbers and counts the total of those that are both positive and less than 1—that is, all that are in the range $0 < X < 1$.

```
10 LET N=0
20 LET I=0
30 READ X
40 IF X<=0 OR X>1 THEN 60
50 LET N=N+1
60 LET I=I+1
70 IF I<10 THEN 30
80 PRINT "TOTAL=" N
90 DATA -1,0,.95, -2,1.8,.55, -1.9,.23,2.6,.01
100 END
```

As each number is read, statement 40 checks to determine if the number is *outside* the range $0 < X < 1$ (that is, if it is either zero, negative, or greater than 1). If so, then the computer branches to statement 60 and I is incremented (increased by 1). I is used to keep track of the number of times that the READ statement is executed. When I reaches 10 we know that all ten numbers have been checked. Statement 70 is the conditional branch statement that checks to determine if I has reached 10. It sends the computer back to statement 30 as long as I is less than 10. (This is an example of a *loop*. We will have more to say about loops in Section 5.3).

If a data value fails to satisfy the condition $X <= 0$ or $X > 1$ in statement 40, it must then be in the range $0 < X < 1$ and therefore must be counted. Thus, if the condition in statement 40 is not satisfied, statement 50 is executed and the variable N is incremented. At the end of program execution, N will therefore equal the total number of numbers that were found to be in the range $0 < X < 1$. Using the data shown in the DATA statement above, execution of the program results in the display

```
TOTAL=4
```

Note that 0.95, 0.55, 0.23, and 0.01 are the only values within the specified range.

Most versions of BASIC permit a conditional branch statement to be used to test the equality of strings and string variables. For example, suppose a program assigns the string "DOG" to the string variable P$. Then the conditional branch statement

IF P$ = "DOG" THEN 150

will cause a branch to statement 150. The following program shows how this kind of branching can be used to terminate an infinite loop:

```
10 INPUT N
20 PRINT N**.5
30 PRINT "DONE? TYPE YES OR NO."
40 INPUT A$
50 IF A$="NO" THEN 10
60 END
```

This program prints the square root of a number N entered by the user and asks if the user is done. If the user types "NO", the program then branches back to statement 10 and prompts for the input of a new number. If the user types "YES" (or anything else, for that matter), the program ends.

EXERCISES

5.5 Write the statement number of the next statement executed after each of the following sequences:

(a)
```
10 LET A=40
20 LET B=20
30 IF A>2*B THEN 70 ELSE 50
40 GOTO 80
50 IF A>=2*B THEN 40 ELSE 70
```

(b)
```
10 LET J=5
20 LET K=15
30 IF (K-J)/5<K/J THEN 40 ELSE 80
40 IF (K+J)/5<K/J THEN 70 ELSE 90
50 GOTO 100
```

(c)
```
10 READ X
20 IF X/2>4 OR 2*X<.5 THEN 10
30 IF X<10 AND X>=0 THEN 80
40 IF X>0 THEN 90
50 DATA 9,0,5
```

(d)
```
10 LET W=4
20 LET Z=2
30 IF 2*(W-3)>2 THEN W=Z
40 IF 2*(W-3)=W/Z THEN Z=W
50 IF W>Z THEN 150 ELSE 70
```

5.6 Show the results that would be displayed if each of the following programs were run:

(a)
```
10 LET R=0
20 LET R=R+5
30 IF R<25 THEN 40 ELSE 60
```

```
    40 PRINT R
    50 GOTO 20
    60 END
(b) 10 LET M=149.5
    20 LET N=73.2
    30 IF M-N>N THEN M=N-M
    40 IF M-N<=N THEN N=M-N
    50 IF M<N THEN PRINT M
    60 IF M>N THEN PRINT N
    70 END
(c) 10 READ X
    20 IF X=0 THEN 70
    30 IF X<10 AND -X<10 THEN 40 ELSE 10
    40 PRINT "ABS VALUE OF" X "IS < 10"
    50 GOTO 10
    60 DATA-5,24,1,-12,0,7
    70 END
(d) 10 LET W$="BIG"
    20 LET X$="BIGGER"
    30 LET Y$="BIGGEST"
    40 LET A=13.9
    50 LET B=6.22
    60 IF (A-B)/2>B THEN Y$=X$
    70 IF A/B>1 THEN W$=X$
    80 IF X$=Y$ THEN 110
    90 PRINT W$,X$,Y$
    100 GOTO 120
    110 PRINT X$ " AND " Y$
    120 END
```

5.7 (a) Use IF-THEN-ELSE statement(s) to write a BASIC program to find and print the larger and smaller of two numbers entered by the user. Each number should be printed with an appropriate identifying string.

(b) Use compound logic expressions to write a BASIC program to print YES if a number N entered by the user is either in the range $-9 \leq N < -3$ *or* in the range $3 \leq N < 9$. If not, print NO.

(c) Write a BASIC program that prints all even numbers between 0 and 10 (inclusive), if the user enters "EVEN", and all odd numbers in that range, if the user enters "ODD".

3 LOOPS

A loop is a set of statements that is executed over and over again. The number of times the loop is executed is controlled by a variable whose value is changed before or after each single execution of the loop. A conditional branch statement is used to test this variable to determine when the desired number of executions have been performed.

This concept is best understood by example. The following program contains a loop that prints the integers 1 through 5.

```
10 LET I=1
20 PRINT I
```

```
30 LET I=I+1
40 IF I<6 THEN 20
50 END
```

The loop in this example consists of statements 20, 30, and 40. Each time the loop is executed, the value of I is printed and I is then increased in value by one ($I = I + 1$). Statement 40 then determines whether all five integers have been printed; if I is less than 6, then all five have *not* been printed, so the program branches back to 20. When $I = 6$, the condition $I < 6$ fails and the program ends.

For our next example, we will write a program to compute and print the sum of the first N positive integers, where N is a number entered by the user. The variable J is used to count the total number of times that the loop is executed.

```
10 PRINT "ENTER A POSITIVE INTEGER"
20 INPUT N
30 LET S=0
40 LET J=1
50 LET S=S+J
60 LET J=J+1
70 IF J<=N THEN 50
80 PRINT "THE SUM OF THE FIRST" N "INTEGERS IS" S
90 END
```

After the user enters a value for N, the program *initializes* the variables S and J—that is, it assigns to these variables the values that they will have during the first execution of the forthcoming loop. The loop in this example consists of statements 50, 60, and 70. We say that one *pass* through the loop is completed each time this set of statements is executed. In the first pass, statement 50 sets S equal to $S + J$, that is, S is updated to the new value $0 + 1 = 1$. J is then updated in statement 60 to $J = J + 1 = 1 + 1 = 2$. If this value of J is less than or equal to the value of N entered previously, the conditional branch at statement 70 causes the computer to go back to statement 50 and begin the second pass through the loop. During each pass through the loop, the sum S is increased by adding the sum accumulated thus far to the current integer value represented by J. This process continues until the last value of J added to S is equal to N. When J equals $N + 1$, the conditional branch at statement 70 fails (the condition $J \leq N$ is no longer satisfied) and so the PRINT statement is executed.

The variable whose value changes in accordance with the number of completed passes through a loop is often called a *counter* variable, because it counts the total number of executions of the loop. In this example, J served as the counter variable. J was also used in the computation performed during each pass through the loop ($S = S + J$), but this need not always be the case: A counter variable is often used only to keep track of the number of passes.

A good way to understand the operation of a program containing a loop is to make a table showing the values that the variables have after each pass through the loop. Table 5.2 shows the values of S and J in the previous program example, when $N = 4$.

Table 5.2

	Initial	Pass 1	Pass 2	Pass 3	Pass 4
S	0	1	1 + 2 = 3	3 + 3 = 6	6 + 4 = 10
J	1	2	3	4	5

You should "step through" the program one pass at a time, while referring to and confirming the values in Table 5.2. Note that each new value of S is obtained by adding the current value of S to the previous value of J.

In many versions of BASIC, the values of all variables are automatically initialized to zero. In these versions, statement 30 of the preceding example would not be required. However, initialization of S by some means is critical to the success of this program. If you are using a version of BASIC that does not automatically initialize variables to 0, the value of a variable before it is assigned some specific value by a program statement will be some unknown, random number, such as $-1.56E15$. In this case, statement 50 (LET S = S + J) would then add 1 to this ridiculous initial value of S, during the first pass, and it is obvious that the final value of S would be incorrect.

Whenever you set up a loop, you must be careful to avoid constructing an infinite loop by improperly expressing the condition in the conditional branch. Consider, for example, the following program:

```
10 INPUT "ENTER A POSITIVE INTEGER"; N
20 LET I=N
30 LET P=1
40 LET P=P*I
50 LET I=I-1
60 IF N>1 THEN 40
70 PRINT "P=" P
80 END
```

In this program, statement 60 will *always* cause a branch back to 40, assuming that the value of N entered by the user is greater than 1. If the condition were expressed correctly as $I > 1$, the program would then compute the value of N! ($N \times N - 1 \times N - 2 \times \ldots 1$.)

A loop can be set up by either *incrementing* (increasing) a counter variable before or after each pass through the loop or else by *decrementing* (decreasing) it before or after each pass. The counter can be tested either at the beginning or at the end of the loop. For example, the following four loops are all equivalent:

```
(1) 10 LET I=0
    20 PRINT I
    30 LET I=I+1
    40 IF I<10 THEN 20
    50 END
(2) 10 LET I=0
    20 IF I>9 THEN 60
    30 PRINT I
    40 LET I=I+1
    50 GOTO 20
    60 END
(3) 10 LET I=10
    20 PRINT 10-I
    30 LET I=I-1
    40 IF I>=1 THEN 20
    50 END
(4) 10 LET I=11
    20 LET I=I-1
    30 PRINT 10-I
    40 IF I> 1 THEN 20
    50 END
```

In loops 1 and 2 above, the counter variable *I* is *incremented*; in loops 3 and 4, I is *decremented*. In loops 1, 2, and 3, the value of *I* is changed at the *end* of the loop (after the PRINT statement), and in loop 4 it is changed at the *beginning* of the loop. Note that a different kind of conditional branch statement is required in each case. "Step through" each of these examples until you are convinced that they each cause the digits 0 through 9 to be printed.

The counter variable does not always have to be incremented or decremented by 1. In the following example, the counter *J* is incremented by 2 after each pass through the loop:

```
10 LET J=1
20 PRINT "ENTER AN ODD NUMBER"
30 INPUT N
40 PRINT J
50 LET J=J+2
60 IF J<=N THEN 40
70 END
```

This program prints all odd integers from 1 through N.

EXERCISES

5.8 In the programming example used to demonstrate a loop by computing the sum of the first *N* positive integers, what would the PRINT statement display if the user entered (a) 0 for *N*? (b) a negative *N*? (c) 7.5?

5.9 Modify the example program described in Exercise 5.8 so it will print the message

```
ILLEGAL ENTRY
```

if *N* is 0 or negative and then prompt the user to enter another *N*.

5.10 Show the display that would be generated if each of the following programs were run. In each case, make a table showing how the values of the variables change during execution of the program.

```
(a) 10 LET T=25
    20 LET K=T**.5
    30 PRINT 2*K
    40 LET K=K-1
    50 IF K>1 THEN 30
    60 END

(b) 10 LET A=20
    20 IF A-12/3*2>A*2/5+4 THEN 40
    30 IF A<>0 THEN 50
    40 PRINT "A="A
    50 LET B=A
    60 PRINT "B="B
    70 LET B=B-5
    80 IF B>=-5 THEN 60
    90 END
```

(c) 10 LET X=10
   20 LET I=10
   30 LET J=10
   40 PRINT X
   50 LET X=X-5
   60 LET J=J-5
   70 IF J>=0 THEN 40
   80 LET I=I-5
   90 IF I<=0 THEN 30
   100 END

(d) 10 LET P=0
   20 LET Q=1
   30 PRINT Q
   40 LET Q=Q+1
   50 IF Q<4 THEN 30
   60 LET P=P+1
   70 IF P<=3 THEN 20
   80 END

5.11 Write a BASIC program that prompts the user to enter two integers M and N (positive, negative, or zero) where $M < N$ and then computes the sum of all integers between M and N, inclusive. For example, If $M = -2$ and $N = +4$, the result 7 should be printed since $(-2) + (-1) + 0 + 1 + 2 + 3 + 4 = 7$. If the user enters a value of M that is greater than or equal to N, the computer should print an appropriate error message and prompt for new values for both M and N.

4 FLOWCHARTS

Programs that contain many conditional branch statements and involve testing numerous conditions can be difficult to write and understand. In these cases, it is not easy to visualize the logical progression or "flow" that program statements must follow to achieve a certain computational goal. This is especially true when that goal requires branches back to earlier statements in the program, loops, or loops within loops. We have shown some examples of program loops, and you might have already experienced some difficulty understanding the logic implemented to construct these segments.

A *flowchart* is a symbolic diagram that shows all the logical decisions, branches, and computations that must be made in the course of a program; it can be a valuable aid in the design of the program.

The most fundamental kind of block found in a flowchart is called a *process* block. This is represented by a *rectangle* and is used to show a computation or value assignment. Figure 5.1 is an example of a process block that corresponds to the BASIC statement

```
LET X = 17*Y**(3/A)
```

Note the arrows entering the block at the top and leaving the block from the bottom. In a flowchart, these would be connected to the previous and next blocks, respectively, showing the operations just prior to and just after the computation $Y = 17X^{3/A}$.

A diamond-shaped *decision* block is used in a flowchart to represent a conditional branch. The condition that is tested in a decision block is expressed in the form of a

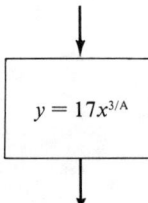

FIGURE 5.1 Example of a process block in a flowchart

question, and arrows are drawn leaving the block to show the paths corresponding to the "yes" and "no" answers to that question. For example, the statement

```
100 IF A>10 THEN 200
```

would be represented in the flowchart by the decision block shown in Figure 5.2.

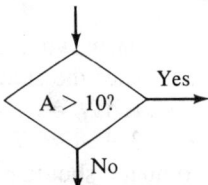

FIGURE 5.2 Example of a decision block in a flowchart

In Figure 5.2, the arrow labeled "yes" would point to the block representing the operation performed at statement 200 and the arrow labeled "no" would point to the block representing the operation performed by the next statement in sequence (the one following statement 100). It makes no difference where the "yes" and "no" exit arrows are drawn on the diamond or where the entry arrow is drawn, but all should be present.

To illustrate how to flowchart a simple loop, consider the following sequence:

```
10 LET B=0
20 LET B=B+1
30 IF B<=25 THEN 20
```

Of course, this sequence of statements does not accomplish anything useful, since it merely increments the variable B 26 times. We use it here only to illustrate how this kind of operation can be diagramed in a flowchart (see Figure 5.3). Note that the loop diagramed in the flowchart of Figure 5.3 could also be diagramed in the equivalent flowchart shown in Figure 5.4. Study these flowcharts until you are convinced that Figures 5.3 and 5.4 are logically equivalent.

An input/output (I/O) operation is represented on a flowchart by a parallelogram with two sides inclined to the right. Examples of operations in this category are those generated by INPUT and PRINT statements. Figure 5.5 illustrates the use of an I/O block to show the printing of data, such as that generated by the statement

```
PRINT "THE VALUE OF X IS:" X
```

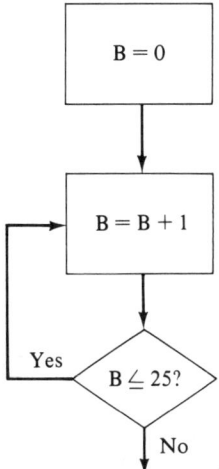

FIGURE 5.3 Flowchart for a loop

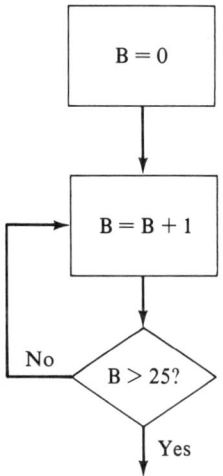

FIGURE 5.4 Flowchart for a loop equivalent to Figure 5.3

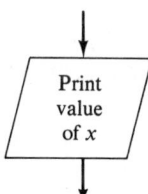

FIGURE 5.5 Example of an I/O block in a flowchart

The beginning and end of a program are indicated on a flowchart by using the special symbols shown in Figure 5.6. Note that no arrow enters the START block and none leaves the STOP block.

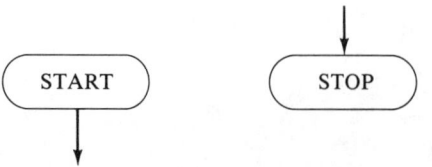

FIGURE 5.6 START and STOP blocks in a flowchart

We will now illustrate how a complete program is designed through construction of a flowchart diagraming all computations and logical decisions that must be made. The goal of this program is to compute and print the square and cube roots of four positive numbers entered by the user. If a negative number is entered, an error message should be printed, and the user should be prompted to enter a new number in its place. A negative number should not be counted as one of the four numbers allowed for input. We shall let the variable I serve as the counter variable to keep track of the total number of positive numbers entered. A first attempt to construct the flowchart might produce a result similar to that shown in Figure 5.7.

Some programmers prefer to show blocks that can have more than one source of entry by drawing junctions at a single arrow entering the block. For example, the first I/O block in Figure 5.7 could also be drawn as shown in Figure 5.8. There is no right or wrong way to show such multiple entries; after all, a flowchart is only a means of helping the programmer to visualize the logical progression of a program, so use whatever system works best for you.

The flowchart in Figure 5.7 shows that the counter variable I is incremented after each new number is entered by the user. The number is then tested to determine whether it is negative ($N < 0$?). If it is, an error message is printed, and the counter variable I is decremented (decreased by one), since a negative number cannot count toward the allowed total of four inputs. The program then branches back to prompt the input of a new number. If the number entered is *not* negative, its roots are then calculated and printed. Next the counter variable I is checked to find out whether all four positive entries have been made ($I = 4$?). If not, a branch is made back to the I/O block where the next number is entered. When all four numbers have been entered, the program ends.

A flowchart is very useful for revealing flaws in logic or inefficient programming techniques. For example, if you study the flowchart in Figure 5.7 carefully, you will observe that the need to decrement I can be eliminated by simply changing the location where I is incremented. If you wait to increment I until a valid number N is entered (that is, until N is found to be positive) it will never be necessary to decrement I. We can further improve the efficiency of the program by combining the computation of the roots with the printing of the results. The revised and improved flowchart is shown in Figure 5.9.

Note again that there are several equivalent ways to express the conditional branches shown by the decision blocks. For example, the decision block posing the question $N < 0$? could be replaced by one asking the question $N >= 0$? instead. In that case, the "yes" and "no" labels on the decision blocks would be interchanged. Also, the question $I = 4$? could be replaced by the question $I >= 4$?, which would have exactly the same effect in this example, or we could ask $I < 4$? and interchange the "yes" and "no" labels.

After developing a satisfactory flowchart, you will be able to implement the logic it represents by writing the BASIC statement(s) corresponding to each block. Match the statement(s) in the following program with the corresponding blocks in Figure 5.9.

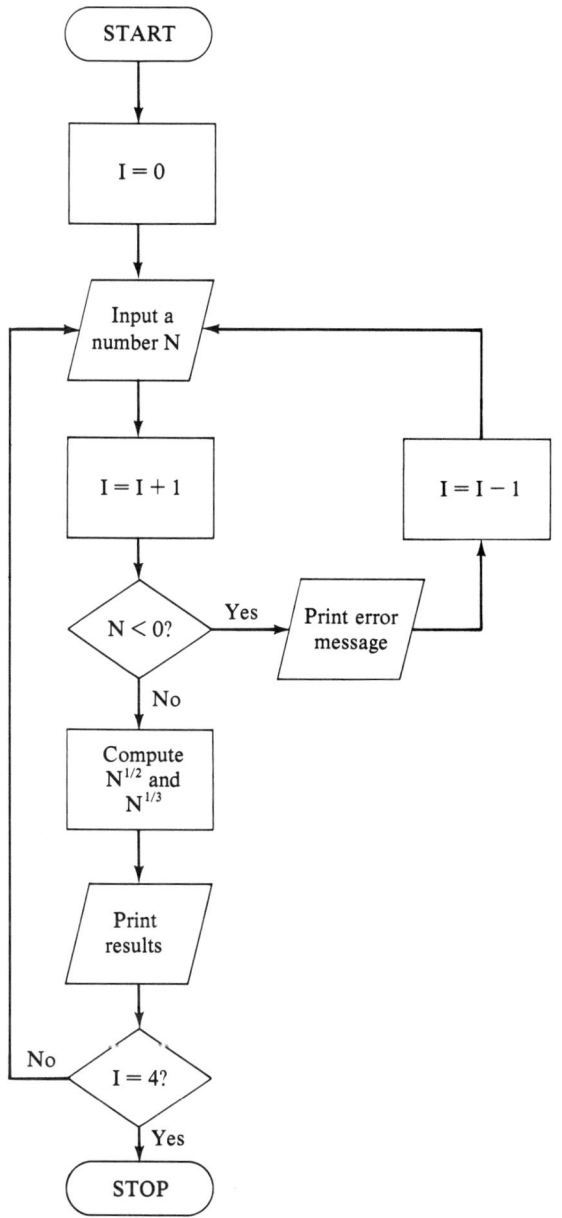

FIGURE 5.7 Flowchart diagraming the computation of square and cube roots of four numbers entered by the user

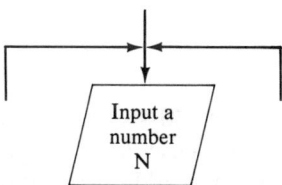

FIGURE 5.8 Alternate method of showing multiple paths entering a block in a flowchart

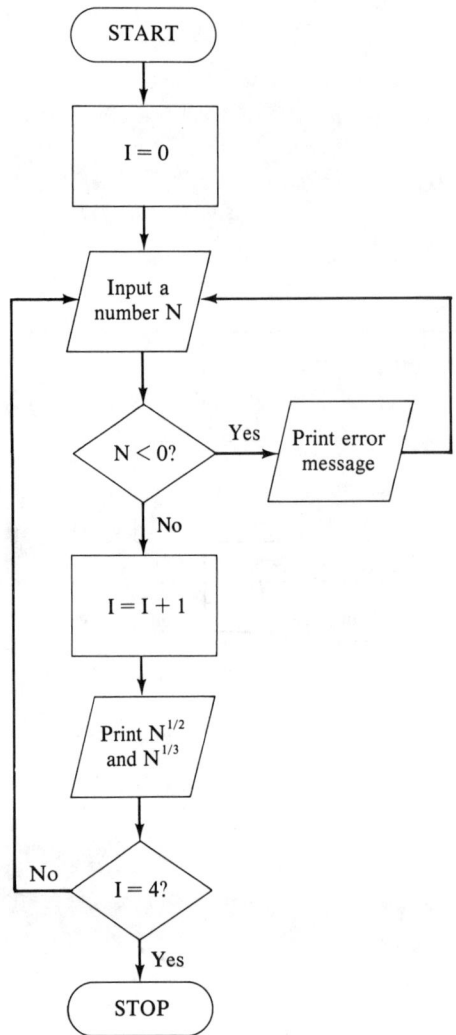

FIGURE 5.9 Flowchart showing a more efficient method of performing the computations diagramed in Figure 5.7

```
10 LET I=0
20 INPUT N
30 IF N>=0 THEN 60
40 PRINT "ERROR: NUMBER MUST BE POSITIVE."
50 GOTO 20
60 LET I=I+1
70 PRINT "THE ROOTS ARE:" N**.5 " AND" N**(1/3)
80 IF I<4 THEN 20
90 END
```

Note that the conditional branch statements 30 and 80 are the logical equivalents of those shown in the flowchart, as previously described. The program can be shortened somewhat by using these equivalents. (You should verify this by rewriting the program using conditional branch statements to express the conditions shown by the decision blocks in the flowchart.) Shorter programs are generally preferable because less memory space is required to store them. Although memory space can be at a premium in small microcomputer systems, it is rarely necessary to expend a great deal of creative energy searching for the shortest possible program that will accomplish a goal.

Some programs contain a great many branches, so their flowcharts show a correspondingly large number of paths between blocks. The presence of numerous lines between and around blocks can clutter a flowchart and detract from its legibility, thus making it difficult to follow any single path from its origin to its destination. One way to avoid this situation is to use *connector* symbols. A connector symbol is simply a circle with a letter or number written inside it. Instead of a line between two blocks, one connector is drawn at the block where the path originates and another at the block where it terminates. The same letter or number is used in both connectors to identity the particular path that they replace. Figure 5.10 shows the flowchart of Figure 5.9 redrawn with a pair of connectors replacing the path originally drawn between the last decision block and the first I/O block. This flowchart is not complex enough to require connector symbols, but we show them here merely to illustrate how those symbols are incorporated into the diagram. Note that the "no" exit from the last decision block is shown entering a connector designated "A", and another connector designated "A" has an arrow entering the I/O block.

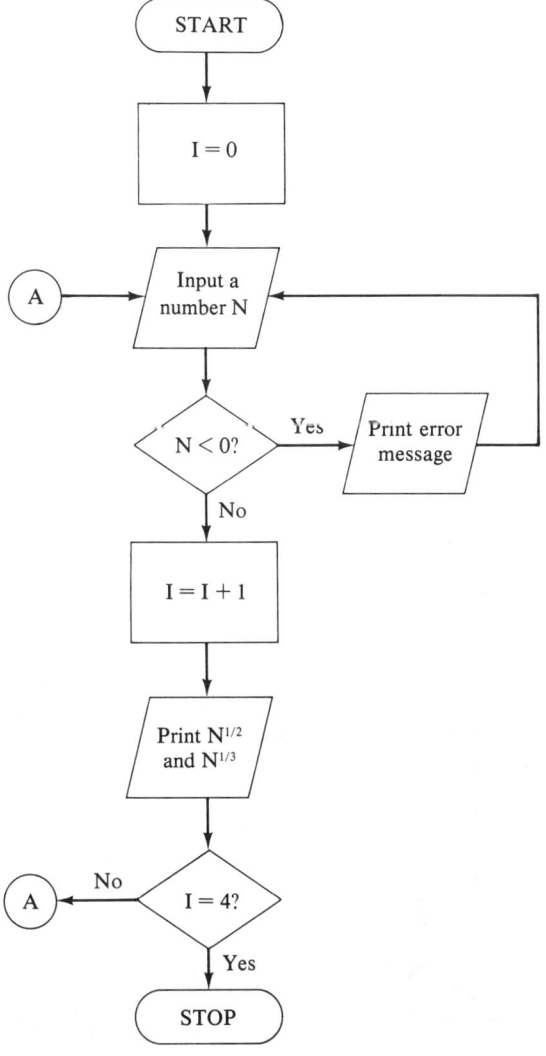

FIGURE 5.10 Flowchart illustrating the use of connector symbols

As another example of the uses of a flowchart, suppose you want to write a program to compute and print the absolute value of the difference between two numbers ($N1$ and $N2$) entered by a user. The logic we will implement is as follows: If $N1 > N2$, then $|N1 - N2| = N1 - N2$; if $N1 < N2$, then $|N1 - N2| = -(N1 - N2)$; and if $N1 = N2$, then $|N1 - N2| = 0$. Figure 5.11 shows the flowchart.

Figure 5.12 is a program incorporating the logic shown in the flowchart of Figure 5.11.

Sometimes a flowchart is too long to fit on a single page. In this case, another kind of connector symbol is used to show paths that run from one page to the other. The *offpage connector* symbol is shown in Figure 5.13.

Example 5.1 In a certain community, the fine for speeding is $2 for each mile per hour over the legal limit, up to and including 20 MPH over the limit. The charge is then $5 for each mile per hour greater than 20 MPH over the limit. The total fine computed this way cannot exceed $100. However, the actual fine is equal to the previously computed fine multiplied by the number of previous speeding convictions, unless there are no previous convictions (in this case the total fine is unchanged).

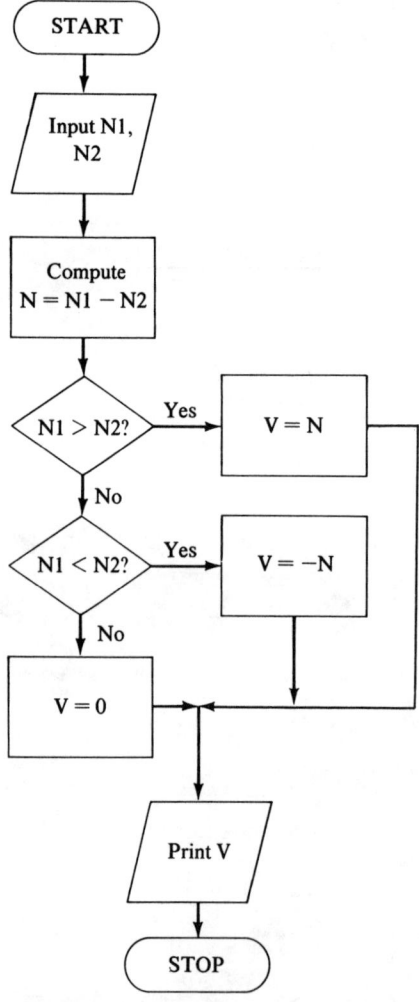

FIGURE 5.11 Flowchart diagraming the computation of the absolute value of the difference between two numbers

```
10 INPUT N1,N2
20 LET N=N1-N2
30 IF N1>N2 THEN 70
40 IF N1<N2 THEN 90
50 LET V=0
60 GOTO 100
70 LET V=N
80 GOTO 100
90 LET V=-N
100 PRINT V
110 END
```

FIGURE 5.12 Program performing the computations diagramed in the flowchart of Figure 5.11

Write a BASIC program that prompts the user to enter the legal limit L, the driver's speed S, and the number of previous convictions C, and then calculates the fine.

A flowchart showing the logical sequence of decisions that must be made to calculate the fine is diagramed in Figure 5.14 on page 100.

As you can see in the flowchart, we first determine whether the driver's speed was greater than 20 MPH over the limit ($S - L > 20$?). If *not,* the fine F is \$2 times the excess speed, $S - L$. If so, the fine is then \$40 (\$2 for each 20 MPH over the limit) plus \$5 times the miles per hour in excess of 20 over the limit. Therefore, in this case, F is calculated as $40 + 5(S - L - 20)$. We then check to determine whether the latter value exceeds the maximum allowable (\$100) and set F to \$100 if it does. Finally, we check to see whether the total number of convictions C is zero. If so, the fine is F; otherwise, it is CF. A BASIC program implementing this logic is shown in Figure 5.15 on page 101. Also shown is the result of a run where $L = 55$ MPH, $S = 80$ MPH, and $C = 2$.

Example 5.2 A manufacturer produces a steel rod with a circular cross-section and a nominal (ideal) cross-sectional area of 1 square inch. The rods are to be classified according to their actual cross-sectional areas. Those having an area of 1 in² ± 0.001 in² have a 0.1% tolerance; those having an area of 1 in² ± 0.005 in² have a 0.5% tolerance; those having an area of 1 in² ± 0.01 in² have a 1.0% tolerance; and those having an area outside the range 1 in² ± 0.01 in² are rejected as being out of tolerance.

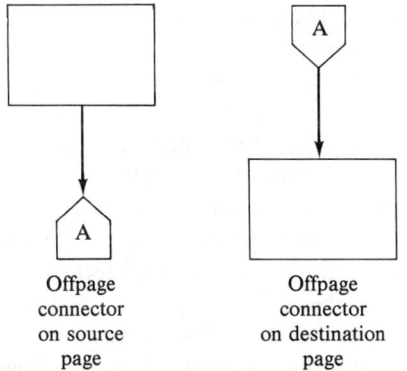

FIGURE 5.13 Examples of offpage connectors used in a flowchart

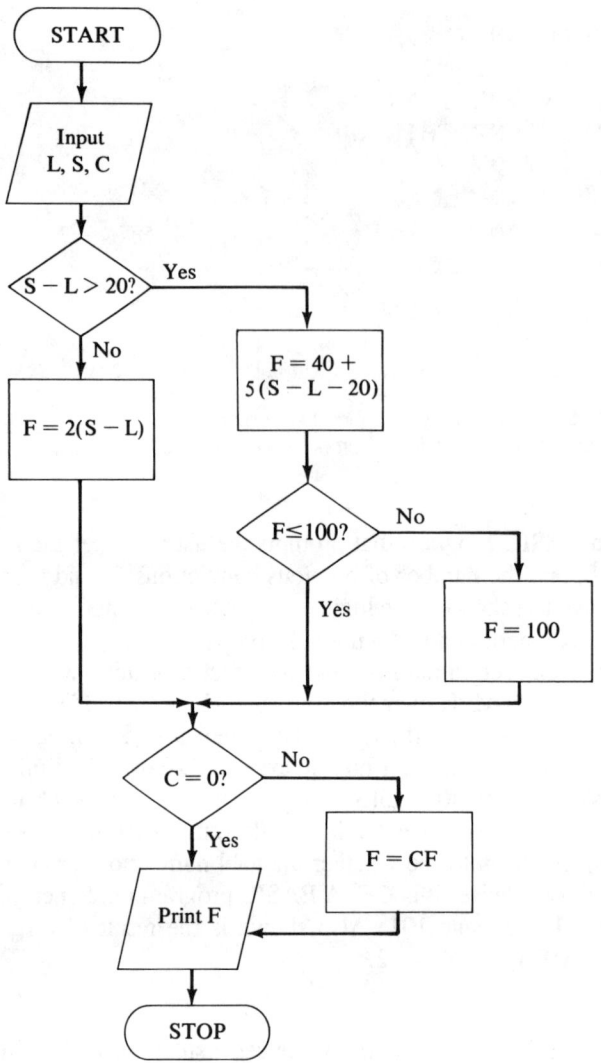

FIGURE 5.14 Flowchart for computing a speeding fine (Example 5.1)

Write a BASIC program to read a set of 10 values of rod diameters and then assign each rod the *best* tolerance for which it qualifies.

The logic that must be implemented to design this program is somewhat complex. The design process requires considerable thought to determine what conditional branch statements must be used to perform the classifications and in what sequence the statements should be executed. In problems like these, a few trial-and-error flowcharts might be needed before a satisfactory solution is found. In this example, a good starting point would be to make a diagram like that of Figure 5.16 showing the exact ranges of cross-sectional area A that correspond to each tolerance classification.

Figure 5.17 on page 102 is a flowchart showing one logical sequence of decision blocks that will result in the proper classification of the rods. The logic represented by the flowchart in Figure 5.17 is as follows: If $A > 1.01$, then A is greater than its maximum permissible value and the unit is therefore out of tolerance; if A is not >1.01 but is >1.005, A must then be in the range from 1.005 (noninclusive) through 1.01 (inclusive) and the unit is therefore classified as 1.0%; if A is not >1.005 but is >1.001, A must then be in the range from 1.001 to 1.005 and is therefore classified as 0.5%. This logic continues until A is finally compared to 0.990, the lower limit of the 1% tolerance range.

```
10 INPUT L,S,C
20 IF S-L>20 THEN 50
30 LET F=2*(S-L)
40 GOTO 80
50 LET F=40+5*(S-L-20)
60 IF F<=100 THEN 80
70 LET F=100
80 IF C=0 THEN 100
90 LET F=C*F
100 PRINT "FINE IS $"F
110 END
>

RUN

?55,80,2
FINE IS $ 130
```

FIGURE 5.15 Program corresponding to the flowchart in Figure 5.14 (Example 5.1)—The fine for driving 80 MPH in a 55-MPH zone, with two previous convictions, is seen to be $130.

Figure 5.18 on page 103 shows the program to implement the logic of Figure 5.17.

Example 5.3 If compound logic expressions are permitted in the version of BASIC being used, then you can approach the programming problem of Example 5.2 more directly. The expression of the test conditions corresponding to each tolerance range can then be written in terms of the minimum and maximum values that A can have to qualify for a particular tolerance. For example, to check a unit's qualifications for a 1% tolerance, we can test directly to determine if either $1.005 \leq A < 1.01$ or $0.990 \leq A < .995$. (See Figure 5.19 on page 103.) This testing could then be accomplished by two successive conditional branches (decision blocks) as shown in Figure 5.19.

Most versions of BASIC that permit the use of compound logic expressions in conditional branch statements also permit more than one such expression to be used in a single statement. In that case, the solution to Example 5.2 is even more straightforward. For example, we can test to determine if a unit qualifies for the 1% tolerance rating by using the conditional branch statement

```
IF (A>1.005 AND A<=1.01) OR (A>=0.990 AND A<0.995) GOTO sn
```

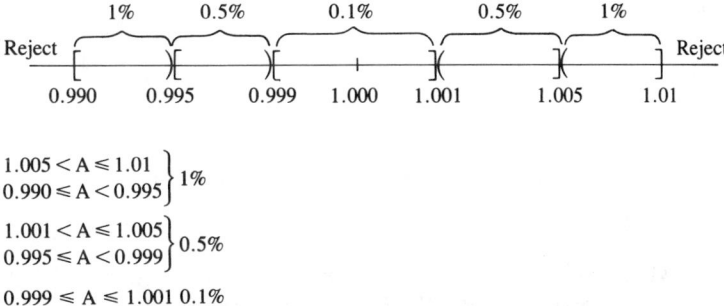

$$1.005 < A \leq 1.01$$
$$0.990 \leq A < 0.995 \Big\} 1\%$$

$$1.001 < A \leq 1.005$$
$$0.995 \leq A < 0.999 \Big\} 0.5\%$$

$$0.999 \leq A \leq 1.001 \quad 0.1\%$$

FIGURE 5.16 Ranges of cross-sectional area corresponding to various tolerance ratings (Example 5.2)

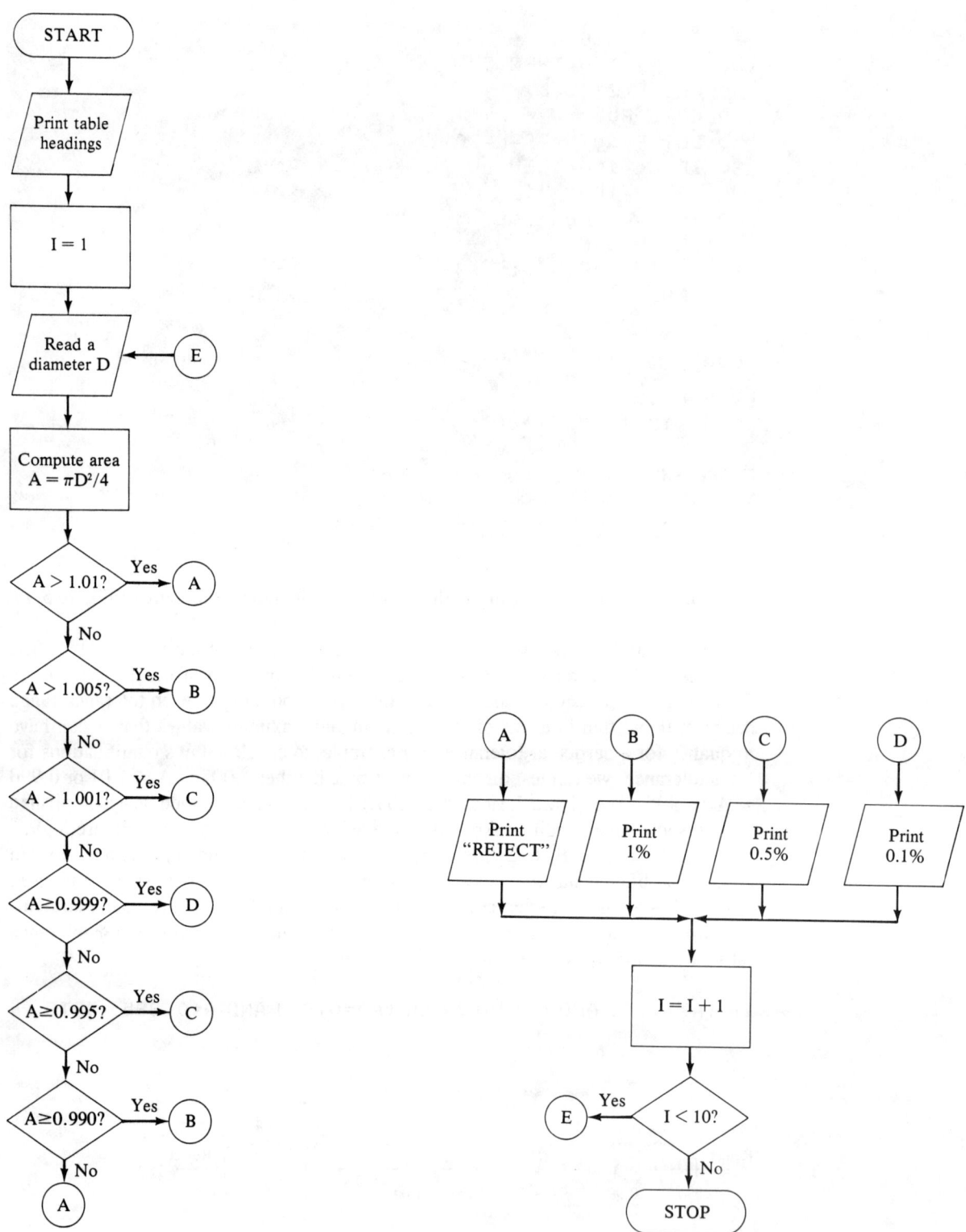

FIGURE 5.17 Flowchart showing the logic for assigning tolerances (Example 5.2).

where *sn* is the statement number where the appropriate PRINT statement is located. Some versions of BASIC impose different requirements on the placement of parentheses in such expressions and on the keyword (THEN or GOTO) following the expression, so consult your programming manual for specifics. Figure 5.20 on page 104 shows a flowchart to construct a program using these kinds of logic expressions.

```
5 PRINT "DIAMETER","AREA","TOLERANCE"
10 LET I=1
20 READ D
30 LET A=3.1416*D**2/4
40 IF A>1.01 THEN 100
50 IF A>1.005 THEN 120
60 IF A>1.001 THEN 140
70 IF A>=.999 THEN 160
80 IF A>=.995 THEN 140
90 IF A>=.990 THEN 120
100 PRINT D,A,"REJECT"
110 GOTO 170
120 PRINT D,A,"1%"
130 GOTO 170
140 PRINT D,A,"0.5%"
150 GOTO 170
160 PRINT D,A,"0.1%"
170 LET I=I+1
180 IF I<=10 THEN 20
190 REM USER SUPPLIES DATA (10 DIAMETER VALUES) IN
    LINE 200
200 DATA
210 END
>
```

FIGURE 5.18 Program corresponding to the flowchart in Figure 5.17 (Example 5.2)

We have shown that a flowchart can be a valuable aid in the design of a program. It is also an important part of program *documentation*. Even if a flowchart is not used in the design process to help define the statements necessary to achieve a certain logical sequence, the programmer should construct one after the program has been written, to help others understand the logic. Also, programmers must often modify or expand programs that they have previously written, and flowcharts are very useful for recalling the thought processes from which the program evolved. A well-documented program, containing a generous number of remark (REM) statements and accompanied by a flowchart, can save a lot of time that might otherwise be spent reinventing the proverbial wheel.

As you become more familiar with the structure of certain well-defined programming techniques such as loops, we will abbreviate our flowcharts so that single blocks

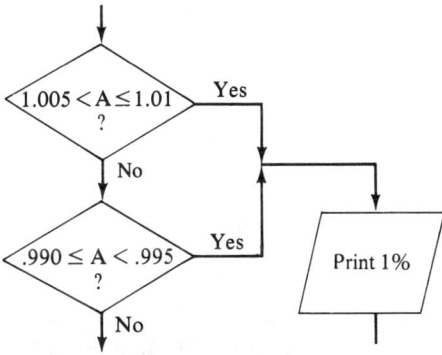

FIGURE 5.19 Compound logic expressions used to simplify the tolerance assignment problem (Example 5.3)

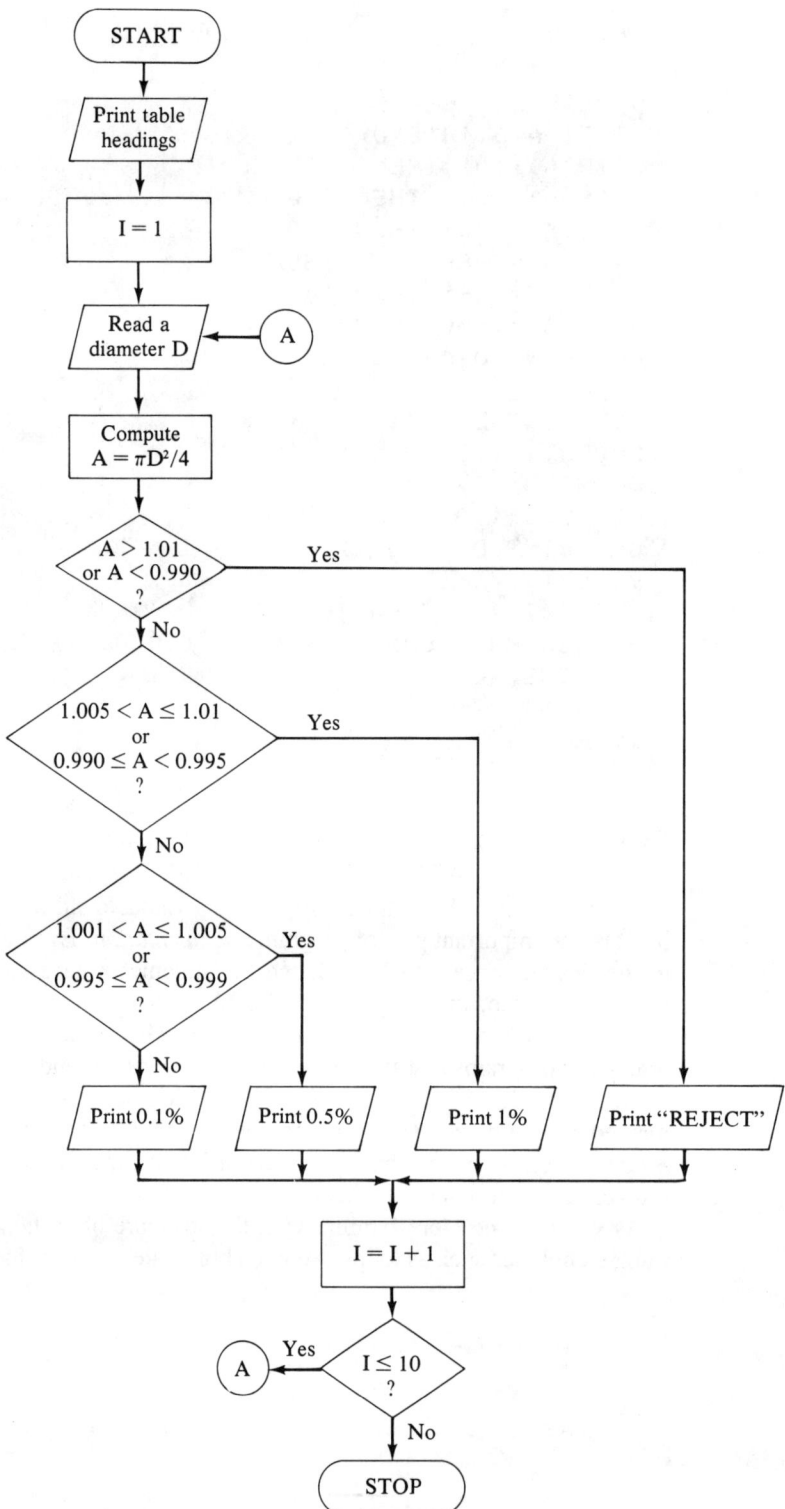

FIGURE 5.20 Flowchart for the tolerance assignment problem, using compound logic expressions (Example 5.3)

are used to show major segments of a program. For example, suppose that we expand the tolerance assignment problem discussed in Examples 5.2 and 5.3 to include provisions for determining the tolerances of ten additional rods having nominal areas of 2 in². An abbreviated flowchart might then be constructed as shown in Figure 5.21. In Figure 5.21, the process block labeled "Determine tolerance" replaces all decision

BRANCHING AND LOOPING 105

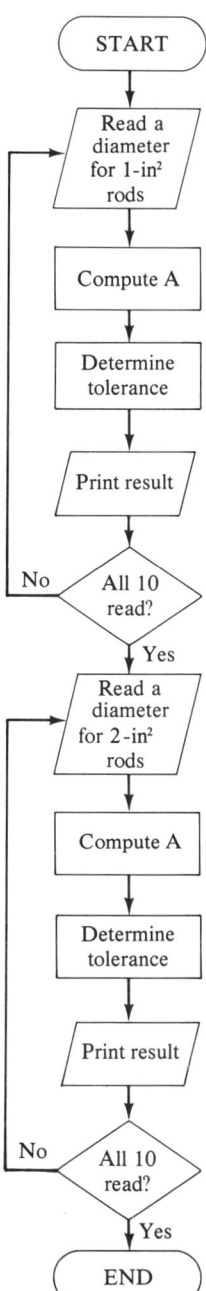

FIGURE 5.21 Abbreviated flowchart for the tolerance assignment problem

blocks that test the value of A in the detailed flowchart shown in Figure 5.17 or 5.20. Documentation of very long, complex programs should include an abbreviated flowchart giving an overview of the sequence in which computations and decisions are made, as well as detailed flowcharts for any intricate segments of the program.

EXERCISES

5.12 Construct detailed flowcharts for each of the following programs:

(a) ```
10 INPUT X1,X2
20 IF X1>X2 THEN 50
```
```

```
           30 PRINT X2
           40 GOTO 60
           50 PRINT X1
           60 END

    (b) 10 INPUT A,B,C
        20 IF B*B>4*A*C THEN 60
        30 IF B*B<4*A*C THEN 80
        40 PRINT "ROOTS ARE EQUAL"
        50 GOTO 90
        60 PRINT "ROOTS ARE REAL, UNEQUAL"
        70 GOTO 90
        80 PRINT "ROOTS ARE COMPLEX"
        90 END

    (c) 10 LET T=0
        20 INPUT M
        30 LET T=T+1
        40 PRINT 2*T
        50 IF T<M THEN 30
        60 END

    (d) 10 LET P=100
        20 INPUT B1,B2
        30 IF B1>B2 THEN 60
        40 LET B=B2
        50 GOTO 70
        60 LET B=B1
        70 LET P=P-2
        80 PRINT P/2
        90 IF P>=B THEN 70
        100 END
```

5.13 Write a BASIC program to implement the logic diagrammed in the flowchart shown in Figure 5.22. What does this program accomplish? (What is the relationship of M to X, Y, and Z?)

5.14 Construct a flowchart and write a BASIC program that prompts the user to enter two integers, M and N, and then does the following: If $M > N$, the program prints the integers counting down from M through N ($M, M - 1, \ldots, N + 1, N$); If $M < N$, it prints the integers counting up from M through N ($M, M + 1, \ldots, N - 1, N$); If $M = N$, it prints M = N.

5.15 A certain state assesses fees for automobile license plates on the basis of automobile weight, age, and horsepower. The minimum fee is $50. If the weight is greater than 3500 lbs, $20 is added to the minimum fee. For each year of age greater than 10, $1 is subtracted from the total fee, but the total fee cannot be less than $50. For each unit of horsepower greater than 150, $1 is added to the total fee. Construct a flowchart and write a BASIC program to compute and print the total license fee after the user enters automobile weight, age, and horsepower.

5.16 Let y_1, y_2, and y_3 be three successive values of a mathematical function. If y_2 is less than both y_1 and y_3, then y_2 is the minimum. If y_2 is greater than both y_1 and y_3, then y_2 is the maximum. If $y_1 < y_2 < y_3$ or $y_1 > y_2 > y_3$, then y_2 is neither minimum nor maximum. Construct a flowchart and write a BASIC program to prompt the user to enter the values of y_1, y_2, and y_3 (in that order) and then determine and print the appropriate message: "Y2 IS MINIMUM", "Y2 IS MAXIMUM", or "Y2 IS NEITHER MIN NOR MAX".

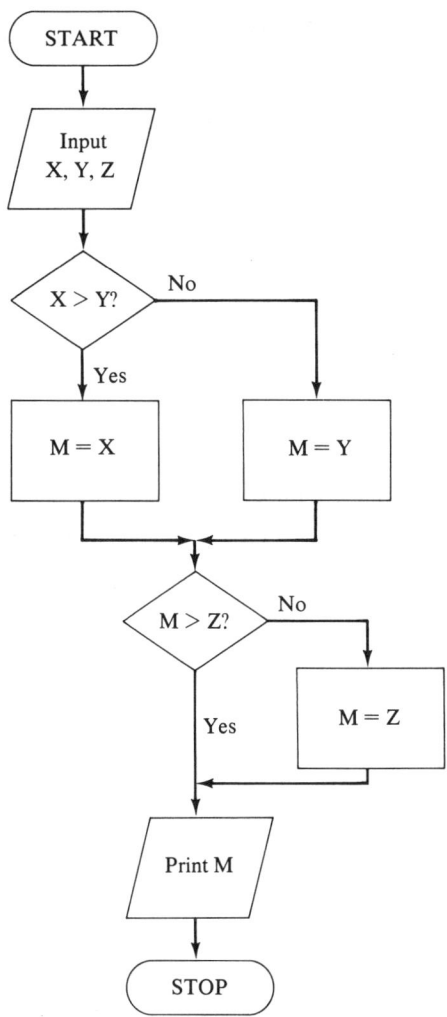

FIGURE 5.22 Exercise 5.13

5 THE FOR-TO AND NEXT STATEMENTS

We have shown that a programming loop consists of a set of statements that is executed over and over again. The number of passes made through the loop is controlled by a counter variable whose value is tested once for each pass. In BASIC there is another way to set up a loop without the explicit testing of a counter variable. The FOR-TO and NEXT statements allow you to define a counter variable and specify the number of times you would like a certain set of statements to be executed. The format of a FOR-TO, NEXT loop is as follows

$$\text{FOR } v = n_1 \text{ TO } n_2$$
$$\ldots$$
$$\ldots$$
$$\ldots$$
$$\text{NEXT } v$$

where v is a variable name and n_1 and n_2 are integer numbers. All statements written between the FOR-TO statement and the NEXT statement constitute the *body* of the loop (the statements that will be repeated). During the first pass through the loop, the variable v has a value of n_1; during the second pass, v is automatically incremented (set equal to $n_1 + 1$); during the next pass, v is incremented again. This process continues until, during the last pass, v has a value of n_2. n_1 is called the *initial* value of v and n_2 the *final* value. (Obviously, n_1 should be less than n_2.) The body of the loop must *always* be terminated by a NEXT statement. Think of the NEXT v statement as telling the computer to repeat the loop using the next value of v—that is, the previous value of v plus 1.

As a simple example, consider how we could use a FOR-TO, NEXT loop to print integers from 1 through 10:

```
10 FOR I=1 TO 10
20 PRINT I
30 NEXT I
```

In this example, the loop consists only of statement 20: PRINT I. In the first pass, $I=1$, so statement 20 causes the number 1 to be printed. In the second pass, $I=2$, so the number 2 is printed, and so forth, until $I=10$ and the number 10 is printed. When all passes have been completed (when the loop is terminated), the computer automatically goes to the statement following the NEXT statement and proceeds from there through the remainder of the program. Programmers usually say that you have "fallen through" the loop when all passes have been completed.

As an example of how a FOR-TO, NEXT loop could be used to set up a table of values, the following program prints positive and negative powers of 10 ($10^{\pm N}$), where N ranges from 0 through 1 in increments of 0.1.

```
10 PRINT "N", "POS. POWER", "NEG. POWER"
20 FOR J=0 TO 10
30 LET N=J/10
40 LET A=10**N
50 LET B=10**(-N)
60 PRINT N,A,B
70 NEXT J
80 END
```

The results of running this program are shown in Figure 5.23.

The quantities n_1 and n_2 identifying the initial and final values of the counter variable v in a FOR-TO, NEXT loop can also be integer-valued expressions. Consider, for example, the following sequence:

```
10 INPUT X,Y
20 FOR B=(Y-X)/3 TO Y/3
30 PRINT B
40 NEXT B
```

Note that $n_1 = (Y-X)/3$ and $n_2 = Y/3$, so the number of passes made through the loop depends on the values entered for X and Y. If the user enters $X=6$ and $Y=12$, then there will be three passes, corresponding to $B=2$, $B=3$, and $B=4$. If the expressions do not produce integer values, then some versions of BASIC will *truncate* the values—that is, drop any fractional parts. In the preceding example, if the user enters $X=1$ and $Y=9$, then $(Y-X)/3 = 2.66667$ and $Y/3 = 3$. The computer will then truncate 2.66667 to produce 2 and perform two passes through the loop, corresponding to $B=2$ and $B=3$. Note that truncation is *not* the same as rounding off a number. In other versions of BASIC, the passes will be performed regardless of the value of n_1, and 1 will be added to v during each pass until v acquires a value larger

```
N              POS. POWER      NEG. POWER
0              1               1
.100000        1.25893         .794328
.200000        1.58489         .630957
.300000        1.99526         .501187
.400000        2.51189         .398107
.500000        3.16228         .316228
.600000        3.98107         .251189
.700000        5.01187         .199526
.800000        6.30957         .158489
.900000        7.94328         .125893
1              10              .100000

>
```

FIGURE 5.23 Table of positive and negative powers of 10, constructed by a FOR-TO, NEXT loop

than n_2. In this case, the preceding example, with $X=1$ and $Y=9$, would result in just the display

2.66667

since a second pass would have $v = 3.66667$, which is greater than $n_2 = 3$.

As a general rule, the value of a counter variable v should not be altered by any statements in the body of the FOR-TO, NEXT loop. For example, the following is *not* correct:

```
10 FOR I=1 TO 5
20 PRINT I
30 LET I=I+1
40 NEXT I
```

Statement 30 illegally alters the value of the counter variable *I*.

As another general rule regarding the use of FOR-TO, NEXT loops, never enter (branch to) a statement within the body of the loop from somewhere outside the loop. Thus, you should never have a sequence such as the following:

```
10 LET X=5
20 FOR I=1 TO 10
30 PRINT I
40 NEXT I
50 IF X>3 THEN 30
```

Statement 50 incorrectly causes a branch to statement 30, which is inside the loop. In some situations, it is necessary to exit a loop (branch to someplace outside the loop) before the loop has terminated, and most versions of BASIC will permit this maneuver. In this case, the counter variable retains the value it had when the loop was exited.

In many programming applications, we want to increment the counter variable v in a FOR-TO, NEXT loop by a value other than 1. For example, we might want $v = 2$ during the first pass, $v = 4$ during the second pass, and so forth. This can be achieved by appending STEP n_3 to the FOR-TO statement, where n_3 is the desired increment size. For example, the statement

```
FOR I=2 TO 10 STEP 2
```

sets up a loop that will be executed for $I = 2, I = 4, I = 6, I = 8$, and $I = 10$. In the following example, we compute the product of the first N odd numbers, where N is an odd number entered by the user.

```
10 LET P=1
20 INPUT N
30 FOR I=3 TO N STEP 2
40 LET P=P*I
50 NEXT I
60 PRINT P
```

The quantity n_3 specifying the size of the step can be an expression, just as n_1 and n_2 can be expressions. Also, most versions of BASIC permit the step size n_3 to be negative. When the step is negative, the value of n_1 should be greater than the value of n_2, since each pass through the loop reduces n_1 by the size of the step until the result is less than n_2. For example, the following program prints the numbers from 10 to 1 in descending order:

```
10 FOR I=10 TO 1 STEP -1
20 PRINT I
30 NEXT I
40 END
```

In many versions of BASIC, a FOR-TO, NEXT loop is equivalent to a loop in which the value of the counter variable is incremented and tested at the bottom of the loop. Therefore, in these versions of BASIC, the loop will always be executed at least once, regardless of the values specified for n_1 and n_2. For example, the following segment would cause the number 3 to be printed, because the computer does not "learn" that n_1 is already greater than n_2, until it reaches the bottom of the loop:

```
10 FOR K=3 TO 1
20 PRINT K
30 NEXT K
```

In other versions of BASIC, nothing would be printed.

The FOR-TO, NEXT loop can be shown in a flowchart as illustrated in Figure 5.24, which diagrams the loop

```
FOR I=1 TO 10
PRINT I
NEXT I
```

Some programmers prefer to enclose a FOR-TO, NEXT loop within dashed lines on the flowchart, or else show its logical equivalent (see Figure 5.25). Again, use whatever system works best for you.

Example 5.4 Temperature measurements were made at nine different times during an experiment designed to test the isothermal (constant temperature) property of a certain heat process. The temperatures in degrees Celsius, were recorded as follows: 112.5, 98.3, 102.9, 100.6, 99.1, 98.9, 108.6, 94.4, and 111.0. Write a BASIC program using a FOR-TO, NEXT loop to compute and print the mean (average) of the temperatures and the standard deviation. The standard deviation is defined as

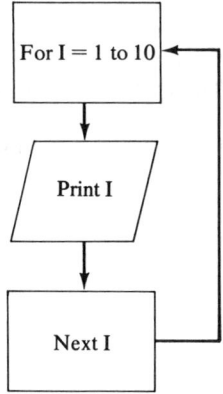

FIGURE 5.24 Diagraming a FOR-TO, NEXT loop in a flowchart

$$s = \sqrt{\sum_{i=1}^{n} \frac{(T_i - \overline{T})^2}{n-1}}$$

where T_i is the ith reading ($i = 1$ to n); $\sum_{i=1}^{n}$ means the sum of n items (as i ranges from 1 to the number n of observations); and $\overline{T}$ is the mean value of the n observations. Note that

$$\overline{T} = \frac{1}{n}\sum_{i=1}^{n} T_i$$

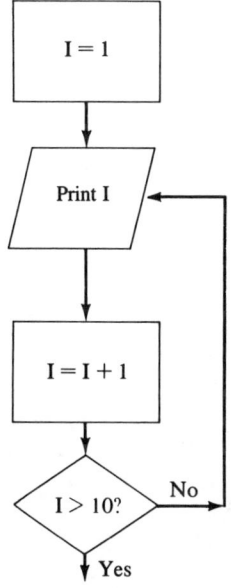

FIGURE 5.25 Logically equivalent method for diagraming a FOR-TO, NEXT loop

A flowchart diagraming the computational procedures necessary to find $\bar{T}$ and s is shown in Figure 5.26.

The program corresponding to the flowchart of Figure 5.26 is shown in Figure 5.27.

Figure 5.28 shows the results of a run of the program in Figure 5.27.

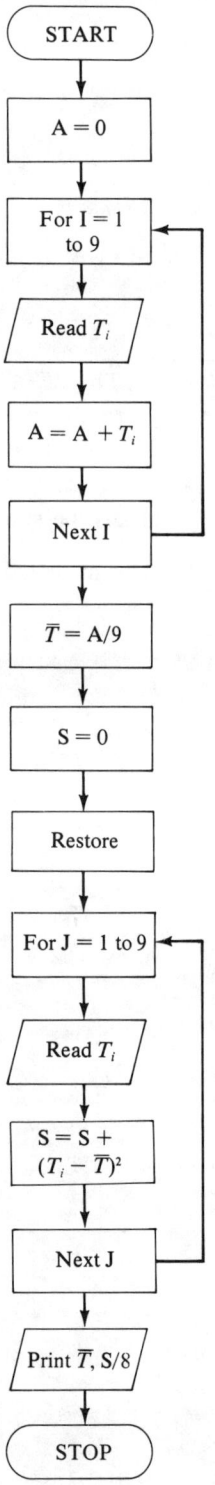

FIGURE 5.26 Flowchart for Example 5.4

BRANCHING AND LOOPING

```
10 REM   COMPUTE MEAN TEMPERATURE
20 LET A=0
30 FOR I=1 TO 9
40 READ T
50 LET A=A+T
60 NEXT I
70 LET M=A/9
80 REM   COMPUTE STD. DEV.
90 LET S=0
100 RESTORE
110 FOR J=1 TO 9
120 READ T
130 LET S=S+(T-M)**2
140 NEXT J
150 PRINT "MEAN="M
160 PRINT "STD. DEV.="(S/8)**.5
170 DATA 112.5,98.3,102.9,100.6,99.1,98.9,108.6,94.4,
    111.0
180 END
>
```

FIGURE 5.27 Program for Example 5.4

```
MEAN=102.922
STD. DEV.=6.31858

>
```

FIGURE 5.28 The results of a run of the program in Figure 5.27

In some programming situations, the body of a loop must contain another loop. When this occurs, we say that one loop is *nested* within the other. The following program is a simple example of this kind of structure:

```
10 FOR I=1 TO 3
20 FOR J=1 TO 2
30 PRINT I "," J
40 NEXT J
50 NEXT I
60 END
```

The flowchart in Figure 5.29 shows logic equivalent to the nested FOR-TO, NEXT loops in the preceding example.

Program execution begins with $I = 1$ and $J = 1$, so the first time the PRINT statement is executed the resulting display is 1,1. While I remains 1, J is incremented to 2, so the next display is 1,2. This completes the "J-loop" (called the inner loop), so I is now incremented to 2. J is then reset to 1. Therefore, the next display is 2,1. This process continues: The inner loop goes through a complete cycle ($J = 1, J = 2$) while the value of I remains fixed; then I is incremented and the inner loop goes through its complete cycle again. The process terminates when I completes its full cycle ($I = 1, I = 2, I = 3$).

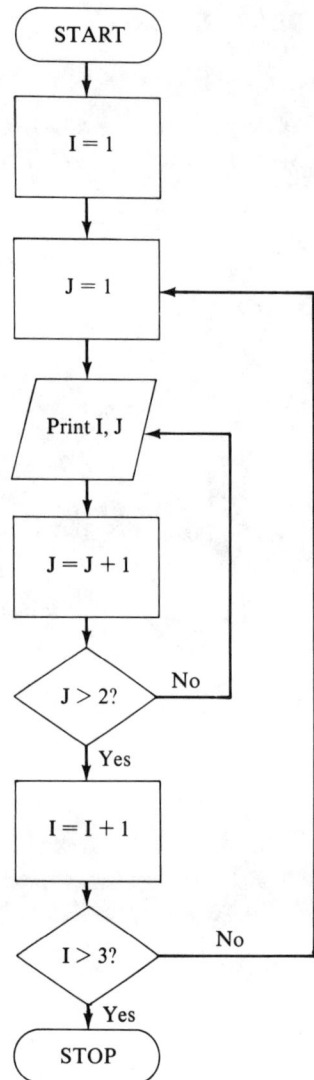

FIGURE 5.29 Flowchart showing a nested loop

You can understand the process we have just described by studying the following diagram and the accompanying sequence of displays generated by the program:

```
START
  ↓
I = 1      I = 2      I = 3           I    J
J = 1      J = 1      J = 1           1    1
J = 2      J = 2      J = 2           1    2
                        ↓             2    1
                      STOP            2    2
                                      3    1
                                      3    2
```

Example 5.5 In a computer programming course attended by eight students, the instructor gave three tests during the semester. The test grades are shown in the following table:

Student	Test 1	Test 2	Test 3
A	82	76	89
B	67	85	85
C	94	100	89
D	38	44	60
E	81	0	77
F	88	74	90
G	68	70	55
H	91	91	100

Write a BASIC program using nested FOR-TO, NEXT loops to compute and print the average grade for each test and the overall test average for all students in the course (the average of the averages). Note the overall average is $(A_1 + A_2 + A_3)/3$, where A_i is the average grade for test i ($i = 1,2,3$).

The flowchart for this program is shown in Figure 5.30. The inner loop in this flowchart is used to compute the sum of the eight grades on one test. The outer loop determines which one of the three tests the inner loop is processing. Thus, when I = 1, the inner loop computes the sum S of the eight grades on test 1 (J = 1,2, . . . ,8). When these eight grades have been summed, the average S/8 is computed and printed. This average is added to the cumulative total T of averages, which will itself ultimately be averaged to produce the overall average (T/3). I is incremented to 2, and the inner loop now sums the grades on test 2. The process continues until all three test averages have been found; the program then falls out of the outer loop and prints the overall average. The program corresponding to the flowchart of Figure 5.30 is shown in Figure 5.31.

Note that the order of the test grades in the DATA statements of Figure 5.31 is critical to the success of the program. The grades must be listed in the same order as they appear in the *columns* of the data table, since the READ statement appears in the inner loop and must therefore read eight successive grades from the same test.

The results of a run of the program in Figure 5.31 are shown in Figure 5.32.

More than one FOR-TO, NEXT loop can be nested within another—that is, an inner loop can itself contain another loop, which may contain another, and so forth. The *depth* of the nesting (the number of loops that can occur within another loop) is limited either by the computer system being used or by the size of its memory.

When FOR-TO, NEXT loops are nested, an inner loop must be closed by its NEXT statement before the outer loop containing the inner one is closed. In other words, the NEXT statements must appear in reverse order from that of their corresponding FOR-TO statements. To illustrate, the following construction is *not* permitted and will cause a run-time error, if attempted:

```
10 FOR I=1 TO 3
20 FOR J=1 TO 2
30 PRINT I,J
40 NEXT I
50 NEXT J
```

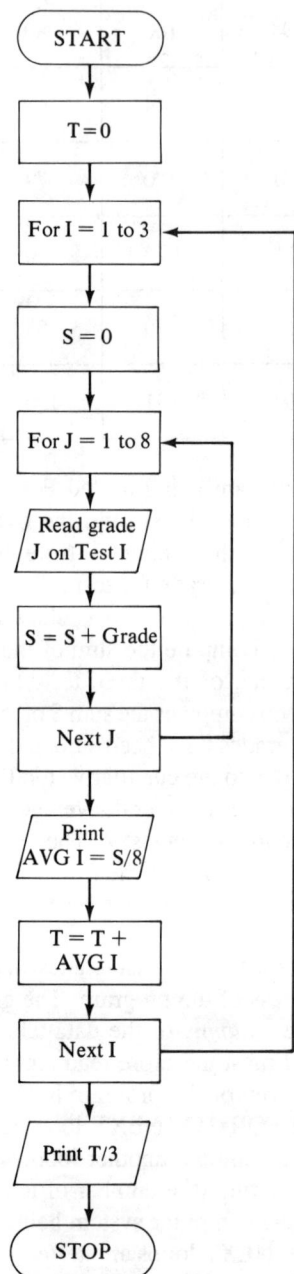

FIGURE 5.30 Flowchart for Example 5.5

```
10 REM   INITIALIZE T (TOTAL OF ALL GRADES)
20 LET T=0
30 FOR I=1 TO 3
40 REM   INITIALIZE S (SUM OF GRADES ON TEST J)
50 LET S=0
60 FOR J=1 TO 8
70 REM   NOW COMPUTE S FOR TEST I
80 READ G
90 LET S=S+G
100 NEXT J
110 REM   PRINT AVERAGE OF GRADES ON TEST I
120 PRINT "AVERAGE OF TEST" I " IS" S/8
130 REM   ADD AVERAGES
140 LET T=T+S/8
150 NEXT I
160 REM   ALL TEST AVERAGES COMPUTED; PRINT AVG. OF
    AVERAGES
170 PRINT "OVERALL AVERAGE IS" T/3
180 DATA 82,67,94,38,81,88,68,91
190 DATA 76,85,100,44,0,74,70,91
200 DATA 89,85,89,60,77,90,55,100
210 END
>
```

FIGURE 5.31 Program for Example 5.5

```
AVERAGE OF TEST 1 IS 76.1250
AVERAGE OF TEST 2 IS 67.5000
AVERAGE OF TEST 3 IS 80.6250
OVERALL AVERAGE IS 74.7500

>
```

FIGURE 5.32 The results of a run of the program in Figure 5.31

EXERCISES

5.17 Show the display that would result if each of the following programs were run:

(a)
```
10 FOR K=3 TO 7
20 LET X=K-3
30 PRINT K-X
40 NEXT K
50 END
```

(b)
```
10 LET P=4
20 FOR R=P**.5 TO P+2
30 PRINT P*R
40 NEXT R
50 END
```

(c)
```
10 LET P=1
20 FOR J=1 TO 6
30 LET P=-P*J
```

```
    40 NEXT J
    50 PRINT P
    60 END
(d) 10 FOR X=1 TO 25 STEP 2
    20 LET Y=X/5
    30 PRINT Y
    40 IF Y>1 THEN 60
    50 NEXT X
    60 PRINT X
    70 END
```

5.18 Write a BASIC program using a FOR-TO, NEXT loop to sum all even numbers from 2 to N, where N is an even number entered by the user. Construct a flowchart for your program.

5.19 Write a BASIC program using a FOR-TO, NEXT loop to generate a table of values of 2^N and $2^{1/N}$, where N ranges from 1 to 10. Construct a flowchart for your program.

5.20 Write a BASIC program using nested FOR-TO, NEXT loops to create the following display:

```
2,5
2,7
2,9
4,5
4,7
4,9
```

Construct a flowchart for your program.

5.21 Write a BASIC program using nested FOR-TO, NEXT loops to compute and print the grade average of each student (using the data from the table in Example 5.5) and the overall grade average for the entire course. The overall grade average is $(\overline{A} + \overline{B} + \ldots + \overline{H})/8$ where $\overline{A}$ is the grade average of student A, $\overline{B}$ is the grade average of student B, and so forth.

6 THE ON-GOTO STATEMENT

Branching in a BASIC program can be achieved by still another method: using the ON-GOTO statement. The format of this statement is

ON v GOTO sn_1, sn_2, . . .

where v is an integer-valued variable or expression, and sn_1, sn_2, . . . are statement numbers. If v has a value of 1, the computer branches to sn_1; if v has a value of 2, it branches to sn_2; and so forth. If there is no statement number sn in the position corresponding to the value of v, the statement following the ON-GOTO is executed next. As an example, consider the statement

```
10 ON T GOTO 120, 150, 80
20
```

BRANCHING AND LOOPING 119

If T has a value of 1, a branch to statement 120 occurs; if $T = 2$, a branch to statement 150 occurs; and if $T = 3$, then a branch to 80 occurs. If $T = 4$, statement 20 is executed next.

If the variable or expression v has a noninteger value, most versions of BASIC will truncate it and respond only to the integer part of the value. For example, if $X = 7$ the statement

```
ON X/4 GOTO 30, 50, 70
```

will cause a branch to statement 30 because $7/4 = 1.75$ is truncated to 1. Some versions of BASIC (Apple, for example) do not allow the variable or expression v to have a negative value.

The following example shows how an ON-GOTO statement is used to print the appropriate message concerning the roots of the quadratic equation $Ax^2 + Bx + C = 0$, after the user enters values for A, B, and C.

```
10  INPUT A,B,C
20  LET D=B*B-4*A*C
30  LET S=3
40  IF D<0 THEN S=1
50  IF D>0 THEN S=2
60  ON S GOTO 70,90,110
70  PRINT "ROOTS ARE IMAGINARY"
80  GOTO 120
90  PRINT "ROOTS ARE REAL AND UNEQUAL"
100 GOTO 120
110 PRINT "ROOTS ARE REAL AND EQUAL"
120 END
```

EXERCISES

5.22 Write the statement number to which the computer branches after executing each of the following program segments:

(a)
```
10 LET C=5
20 LET N=C-C/5
30 ON N GOTO 1,5,50,60,100
40 GOTO 80
```

(b)
```
10 FOR I=1 TO 10
20 LET X=I+1
30 NEXT I
40 ON X GOTO 40,200,80
50 IF I>10 THEN 100
60 GOTO 150
```

5.23 Write a BASIC program using an ON-GOTO statement that

(a) prompts the user to enter a number N,
(b) computes and prints the value of N! if $N = 4$ or 5,
(c) computes and prints the value of $(N^2)!$ if $N = 2$ or 3, and
(d) prints N = 1 if $N = 1$.

Note that $N! = (N)(N - 1) \ldots (1)$.

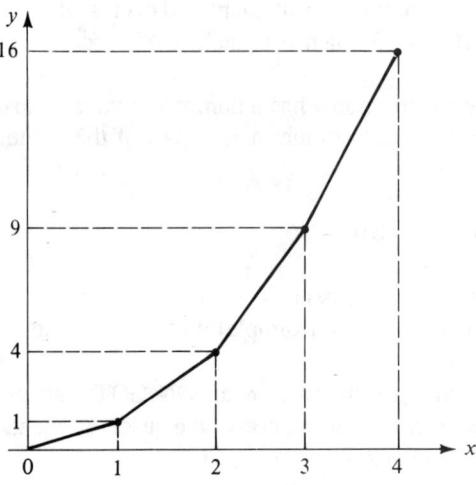

FIGURE 5.33 Exercise 5.24

GENERAL EXERCISES

5.24 Figure 5.33 shows a *straight-line segment approximation* for the parabola $y = x^2$, when x is in the range $0 \leq x \leq 4$.
According to this approximation,

$$\begin{aligned} y &= x & \text{if } 0 \leq x \leq 1 \\ y &= 3x - 2 & \text{if } 1 \leq x \leq 2 \\ y &= 5x - 6 & \text{if } 2 \leq x \leq 3 \\ y &= 7x - 12 & \text{if } 3 \leq x \leq 4 \end{aligned}$$

Write a BASIC program to find the value of y, given by the straight-line approximation, after the user enters a value of x between 0 and 4, inclusive. Print the message X OUT OF RANGE if $x < 0$ or $x > 4$. Run your program for $x = 0.5$, $x = 2$, and $x = 3.75$.

5.25 Six astronomers participated in an experiment in which each measured the distance from the earth to the sun, using three different measurement techniques. The results of these measurements (in millions of miles) are tabulated below:

Astronomer	Method A	Method B	Method C
1	93.07	92.89	94.60
2	93.75	93.01	93.84
3	92.99	92.76	93.57
4	93.52	92.98	94.12
5	91.89	91.60	93.04
6	94.02	94.12	95.09

Write a BASIC program using FOR-TO, NEXT loops to compute and print the average distance measured by each astronomer, the average distance obtained by each measurement technique, and the overall average of all measurements. Construct a flowchart for your program.

EXERCISES FOR ARCHITECTURAL/CIVIL/CONSTRUCTION TECHNOLOGY

5.26 A group of architectural and engineering consultants is preparing a study on the air conditioning requirements of a large office building. They wish to perform a computer simulation that will allow them to input certain environmental conditions and determine the response of the cooling system. The system is designed to turn on under any of the following conditions:

(1) The temperature exceeds 80° F and the relative humidity exceeds 50%, provided the time of day is between 8:00 AM and 5:00 PM.
(2) The temperature exceeds 75° F, and the relative humidity exceeds 90%, provided the time of day is between 8:00 AM and 5:00 PM.
(3) The temperature exceeds 85° F, and the time of day is after 5:00 PM and before 8:00 AM.
(4) The temperature exceeds 90° F.

Write a BASIC program that prompts the user to enter temperature, humidity, and time of day and then prints A/C IS ON or A/C IS OFF. Assume that the time is entered using a 24-hour clock (12 noon to 12 midnight is 1200 to 2400). Construct a flowchart for your program, and run your program for each of the following conditions:

(a) $T = 82°F, H = 65\%$, time = 11:00 AM
(b) $T = 80°F, H = 95\%$, time = 2:30 PM
(c) $T = 72°F, H = 99\%$, time = 12:00 noon
(d) $T = 93°F, H = 70\%$, time = 5:00 PM
(e) $T = 86°F, H = 50\%$, time = 4:00 AM

5.27 The number of bricks required to construct a set of brick steps depends on the width W and number N of steps. For a certain design, the first (lowest) step requires $6W$ bricks, where W is the width in feet; the second step requires $6W + 6W = 12W$ bricks; the third step requires $6W + 6W + 6W = 18W$ bricks, and so forth. The top (Nth) step is twice as deep as the others and therefore requires twice as many bricks as it would if it had the same depth as the others. Figure 5.34 shows an example in which the number of steps is $N = 4$. The total number of bricks in this example is $6W + 12W + 18W + 2(24W) = 84W$.

Write a BASIC program using a FOR-TO, NEXT loop to determine the total number of bricks required, after the user enters the number of steps N and the width W in feet. Construct a flowchart for your program. Run your program for $N = 6$ steps and $W = 6$ feet and for $N = 10$ steps and $W = 4$ feet.

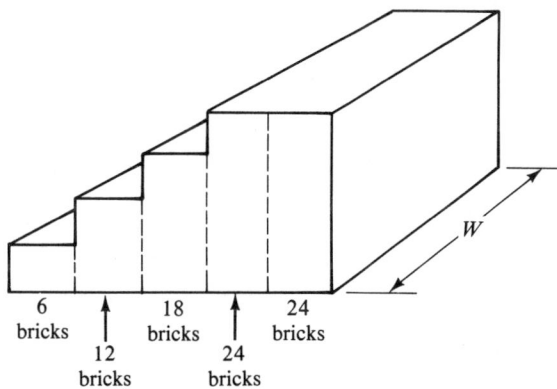

FIGURE 5.34 Exercise 5.27

5.28 A survey of building costs was made in three different regions of the country. Four types of buildings were considered in the survey. The following table shows the area of each building (in thousands of square feet) and the total construction cost (in thousands of dollars).

	REGION 1		REGION 2		REGION 3	
	Area	Cost	Area	Cost	Area	Cost
Type A	24.5	87.5	19.8	65.3	21.4	72.1
Type B	12.9	43.2	15.1	50.4	13.7	60.0
Type C	63.6	181.0	54.2	148.5	48.5	137.7
Type D	185.0	483.9	170.7	311.9	201.3	568.0

Write a BASIC program using FOR-TO, NEXT loops to compute and print the average building cost (in dollars per square foot) for each type of building, the average building cost in each region of the country, and the overall average building cost (for all types of buildings in all regions of the country). Construct a flowchart for your program.

EXERCISES FOR ELECTRICAL/ELECTRONICS/COMPUTER TECHNOLOGY

5.29 The nominal value of a carbon resistor is 47 kΩ. Standard tolerance ratings are 5%, 10%, and 20%, so the actual value of the resistance is $47 \times 10^3 \pm 0.05(47 \times 10^3)$ or $\pm 0.1(47 \times 10^3)$ or $\pm 0.2(47 \times 10^3)$ ohms. Write a BASIC program that prompts the user to enter a resistance value and then assigns the best tolerance for which it qualifies. If the resistance is outside the range $47 \times 10^3 \pm 20\%$, print the message REJECT. Construct a flowchart for your program, and run your program for each of the following resistance values:

(a) 51.6 kΩ (c) 37.2 kΩ
(b) 44.0 kΩ (d) 56.1 kΩ

5.30 When N resistors are connected in parallel, their total equivalent resistance is

$$R = \frac{1}{\frac{1}{R_1} + \frac{1}{R_2} + \ldots + \frac{1}{R_N}}$$

(See Figure 5.35.) Write a BASIC program using a loop to compute the total equivalent

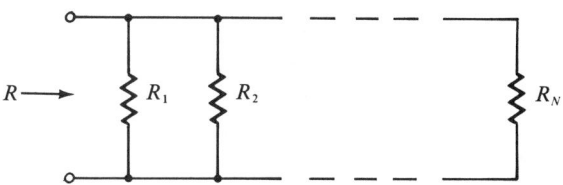

FIGURE 5.35 Exercise 5.30

resistance of 10 resistors connected in parallel. Construct a flowchart for your program. Run your program using the following 10 resistance values: 1.5K, 2.2K, 100K, 4.7K, 1.5K, 3.3K, 10K, 15K, 22K, 47K.

5.31 A survey was conducted to determine electric power costs in four different regions of the country. Data was obtained for three kinds of customers: business, industrial, and residential. The energy consumption, in kilowatt-hours, and the total cost in dollars was reported for each combination of region and customer type, as shown in the following table:

	REGION 1		REGION 2		REGION 3		REGION 4	
	kw-hrs	cost	kw-hrs	cost	kw-hrs	cost	kw-hrs	cost
Business	2,841	$142.05	12,793	$550.09	8,170	$449.35	4,350	$104.40
Industrial	64,800	2916.32	114,080	6274.40	78,230	2738.05	58,900	1590.30
Residential	900	63.50	1,150	68.20	988	74.10	1,764	89.96

Write a BASIC program using FOR-TO, NEXT loops to compute and print the average energy cost (in cents per kilowatt-hour) for each region, the average cost for each type of customer, and the overall average for all customers in all parts of the country. Construct a flowchart for your program.

EXERCISES FOR INDUSTRIAL/MANUFACTURING/PRODUCTION TECHNOLOGY

5.32 An industrial robot is used to perform certain welding operations on an assembly passing in front of it on a conveyor belt. An operator monitors the process and disconnects the power from the robot whenever certain combinations of conditions occur. Power is removed under any of the following conditions:

(a) The conveyor speed falls below 0.5 ft/min and more than 8 minutes have elapsed since the last unit was welded, provided at least 100 units have already been welded.

(b) The conveyor speed is bewteen 0.75 ft/min and 1.25 ft/min, inclusive, and more than 5 minutes have elapsed since the last unit was welded—provided at least 60 units have already been welded.

(c) More than 150 units have been welded and more than 2 minutes have elapsed since the last unit was welded.
(d) The conveyor speed exceeds 2.0 ft/min.

Write a BASIC program that prompts the user to enter the conveyor speed in ft/min, the time that has elapsed since the last unit was welded (in minutes), and the total number of units welded and then prints ON or OFF, depending on whether the robot should be shut down or allowed to continue. Construct a flowchart for your program, and run your program for each of the following combinations of input:

Conveyor speed	0.4	1.0	1.25	2.3	1.1
Time elapsed	9.1	3.6	2.1	1.7	1.4
Units welded	120	75	210	52	85

5.33 A certain manufacturing process consists of ten major components, and the failure of any of these components will cause the entire process to fail. The probability that the system will fail is $P = 1 - P_1 P_2 \ldots P_{10}$, where P_i is the probability that the ith component will operate without failure. Write a BASIC program using a loop to compute and print the probability of process failure when $P_1 = 0.990$, $P_2 = 0.995$, $P_3 = 0.989$, $P_4 = 0.999$, $P_5 = 0.994$, $P_6 = 0.990$, $P_7 = 0.987$, $P_8 = 0.999$, $P_9 = 0.985$, and $P_{10} = 0.995$. Construct a flowchart for your program.

5.34 A survey was conducted to determine the rate at which each of three machines could produce a certain part. Each machine was operated by four different machinists. The number of parts produced and the time required to produce them (in minutes) for each combination of machine and machinist are shown in the following table:

	MACHINE 1		MACHINE 2		MACHINE 3	
	Parts	Time	Parts	Time	Parts	Time
Machinist A	12	38.4	14	40.0	14	39.1
Machinist B	11	37.6	13	39.7	16	52.2
Machinist C	11	39.9	14	42.1	15	45.6
Machinist D	14	49.5	14	48.0	17	68.3

Write a BASIC program to compute and print the average number of parts per minute produced by each machine, the average number of parts per minute produced by each machinist, and the overall average number of parts per minute produced by all machinists on all machines. Construct a flowchart for your program.

EXERCISES FOR MECHANICAL TECHNOLOGY

5.35 The autopilot on a jet aircraft controls its altitude, heading (its direction in magnetic degrees), and its speed. If certain combinations of actual altitude, heading, and speed deviate from those that the autopilot is supposed to maintain, an alarm is sounded. The alarm in a certain autopilot is sounded under any of the following conditions:

(a) The deviation in speed exceeds 20 MPH, and the deviation in altitude exceeds 100 ft, provided the deviation in bearing exceeds 5°.
(b) The deviation in speed exceeds 40 MPH, and the deviation in bearing exceeds 5°, provided the deviation in altitude exceeds 75 ft.
(c) The deviation in speed exceeds 50 MPH, and the deviation in altitude exceeds 50 ft.
(d) The deviation in altitude exceeds 200 ft.

Write a BASIC program that prompts the user to enter the deviations in speed, altitude, and bearing and then prints either ON or OFF to indicate whether the alarm should be sounded. Construct a flowchart for your program, and run your program for each of the following combinations of inputs:

Speed deviation (MPH)	33	50	60	10	40
Bearing deviation (degrees)	6.5	4.2	8.0	2.3	7.8
Altitude deviation (feet)	110	25	60	225	25

5.36 The inertia J reflected from the load side of a gear train to the driving shaft is

$$J = \left(\frac{N_1}{N_2}\right)^2 \left(\frac{N_3}{N_4}\right)^2 \cdots \left(\frac{N_m}{N_n}\right)^2 (J_L)$$

where N_i/N_j is the gear ratio of each pair of gears in the system and J_L is the load inertia. N_i is the number of teeth on the driving gear, and N_j is the number of teeth on the driven gear. Figure 5.36 illustrates a system containing three sets of gears. Write a BASIC program using a loop to compute and print the reflected inertia J in a system containing six sets of gears. The user enters the value of load inertia J_L in N-

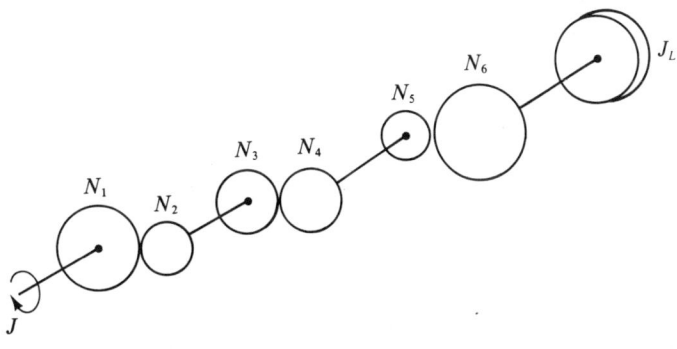

FIGURE 5.36 Exercise 5.36

m-sec^2/rad. Run your program using the following gear data: $N_1 = 32$, $N_2 = 48$, $N_3 = 20$, $N_4 = 30$, $N_5 = 12$, $N_6 = 24$, $N_7 = 48$, $N_8 = 16$, $N_9 = 32$, $N_{10} = 22$, $N_{11} = 15$, $N_{12} = 15$. Use $J_L = 160$ N-m-sec^2/rad. Construct a flowchart for your program.

5.37 The yield strength of a material is the stress (force per unit area) that will cause the material to undergo a specified amount of permanent strain (deformation) after the stress is removed. In an investigation conducted to determine the yield strength of a certain type of steel, 3 different testing machines were used by 4 operators on 12 different test samples of the material. The following table shows the measured force (in thousands of pounds) that caused yielding and the cross-sectional area of each test sample, in square inches, obtained by each operator using each testing machine. (Yield strength = force/area.)

	TEST MACHINE 1		TEST MACHINE 2		TEST MACHINE 3	
	Force	Area	Force	Area	Force	Area
Operator A	52.53	1.04	49.65	0.95	50.18	1.00
Operator B	25.85	0.55	78.02	1.86	47.27	0.88
Operator C	84.61	1.59	63.34	1.25	50.22	1.14
Operator D	39.75	0.80	34.70	0.75	83.03	1.52

Write a BASIC program using FOR-TO, NEXT loops to compute and print the average yield strength (in pounds per square inch) obtained from each test machine, the average yield strength obtained by each operator, and the overall average yield strength obtained from all operators using all machines. Construct a flowchart for your program.

CHAPTER 6

ARRAYS

1 THE NATURE OF AN ARRAY

An *array* is a set of data arranged like a list or a table. Each entry in the array occupies a specific position, so we can refer to the "first" entry (in position number 1), the "second" entry, and so forth. Every array has a name, and any data item contained in an array is specified by stating the name of the array and the position that the data occupies within the array. This specification is accomplished by writing the array name, immediately followed by the position number, the latter enclosed by parentheses. For example, A(3) means the entry in position 3 of the array named A.

In BASIC, the entries in an array are treated in many of the same ways that BASIC variables are treated. In fact, the contents of arrays are referred to as "array variables," and array names are called array variable names. Thus, for example, we can assign values to array variables using the LET statement, as in the following statement:

```
LET B(2)=17
```

This statement assigns the value 17 to entry number 2 in the array named B. Furthermore, array variables can appear in expressions, as in the following conditional branch statement:

```
IF X>2*A(4)/B(1) THEN 30
```

In this example, a branch to statement 30 occurs if the value of X is greater than the value of two times A(4) divided by B(1). In other words, once an array variable has been assigned a value, BASIC treats it in exactly the same way that it would if the value itself were specified. As another example, if M(8) = 10, the statements

```
PRINT 2*M(8)
```

and

```
PRINT 2*10
```

would both result in a display of the number 20.

The choice of an array name must conform to the same rules used to create other variable names in BASIC (although some versions are more restrictive with array names than they are with other variable names). In this book, we will always use a single letter for an array name, since this is permissible in all versions of BASIC. Arrays can also contain string data; in this case, the array name must be followed by a $ sign. For example, we might write

```
LET R$(3)="YES"
```

However, string and numeric data cannot be stored in the same array.

Array variables with position numbers specified in parentheses are much like *subscripted* variables in algebra. For example, suppose that x_1, x_2, and x_3 represent three experimentally determined values of a variable x. These three values could be assigned to an array X and thereafter be referred to in BASIC as X(1), X(2), and X(3). The number appearing in parentheses after the array name is called an *index*, and this index can also be an integer-valued *variable*. The following program segment shows how an array could be assigned ten values by a FOR-TO, NEXT loop using the counter variable I as the index of the array.

```
10 FOR I=1 TO 10
20 PRINT "ENTER A("I")"
30 INPUT A(I)
40 NEXT I
```

During each pass through this loop, the input prompting message

```
ENTER A(i)
```

is printed, where i is the current value of I, corresponding to the number of the pass. Thus, ENTER A(1) is printed during the first pass. After this, the statement INPUT A(1) is executed and the user enters a value for A(1); then "ENTER A(2)" is printed, and the user enters a value for A(2) in response to the INPUT A(2) statement, and so forth.

Arrays are useful in many programming situations. Besides providing a convenient means of storing and retrieving data simply by specifying an index, they also allow us to perform the many different types of algebraic computations that involve subscripted quantities. For example, if you want to calculate the mean and standard deviation of a set of data values x_i, you could first assign these values to an array X, and then proceed to calculate

$$M = \frac{1}{N}\sum_{I=1}^{N} X(I) \quad \text{and} \quad S = \sqrt{\sum_{I=1}^{N} \frac{\{X(I) - M\}^2}{N-1}}$$

Since the data values are stored in array X, the computations of *M* and *S* can be performed without having to READ (or READ and RESTORE) data values from a DATA statement.

2 DIMENSIONING ARRAYS

The word *dimension* is used in two different contexts in connection with arrays. When we think of an array as a list of numbers, its dimension is the same as its capacity (the total number of numbers it can store). To ensure that the computer will reserve enough memory space for storing all the data that a particular array must hold, the computer must be informed of the dimension of that array. This information is conveyed in the program by a DIM statement. The format of a DIM statement is

$$\text{DIM } a(n)$$

where *a* is an array name and *n* is the required dimension. The variable *n* must be a positive integer; it cannot be an expression. More than one array can be dimensioned by a single DIM statement, simply by inserting commas between dimension specifications. For example, the statement

```
DIM A(13),X(20),T$(30)
```

assigns dimension 13 to array A, dimension 20 to array X, and dimension 30 to array T$. Once an array has been dimensioned, any use of an index value outside the range of the dimension will cause an error message to be generated. For example, the following sequence would generate an error message when executed:

```
10 DIM V(20)
20 FOR I=1 TO 21
30 LET V(I)=I/2
40 NEXT I
```

This program segment attempts to assign 11.5 to V(21), which is outside the range (20) for which V has been dimensioned.

Many versions of BASIC treat zero as the "first" position in an array. In these versions, the statement DIM A(13) actually reserves memory spaces for 14 data values, corresponding to A(0), A(1), ..., A(13). The programmer is not required to use index 0 in any array references and, as far as computations or data handling operations are concerned, the programmer is always free to interpret A(1) as the "first" data item in the array. Similarly, it is not necessary to assign values to any particular position in an array. In other words, not all locations for which an array is dimensioned need to be filled. However, it is wasteful of memory space to dimension an array for a greater capacity than will actually be used.

In most versions of BASIC, the DIM statement is optional for an array with 11 or fewer entries (corresponding to subscripts 0 through 10). When the computer encounters a quantity such as X(1), it automatically reserves 11 spaces for an array named X. In some versions of BASIC, including Atari and the proposed ANSI standard BASIC, *all* arrays must be dimensioned, regardless of size. In subsequent programs in this book, we will dimension all arrays used.

To illustrate a simple application of an array, suppose we have two sets of data, one consisting of four values of the variable *A* and the other consisting of four values of *B*. We want to produce a third set of data representing the sums of the corresponding pairs of values of $A + B$ ($A_1 + B_1$, $A_2 + B_2$, ..., etc.) and a fourth set representing products of corresponding pairs (A_1B_1, A_2B_2, ..., etc.). Suppose the data values are $A = 1,4,-12,0$ and $B = 6,2,1,5$. We proceed as follows:

```
10 DIM A(4),B(4),S(4),P(4)
20 FOR I=1 TO 4
30 READ A(I),B(I)
40 LET S(I)=A(I)+B(I)
50 LET P(I)=A(I)*B(I)
60 NEXT I
70 DATA 1,6,4,2,-12,1,0,5
80 END
```

Note that the DATA statement lists values of A and B in an alternating sequence, as required by the program's structure: READ A(I),B(I). Values are read in the sequence A(1), B(1), A(2), B(2), A(3), ...

The word "dimension" is also used to describe another, completely different characteristic of an array. In this new context, the arrays we have described so far are said to be *one-dimensional,* because a single index value is sufficient to identify any one data item in the array. For example, to recover the number 8 from an array T containing $-4,3,8$, and -2, you need only to specify the single index value 3—that is, to write T(3). On the other hand, in a *two-dimensional* array, a *pair* of index numbers is associated with every data value in the array. Thus, A(1,2) can represent one value in the two-dimensional array A, while A(2,1) and A(1,3) represent other values. A two-dimensional array is like a set of doubly subscripted variables in algebra: a_{11}, $a_{12}, \ldots, a_{21}, a_{22}, \ldots$, and so forth. It is convenient to think of the subscripts, or indices, as representing the *row* and *column* in which a given data item is located in a *matrix,* (a table of data). The first subscript represents the row and the second represents the column in which the data is located. Consider, for example, the following matrix M. This matrix consists of three rows and four columns and thus contains $3 \times 4 = 12$ data values.

		Column			
		1	2	3	4
	1	1	3	5	7
Row	2	9	11	13	15
	3	17	19	21	23

In the two-dimensional array M, we see, for example, that

$$M(1,2) = 3, M(3,2) = 19, M(2,2) = 11, \text{ and } M(3,4) = 23.$$

Arrays can also have three or more dimensions, with an additional subscript added for each new dimension. Different versions of BASIC impose different limits on the total number of dimensions permissible, but in this book we will not be concerned with arrays having dimensions greater than 2. One-dimensional arrays are often called *vectors,* and a two-dimensional matrix can therefore be considered to be composed of a number of one-dimensional vectors. Each vector is like a single *column* in a two-dimensional array so, for example, the matrix M that we just described could be regarded as three vectors, or 3 one-dimensional arrays, with four data values each. Two-dimensional arrays are referred to as being $m \times n$, where m is the number of rows and n is the number of columns. So the 3×4 matrix M consists of three 4×1 vectors.

In some versions of BASIC, two-dimensional arrays must always be dimensioned; other versions do not require dimensioning of arrays smaller than 11×11. One or more two-dimensional arrays can be dimensioned in the same DIM statement used to dimension other arrays. For example, the statement DIM X(12), A(3,20), Y(12), B(4,4) establishes X and Y as one-dimensional arrays, each having "dimension" 12 (in the context of size); A as a 3×20, two-dimensional array, and B as a 4×4, two-dimensional array.

As a simple illustration of the formation of a two-dimensional array, we will write a program to compute the products of every possible pair of integers between 1 and 9 and store them in a two-dimensional product matrix P. The matrix P is like a multiplication table in which each entry P(I,J) is equal to I * J.

```
10 DIM P(9,9)
20 FOR I=1 TO 9
30 FOR J=1 TO 9
40 LET P(I,J)=I*J
50 NEXT J
60 NEXT I
70 END
```

Note that the entries in P are assigned in the following order: P(1,1), P(1,2), . . . , P(1,9), P(2,1), P(2,2), . . . , and so forth.

Example 6.1 The gasoline consumption (mileage) of each of four different automobiles was measured at each of three different speeds: 20 MPH, 40 MPH, and 55 MPH. The results (in miles per gallon) were tabulated as follows:

	20 MPH	40 MPH	55 MPH
Auto 1	38	32	26
Auto 2	41	37	30
Auto 3	26	25	21
Auto 4	34	31	29

Write a BASIC program using a two-dimensional array to compute and print the average mileage of each type of automobile, the average mileage at each speed, and the overall average mileage.

A flowchart for the program is shown in Figure 6.1 (page 132).

The program and the results of a run are shown in Figure 6.2 (page 133).

Example 6.2 The Transportation Engineering Department at a certain university collected data on the circumstances surrounding ten serious traffic accidents, each involving several vehicles. The data collected included an identifying number for each

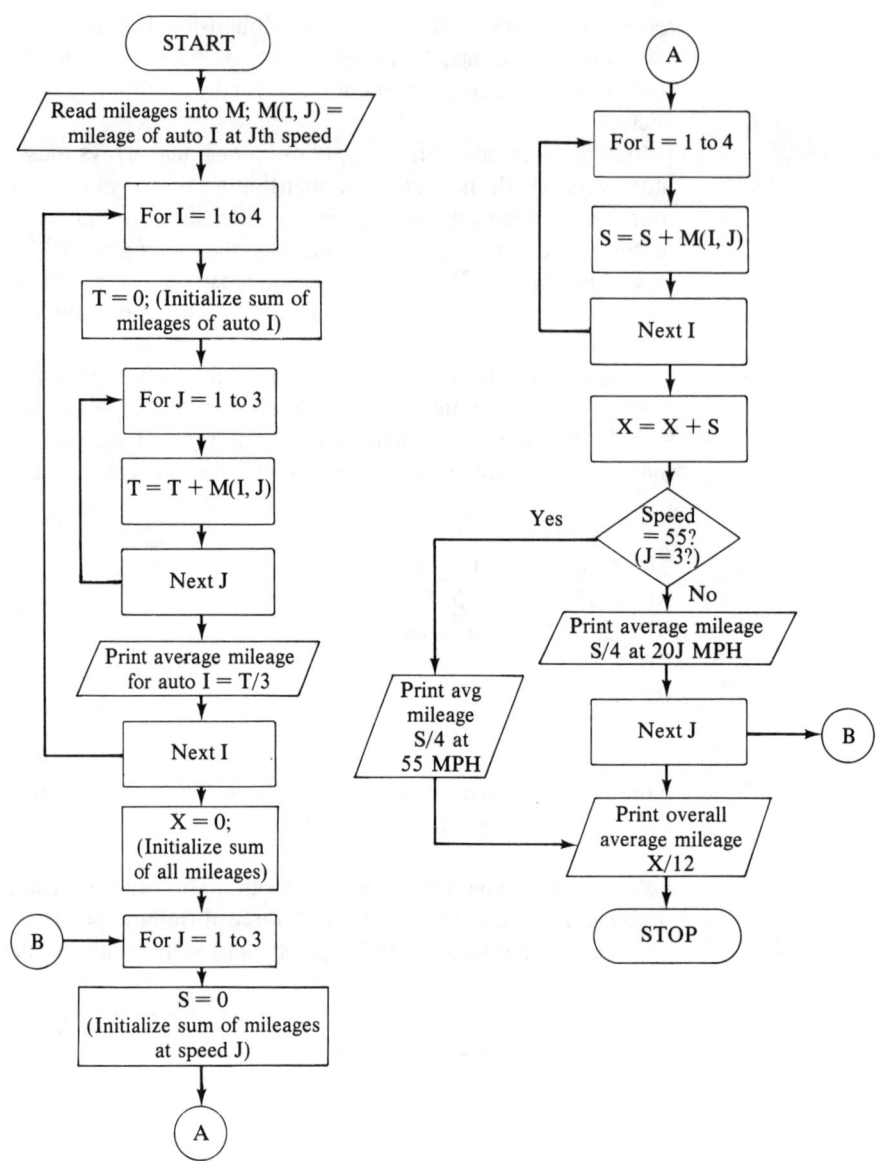

FIGURE 6.1 Flowchart diagraming the computation of the average mileages of four automobiles at three speeds (Example 6.1)

accident (1 through 10), the number of vehicles involved, the condition of the road surface at the time (dry, wet, icy, or under repair), the lighting conditions (day or night), and whether any 18-wheel trucks were involved. Write a BASIC program to read all the data into an array and then permit the user to enter the identifying number of an accident to obtain a display of the percent of the total number of accidents that occurred at night and the percent involving 18-wheel trucks.

Since we cannot store string and numeric data in the same array, we must assign *code* numbers to some of the data and store these codes in the array. We decide to code the road conditions as follows: 1 = dry, 2 = wet, 3 = icy, and 4 = under repair. Similarly, for lighting conditions we let 0 = day and 1 = night, and for truck involvement we let 1 = yes and 0 = no. (These codes are chosen to facilitate the computations of percent of night accidents and of truck involvement.) Thus, we will need a 10 × 4 array to store all the data. We will let the first column contain the number of vehicles; the second column contain the coded road data; the third, the

```
10 DIM M(4,3)
20 REM   ASSIGN MILEAGE DATA TO ARRAY M:
30 FOR I=1 TO 4
40 FOR J=1 TO 3
50 READ M(I,J)
60 NEXT J
70 NEXT I
80 REM   COMPUTE TOTAL T OF MILEAGES FOR EACH AUTO
90 FOR I=1 TO 4
100 LET T=0
110 FOR J=1 TO 3
120 LET T=T+M(I,J)
130 NEXT J
140 PRINT "AVERAGE MILEAGE FOR AUTO" I " IS" T/3
150 NEXT I
160 REM   COMPUTE SUM S OF MILEAGES AT EACH SPEED
170 REM   X=OVERALL SUM, USED TO COMPUTE OVERALL
          AVERAGE
180 LET X=0
190 FOR J=1 TO 3
200 LET S=0
210 FOR I=1 TO 4
220 LET S=S+M(I,J)
230 NEXT I
240 LET X=X+S
250 REM   BRANCH IF SPEED IS 55:
260 IF J>2 THEN 290
270 PRINT "MILEAGE AT" J*20 " MPH IS" S/4
280 NEXT J
290 PRINT "MILEAGE AT 55 MPH IS" S/4
300 PRINT "OVERALL MILEAGE IS" X/12
310 DATA 38,32,26,41,37,30,26,25,21,34,31,29
320 END
>

AVERAGE MILEAGE FOR AUTO 1 IS 32
AVERAGE MILEAGE FOR AUTO 2 IS 36
AVERAGE MILEAGE FOR AUTO 3 IS 24
AVERAGE MILEAGE FOR AUTO 4 IS 31.3333
MILEAGE AT 20 MPH IS 34.7500
MILEAGE AT 40 MPH IS 31.2500
MILEAGE AT 55 MPH IS 26.5000
OVERALL MILEAGE IS 30.8333
```

FIGURE 6.2 Program for Example 6.1 and the results of a program run

coded lighting conditions; and the fourth, the coded truck involvement. For example, if the sixth row of the array is

$$3 \quad 2 \quad 1 \quad 1$$

then it signifies that accident number 6 involved three vehicles on a wet road surface at night, and an 18-wheel truck was involved.

To facilitate printing the output, we will also construct a string array containing strings identifying the road conditions. The index for this array will be the code number identifying the road condition.

Figure 6.3 shows a BASIC program that performs the required functions using the following data:

Accident Number	Number of Vehicles	Road Condition	Day or Night	18-Wheel Truck Involved?
1	2	dry	day	yes
2	2	dry	day	no
3	3	icy	night	yes
4	5	dry	day	yes
5	2	wet	night	yes
6	4	in repair	night	no
7	3	icy	night	yes
8	5	wet	day	no
9	2	wet	night	yes
10	3	in repair	day	yes

Note statement 210 in the program:

210 PRINT "ROAD CONDITIONS: " B$(A(M,2))

The value of the index for the string array B$ is A(M,2). But A(M,2) is the entry in the second column of the Mth row of A; this is the code number for the road condition in the Mth accident. Thus, the string in B$ corresponding to this code number is printed.

Figure 6.4 on page 136 shows the results of one run of the program for accident number 7.

EXERCISES

6.1 Show the display that would result if each of the following programs were run:

(a)
```
10 DIM X(3)
20 FOR N=1 TO 3
30 LET X(N)=2*N
40 NEXT N
50 PRINT X(1)+X(2)*X(3)
60 END
```

(b)
```
10 DIM A(3),B(3)
20 FOR I=1 TO 3
30 LET A(I)=I**2+2
40 LET B(I)=A(I)-1
50 PRINT A(I),B(I),B(I)-A(I)
60 NEXT I
70 END
```

(c)
```
10 DIM R(3)
20 LET R(1)=4
```

```
10 DIM A(10,4), B$(4)
20 REM LOAD DATA ARRAY A
30 FOR I=1 TO 10
40 READ A(I,1), A(I,2), A(I,3), A(I,4)
50 NEXT I
60 REM LOAD STRING ARRAY B
70 FOR I=1 TO 4
80 READ B$(I)
90 NEXT I
100 REM COMPUTE NUMBER N OF NIGHT ACCIDENTS AND T OF
    TRUCK INVOLVEMENTS
110 LET N=0
120 LET T=0
130 FOR I=1 TO 10
140 LET N=N+A(I,3)
150 LET T=T+A(I,4)
160 NEXT I
170 PRINT "ENTER ACCIDENT NUMBER"
180 INPUT M
190 PRINT "ACCIDENT NUMBER" M
200 PRINT A(M,1) " VEHICLES INVOLVED"
210 PRINT "ROAD CONDITIONS: "B$(A(M,2))
220 IF A(M,3)=0 THEN 250
230 PRINT "ACCIDENT OCCURRED AT NIGHT"
240 GOTO 260
250 PRINT "ACCIDENT OCCURRED IN DAYLIGHT"
260 IF A(M,4)=0 THEN 290
270 PRINT "TRUCK INVOLVED"
280 GOTO 300
290 PRINT "NO TRUCK INVOLVED"
300 PRINT "PERCENTS DESIRED? TYPE Y OR N."
310 INPUT C$
320 IF C$="N" THEN 350
330 PRINT N*10 " PERCENT NIGHT ACCIDENTS"
340 PRINT T*10 " PERCENT TRUCK INVOLVEMENTS"
350 DATA 2,1,0,1,2,1,0,0,3,3,1,1
360 DATA 5,1,0,1,2,2,1,1,4,4,1,0
370 DATA 3,3,1,1,5,2,0,0,2,2,1,1
380 DATA 3,4,0,1, "DRY", "WET"
390 DATA "ICY", "UNDER REPAIR"
400 END
>
```

FIGURE 6.3 Program for Example 6.2

```
30 LET R(2)=(R(1)-2)/R(1)
40 IF R(2)**2<R(2) THEN 70
50 LET R(3)=R(2)
60 GOTO 80
70 LET R(3)=R(1)+R(2)
80 FOR J=1 TO 3
90 PRINT "R("J")="R(J)
100 NEXT J
110 END
```

```
ENTER ACCIDENT NUMBER
?7
ACCIDENT NUMBER 7
 3 VEHICLES INVOLVED
ROAD CONDITIONS: ICY
ACCIDENT OCCURRED AT NIGHT
TRUCK INVOLVED
PERCENTS DESIRED? TYPE Y OR N.
?Y
 50 PERCENT NIGHT ACCIDENTS
 70 PERCENT TRUCK INVOLVEMENTS

>
```

FIGURE 6.4 The results of a run of the program in Figure 6.3

(d)
```
10 DIM T(25)
20 LET T(0)=101
30 FOR X=2 TO 22 STEP 2
40 READ T(X)
50 IF T(X)>T(X-2) THEN 70
60 NEXT X
70 FOR K=10 TO 2 STEP -2
80 PRINT T(K)
90 NEXT K
100 DATA 100,99,98,97,96,100
110 END
```

(e)
```
10 DIM L$(5)
20 FOR P=1 TO 5
30 READ L$(P)
40 NEXT P
50 FOR N=1 TO 5
60 PRINT L$(6-N)
70 NEXT N
80 DATA "BANANAS","NO","HAVE","WE","YES"
90 END
```

6.2 Show the display that would result if each of the following programs were run:

(a)
```
10 DIM A(3),B(3),C(3,3)
20 FOR I=1 TO 3
30 LET A(I)=4*I
40 LET B(I)=10-A(I)/2
50 NEXT I
60 FOR I=1 TO 3
70 FOR J=1 TO 3
80 LET C(I,J)=A(I)*B(J)
90 PRINT A(I),B(J),C(I,J)
100 NEXT J
110 NEXT I
120 END
```

(b)
```
10 DIM W(4,2)
20 FOR I=1 TO 2
30 FOR J=1 TO 4
40 READ W(J,I)
50 NEXT J
60 NEXT I
70 FOR J=1 TO 4
80 FOR I=1 TO 2
90 IF I>=J THEN 110
100 GOTO 120
110 LET W(I,J)=W(J,I)
120 PRINT J,I,W(J,I)
130 NEXT I
140 NEXT J
150 DATA 1,2,3,4,5,6,7,8
160 END
```

6.3 Write a BASIC program that assigns the first ten odd numbers in succession (beginning with 1) to array A, the first ten even numbers in succession (beginning with 2) to array B, and then assigns the product of each odd-even pair to array C. When the program is finished, array C should contain $C(1) = 1 \times 2 = 2, C(2) = 3 \times 4 = 12$, and so forth. Print the contents of array C.

6.4 Write a BASIC program that prompts the user to enter ten pairs of numbers and stores the first of each pair in array M and the second of each pair in array N (in the same order in which they are entered). The program should also create an array L that contains the larger number of each pair. For example, if the first pair entered is 3,8, then $M(1) = 3$, $N(1) = 8$, and $L(1) = 8$. Print the contents of each array in a table showing $M(i)$, $N(i)$, and $L(i)$ in each of ten rows. For example, if the first pair entered is 3,8, the first row would then be displayed as

3 8 8

6.5 Write a BASIC program using FOR-TO, NEXT loops to assign the following values to a matrix M:

$$M(i,j) = i \quad \text{if } i \leq j$$
$$M(i,j) = j \quad \text{if } i > j$$

Print a display of M in a 5×5 table.

6.6 Write a BASIC program that prompts the user to enter a pair of numbers (m,n) and then displays the rest of a line in the following text (including blanks), beginning with the character in the mth line and nth column:

```
THE QUICK BROWN
FOX JUMPED OVER
THE LAZY DOG!!!
```

(For example, if the user enters 2,7, the display should be MPED OVER.) If the user enters a value of m greater than 3 or a value of n greater than 15, the computer should print an error message and halt.

3 SORTING AND ORDERING NUMERIC DATA

A set of numeric data must often be rearranged so that it appears in increasing or decreasing order. A number of standard algorithms can be used for this purpose, and we will examine one of these in detail.

Let us first consider the problem of extracting the maximum value from a set of numeric data contained in an array. An algorithm that can be used to determine a maximum value is outlined in the following steps:

(1) Assume that the first data value in the set is the maximum of all values.
(2) Compare this assumed maximum value to the next data value. If the latter is greater, then it becomes the new assumed maximum. Otherwise, the previously assumed maximum remains the assumed maximum.
(3) Repeat Step 2 until all data values have been examined. The last assumed maximum is the true maximum of all values.

A flowchart diagraming the logic of the preceding algorithm is shown in Figure 6.5. The flowchart assumes that an array A contains ten values from which the maximum is to be extracted and printed.

To illustrate the sequence of logical operations diagramed in the flowchart, suppose that $A(1) = 5$, $A(2) = 8$, and $A(3) = 6$. Thus, M is initially set to 5, and in the first pass through the loop M is compared to $A(2) = 8$. Since $8 > 5$, M is assigned the new value of 8. In the next pass, this new value of M is compared to $A(3) = 6$. Since $6 < 8$, M retains its value of 8—that is, 8 is the largest of all values examined so far. If the program ended here, M would be the maximum of all values, but the process actually continues until all remaining data values have been checked against M.

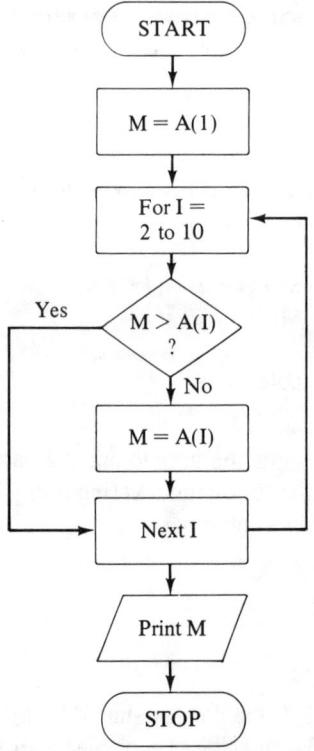

FIGURE 6.5 Flowchart diagraming the algorithm for finding the maximum value in an array

The following program assigns values to the array A and then implements the logic of the flowchart to find the maximum value (25).

```
10  DIM A(10)
20  FOR I=1 TO 10
30  READ A(I)
40  NEXT I
50  LET M=A(1)
60  FOR I=2 TO 10
70  IF M>A(I) THEN 90
80  LET M=A(I)
90  NEXT I
100 PRINT "THE MAXIMUM IS" M
110 DATA 6,-2,14,0,25,3,-18,1,5,11,-30
120 END
```

Here is one way to *order* the numeric data in an array: First find the maximum value, exchange its position with that of the first entry in the array, then find the maximum of the remaining values and exchange its position with the second entry in the array, and so forth. However, to perform the position exchanges this approach requires keeping track of the position occupied by each entry. A more efficient approach is to use an algorithm known as the *bubble sort*. The essence of this technique is to compare the magnitudes of each value in every succeeding *pair* of adjacent values and interchange them, if the first value is smaller than the second. If this procedure is repeated enough times (that is, if smaller values are continually exchanged with adjacent larger ones) the largest must then eventually filter through ("bubble up") to the top, while the smaller values end up at the bottom. The sort process is finished when it is no longer necessary to interchange the positions of any adjacent values.

The bubble sort algorithm can be outlined as follows:

(1) Set $i = 1$.
(2) Compare the ith value with the $(i + 1)$th value. Exchange their positions if the ith value is smaller than the $(i + 1)$th value.
(3) Increase i by 1. If i is now greater than the total number of values in the set, and

   (a) at least one position exchange has been made since i was last equal to 1, go to Step 1.
   (b) no position exchanges have been made since i was last equal to 1, go to Step 4.

   If i is less than the total number of values in the set, go to Step 2.
(4) The sort is complete.

A flowchart diagraming the logic of the bubble sort algorithm is shown in Figure 6.6. We assume that the values to be sorted are contained in an array A of dimension 10.

In the flowchart of Figure 6.6, the variable E is used to count the total number of exchanges that occur during each complete execution of the FOR-TO, NEXT statement—that is, during nine passes through the loop. If E = 0 after nine such passes, this means that all adjacent pairs of values have been examined, without having to interchange any of them, and the sorting is therefore complete.

To illustrate the logic diagramed in this flowchart, suppose that the array A consists of only the three values $A(1) = 4$, $A(2) = 6$, and $A(3) = 8$. The following table illustrates the exchanges that take place during each of the passes (in this case 2) constituting each FOR-TO, NEXT loop.

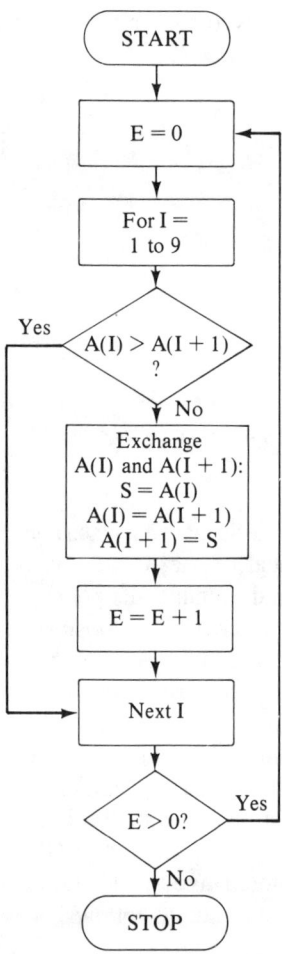

FIGURE 6.6 Flowchart diagraming the bubble sort algorithm

	Initial	I = 1	I = 2	I = 1	I = 2	I = 1	I = 2
A(1)	4	6	6	8	8	8	8
A(2)	6	4	8	6	6	6	6
A(3)	8	8	4	4	4	4	4
E	0	1	2	1	1	0	0

As shown in this table, the first execution of the FOR-TO, NEXT loop results in E = 2 exchanges, the second execution results in E = 1 exchange, and the last execution requires no exchanges. Hence the data are sorted after three complete executions of the FOR-TO, NEXT statement.

The following program assigns values to the array A, implements the bubble sort algorithm, and prints the values of the sorted array.

```
10 DIM A(10)
20 FOR I=1 TO 10
```

ARRAYS

```
30 READ A(I)
40 NEXT I
50 LET E=0
60 FOR I=1 TO 9
70 IF A(I)>A(I+1) THEN 120
80 LET S=A(I)
90 LET A(I)=A(I+1)
100 LET A(I+1)=S
110 LET E=E+1
120 NEXT I
130 IF E>0 THEN 50
140 FOR I=1 TO 10
150 PRINT A(I)
160 NEXT I
170 DATA 6,-2,14,0,25,3,-18,1.5,11,-30
180 END
```

When this program is run, the following values are displayed in the order shown:

```
 25
 14
 11
  6
  3
  1.5
  0
 -2
-18
-30
```

Example 6.3 Five traveling salespeople each enter their total sales and total expenses into a computer terminal at the end of each week. This information is entered by typing three numbers: an identifying digit (1 through 5); total sales, in dollars; and total expenses, in dollars. Write a BASIC program that waits until all five salespeople have entered their data and then determines which one had the greatest net sales (sales minus expenses). The computer should print the identifying digit of the person with the greatest net sales and the value of that person's net sales.

We will solve this programming problem using a method similar to that previously described for finding the maximum value in an array. The only modification we need to make is a provision to determine which salesperson is responsible for the maximum value (net sales) in the array. We can do this by using each salesperson's identification number (ID) as an index for the array containing net sales values. Then, as we select progressively larger values for the maximum (in the process described by the algorithm), we can also keep track of the index values corresponding to these maximums. Thus, upon completion of the program, we will have the true maximum of all values and also the index (ID) corresponding to it.

A program using the procedure we have just described is shown in Figure 6.7 on page 142. The ID number N is used to index an array R that stores net sales values $S-E$. The variable K is used to keep track of the index number corresponding to the maximum M.

One run of the program of Figure 6.7 is shown in Figure 6.8 on page 143.

You might have noticed that there is no provision in the program written for Example 6.3 to cover a situation when there is a tie for maximum net sales. Study the program and determine which salesperson would be credited with the greatest sales if two were actually tied for that honor. Convince yourself that one salesperson would lodge a

```
10 DIM R(5)
20 FOR I=1 TO 5
30 PRINT "ENTER ID, SALES, EXPENSES"
40 INPUT N,S,E
50 LET R(N)=S-E
60 NEXT I
70 LET M=R(1)
80 LET K=1
90 FOR I=2 TO 5
100 IF M>R(I) THEN 130
110 LET M=R(I)
120 LET K=I
130 NEXT I
140 PRINT "SALESMAN"K" HAS MAXIMUM NET SALES OF $"M"
    THIS WEEK"
150 END
>
```

FIGURE 6.7 Program for Example 6.3

protest with the accounting department (or its programmer). The modification of the program to accommodate this situation is given as an exercise at the end of this chapter.

Example 6.4 Figure 6.9 shows a plot of experimental data points (x_i, y_i), where x_i is the x coordinate of the ith point and y_i is the y coordinate of the ith point. Also shown is the "best" straight line that can be drawn through these points. The equation of this line is $y = mx + b$, where m is the slope of the line and b is the value at the point where the line intercepts the y axis. The *deviation* of any point from the line is the vertical distance from the point to the line. As you can see in the figure, the deviation D_i of a point (x_i, y_i) is equal to $(mx_i + b - y_i)$.

For the set of data points shown in the following table, the best straight line has the equation $y = 4.207152x + 7.179604$. One property of the best straight line drawn through any set of data points is that the sum of all the deviations D_i of the data points equals zero.

Point	x_i	y_i
1	1.2	12.0
2	3.5	20.5
3	2.3	19.1
4	6.0	31.5
5	0.7	9.5
6	5.2	28.7
7	10.0	51.1
8	8.6	41.9
9	3.1	20.4
10	1.9	15.9

Write a BASIC program to print a table showing the coordinates of the points and their deviations, in descending order of deviations. Also compute and print the sum of all deviations.

You should create a 10 × 3 array that contains the values of x_i, y_i and $D_i = mx_i + b - y_i$ for each i from 1 through 10. Then use the bubble sort algorithm

```
ENTER ID, SALES, EXPENSES
?4,1465,250
ENTER ID, SALES, EXPENSES
?1,3990,630
ENTER ID, SALES, EXPENSES
?3,8760,385
ENTER ID, SALES, EXPENSES
?5,2910,170
ENTER ID, SALES, EXPENSES
?2,4375,209
SALESMAN 3 HAS MAXIMUM NET SALES OF $ 8375 THIS WEEK

>
```

FIGURE 6.8 The results of a run of the program in Figure 6.7

to order the values of D_i. Each time an exchange of values of D_i occurs, interchange the corresponding x_i and y_i values. Finally, sum the values of D_i.

Figure 6.10 on page 144 shows a BASIC program that performs these tasks.

The results of a run of this program are shown in Figure 6.11 (page 145). Note that the data are listed in descending order of the deviations D, as required. Also note that the sum of the deviations is the very small number $-1.24345E-14$, which except for round-off error would be zero.

It is occasionally necessary to *search* an array for a specific entry. This is easily accomplished by writing a FOR-TO, NEXT loop that uses a conditional branch statement to check each entry for the one desired. We will usually want to know the position of the entry in the array, and we can use the value of the counter variable at the time the entry is found for this purpose. For example, suppose the array A$ contains 12 strings, one of which is "JUNE", and we wish to find the location of that entry. We can use the following program segment:

```
100 FOR I=1 TO 12
110 IF A$(I)="JUNE" THEN 130
120 NEXT I
130
```

When the computer exits the loop at 130, the value of I is the position of "JUNE" in the array A$.

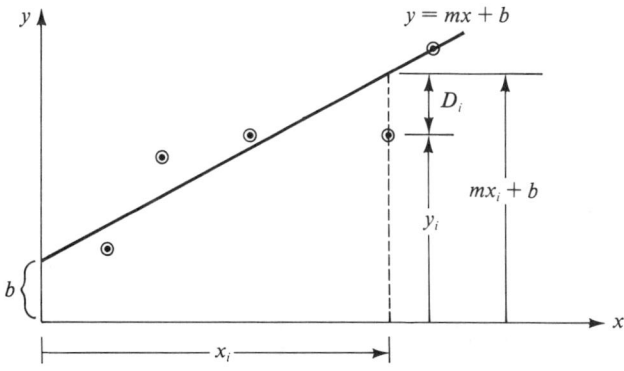

FIGURE 6.9 Deviations of data points from the "best" straight line that can be drawn through them (Example 6.4)

```
10 DIM M(10,3)
20 FOR I=1 TO 10
30 REM   READ X(I)
40 READ M(I,1)
50 REM   READ Y(I)
60 READ M(I,2)
70 REM   COMPUTE AND STORE D(I)
80 LET M(I,3)=4.207152*M(I,1)+7.179604-M(I,2)
90 NEXT I
100 LET E=0
110 FOR I=1 TO 9
120 IF M(I,3)>M(I+1,3) THEN 240
130 REM   INTERCHANGE X,Y, AND D
140 LET A=M(I,1)
150 LET B=M(I,2)
160 LET C=M(I,3)
170 LET M(I,1)=M(I+1,1)
180 LET M(I,2)=M(I+1,2)
190 LET M(I,3)=M(I+1,3)
200 LET M(I+1,1)=A
210 LET M(I+1,2)=B
220 LET M(I+1,3)=C
230 LET E=E+1
240 NEXT I
250 REM   IF ANY EXCHANGES OCCURRED, GO BACK AND SORT
    AGAIN.
260 IF E>0 THEN 100
270 LET S=0
280 REM   PRINT TABLE AND COMPUTE SUM S OF DEVIATIONS.
290 PRINT "    X","    Y","    D"
300 FOR I=1 TO 10
310 LET S=S+M(I,3)
320 PRINT M(I,1),M(I,2),M(I,3)
330 NEXT I
340 PRINT
350 PRINT "SUM OF DEVIATIONS IS " S
360 DATA 1.2,12,3.5,20.5,2.3,19.1,6.0,31.5,.7,9.5
370 DATA 5.2,28.7,10,51.1,8.6,41.9,3.1,20.4,1.9,15.9
380 END
>
```

FIGURE 6.10 Program for Example 6.4

GENERAL EXERCISES

6.7 If the vectors X and Y are given by

$$X = \begin{bmatrix} x_1 \\ x_2 \\ \cdot \\ \cdot \\ \cdot \\ x_n \end{bmatrix} \quad Y = \begin{bmatrix} y_1 \\ y_2 \\ \cdot \\ \cdot \\ \cdot \\ y_n \end{bmatrix}$$

```
     X              Y              D
  8.60000        41.9000        1.46111
  3.50000        20.5000        1.40464
  6              31.5000         .922516
   .700000        9.50000        .624610
  5.20000        28.7000         .356794
  1.20000       12               .228186
  3.10000       20.4000         -.178225
  1.90000       15.9000         -.726807
 10             51.1000        -1.84888
  2.30000       19.1000        -2.24395

SUM OF DEVIATIONS IS -1.24345E-14
```

FIGURE 6.11 The results of a run of the program in Figure 6.10

then the *inner product* of X and Y is defined by

$$X \cdot Y = \sum_{i=1}^{n} x_i y_i$$

Write a BASIC program to read five values into an array X and five values into an array Y and then compute the inner product $X \cdot Y$. Run your program for the following data:

$$X = \begin{bmatrix} 2.3 \\ 1.7 \\ -0.9 \\ 4.6 \\ 0 \end{bmatrix} \qquad Y = \begin{bmatrix} 12.7 \\ -14.0 \\ 6.8 \\ 11.9 \\ 15.5 \end{bmatrix}$$

6.8 If a 3 × 3 matrix M is defined by

$$M = \begin{bmatrix} a_{11} & a_{12} & a_{13} \\ a_{21} & a_{22} & a_{23} \\ a_{31} & a_{32} & a_{33} \end{bmatrix}$$

then the *transpose* M^T of M is defined by

$$M^T = \begin{bmatrix} a_{11} & a_{21} & a_{31} \\ a_{12} & a_{22} & a_{32} \\ a_{13} & a_{23} & a_{33} \end{bmatrix}$$

Note that M^T is formed by making the rows of M equal to the columns of M^T. Write a BASIC program to read the entries for a 3 × 3 matrix M and then compute the matrix equal to the sum $M + M^T$. The sum of two 3 × 3 matrices is a 3 × 3 matrix whose entries are the sums of corresponding entries in the other two matrices. For example, the entry in the first row and second column of the sum matrix is the sum of entries in the first row and second column of each of the other matrices. Run your program and print a table showing $M + M^T$ for

$$M = \begin{bmatrix} 1 & 2 & 3 \\ 4 & 5 & 6 \\ 7 & 8 & 9 \end{bmatrix}$$

6.9 Write a BASIC program to find and print the minimum value in an array with a dimension of 10. Run your program using the following data:

$$A(1) = 12, A(2) = -8, A(3) = 6.5$$
$$A(4) = -0.4, A(5) = 17.1, A(6) = -11.9$$
$$A(7) = 2.4, A(8) = 5.2, A(9) = 1.8, A(10) = 0$$

6.10 Write a BASIC program to list ten data values in ascending order (smallest value first and largest value last). Run your program using the data given in Exercise 6.9.

6.11 Modify the program given in Example 6.3 so that it prints the ID numbers and sales of both salespeople whenever there is a two-way tie for the largest net sales. Run your program with the following input:

$$N = 4, \quad S = 10{,}050, \quad E = 810$$
$$N = 1, \quad S = 6{,}580, \quad E = 866$$
$$N = 2, \quad S = 15{,}790, \quad E = 1{,}090$$
$$N = 3, \quad S = 1{,}800, \quad E = 2{,}000$$
$$N = 5, \quad S = 15{,}010, \quad E = 310$$

EXERCISES FOR ARCHITECTURAL/CIVIL/CONSTRUCTION TECHNOLOGY

6.12 A contractor maintains records of the number of feet of pipe laid each day of the week during a major pipeline construction project. For each day (except Sunday), the contractor records the number of feet of pipe laid, the weather conditions that day (rainy or fair), and the type of terrain (flat, hilly, or mountainous). Assign numeric codes to the data (Monday = 1, Tuesday = 2, etc.) and write a BASIC program to read and store all data in an array. The program should then prompt the user to enter a day of the week, to obtain a display of all data pertaining to that day. Run your program using the following data:

Day	Feet	Weather	Terrain
Monday	2260	fair	flat
Tuesday	2430	fair	flat
Wednesday	1740	rain	hilly
Thursday	1890	fair	hilly
Friday	1070	rain	mountainous
Saturday	1160	fair	mountainous

Obtain a display for each day of the week.

6.13 Figure 6.12 shows a beam with supports at each end and a uniformly distributed load. The load is q pounds per foot, so the total load at a distance of x feet from the left support is qx pounds. The length of the beam is L feet.

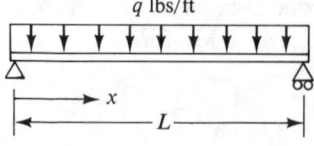

FIGURE 6.12 Exercise 6.13

The *bending moment M* at distance x is

$$M = \frac{-qx^2}{2} + \frac{qLx}{2} \text{ ft-lb}$$

Assuming that the beam is 12 feet long, and $q = 500$ lb/ft, write a BASIC program to compute the bending moment at 1-foot intervals from 0 through 12 feet. The program should then print the value of x at which the bending moment is the maximum and the value of that maximum moment.

6.14 A certain type of lumber is graded according to the number of blemishes, knotholes, and other defects found in samples taken from various production lots. In a production run made to supply a certain contractor, the lumber from 12 production lots was graded as follows: The grade "premium" was assigned to the four lots containing the least number of defects; the grade "good," to the four lots having the next fewest defects; and the grade "utility," to the remaining lots. The following table shows the total number of defects in six samples taken from each lot:

Lot Number	Defects
1	24
2	54
3	6
4	2
5	13
6	31
7	20
8	15
9	8
10	41
11	17
12	12

Write a BASIC program to read the data (in the order given) and print a table showing the lot numbers assigned the grade "premium," the lot numbers assigned "good," and those assigned "utility." The program should also print the average number of defects per sample in each of the three grades. (There are $6 \times 4 = 24$ samples from each grade.)

EXERCISES FOR ELECTRONICS/ELECTRICAL/COMPUTER TECHNOLOGY

6.15 According to the *maximum power transfer theorem,* maximum power is delivered from a source to a load when the load resistance R_L is equal to the source resistance R_S. (See Figure 6.13.)

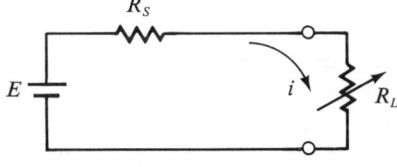

FIGURE 6.13 Exercise 6.15

The power delivered to R_L can be computed from

$$P = \left(\frac{E}{R_S + R_L}\right)^2 R_L \text{ watts.}$$

To verify the theorem when E = 12V and R_S = 4 ohms, write a BASIC program to compute the power delivered to R_L for values of R_L from 0 through 10 ohms in 1-ohm increments and print the value of R that results in maximum power.

6.16 One measure of the quality of a transistor is its *current amplification factor* β. There is typically a large variation in the value of β among transistors of the same type. A certain manufacturer markets transistors as "high gain," "medium gain," or "low gain," based upon measurements made on samples taken from different production lots. In one week's production, 12 lots are manufactured and 40 samples are taken from each lot. The production lots having the four highest average values of β are classified as "high gain," those having the next four highest averages are classified as "medium gain," and the rest are classified as "low gain." The following table shows the average value of β for the transistors in each lot.

Lot Number	Average β
1	80
2	120
3	55
4	38
5	115
6	99
7	195
8	75
9	211
10	66
11	84
12	148

Write a BASIC program to read the data (in the order given) and print the lot numbers of the "high gain," "medium gain," and "low gain" transistors. The computer should also print the average β of the transistors in each lot (the average of the averages within each lot).

6.17 An ac voltage can be represented by the *complex number* $a + jb$, where a is the *real part*, b is the *imaginary part*, and $j = \sqrt{-1}$. The *amplitude* of the ac voltage is found from

$$A = \sqrt{a^2 + b^2}$$

In one experiment, 12 ac voltages were measured and their values were reported in complex number form. Each measurement was made using one of four types of instruments: (a) a digital voltmeter, (b) a VOM, (c) an oscilloscope, *or* (d) a spectrum analyzer. (All phase angle measurements were made with a phase meter.)

The results of these measurements are summarized below:

AC Voltage	Instrument Used
40 + j20	oscilloscope
30 − j50	VOM
10 + j90	digital voltmeter
50 + j50	spectrum analyzer
40 − j30	oscilloscope
10 + j60	digital voltmeter
20 − j80	oscilloscope
100 + j0	digital voltmeter
70 + j30	VOM
0 − j100	spectrum analyzer
80 + j80	spectrum analyzer
10 + j10	oscilloscope

Write a BASIC program to print a table arranged in ascending order of amplitude (smallest amplitude first and largest amplitude last) containing the amplitude, real part, imaginary part, and the measuring instrument used, for each ac voltage.

6.18 The *color code* for resistors is used to identify values of resistance in ohms. Each digit from 0 through 9 is represented by a different color, and three colored bands on the body of the resistor are then used to encode its resistance. The standard color code is as follows:

Color	Digit	Color	Digit
black	0	green	5
brown	1	blue	6
red	2	violet	7
orange	3	gray	8
yellow	4	white	9

The first two color bands (in left-to-right order) represent the first two digits of the resistance value, and the third band represents the power of 10 by which the first two digits are multiplied. Following are two examples:

$$\text{yellow–violet–orange} = 47 \times 10^3 = 47\,\text{k}\Omega$$
$$\text{red–red–black} = 22 \times 10^0 = 22\,\Omega$$

Write a BASIC program that prompts the user to enter three colors (in the left-to-right order in which they appear on a resistor) and then prints the resistance value in ohms. Use an array to store the standard color code.

EXERCISES FOR INDUSTRIAL/MANUFACTURING/PRODUCTION TECHNOLOGY

6.19 A survey was conducted in a manufacturing plant to study the productivity of a certain machine tool, in relation to its age and the manufacturer who produced it. Data was gathered on machines produced by five manufacturers: A, B, C, D, and E. For each machine surveyed, the investigators recorded the total number of units it produced and the total operating time required to produce those units. Also recorded was the skill level of the machine operator: apprentice, journeyman, or master machinist. The results of the survey are shown in the following table:

Machine Type	Age (Years)	Units Produced	Operating Time (Hours)	Operator Skill
A	7	1473	12.4	journeyman
B	2	2060	11.1	apprentice
C	15	1389	10.9	journeyman
D	9	2552	14.7	master
E	5	989	9.6	apprentice

After assigning numeric codes to the data, as necessary, write a BASIC program to read and store all the data in an array. The program should then prompt the user to enter a machine type (A, B, C, D, or E) to obtain a display of (a) the machine type, (b) its age, (c) its productivity (in units produced per hour), and (d) the skill level of the operator. Run your program and obtain a display for each machine type.

6.20 The total annual cost Y of inventory is given by

$$Y = \frac{CB}{X} + \frac{XE}{2}$$

where C = consumption rate in units per year,
B = ordering cost in dollars per order,
E = carrying cost in dollars per unit per year, and
X = lot size (number of units per order).

According to the *economic lot-size formula,* the cost of inventory is at the minimum when the lot size is

$$X = \sqrt{2CB/E}$$

To verify the economic lot size for C = 40,000 units/year, B = $40/order, and E = $0.05/unit, write a BASIC program to compute the total annual cost for lot sizes from 2000 through 20,000, in increments of 2000. The program should then find the lot size corresponding to the minimum of the computed annual costs and print the value of that lot size with its annual cost. Finally, for comparison, compute the lot size using the economic lot-size formula.

6.21 The quality control department of a food processing company takes samples from various batches of roasted peanuts, to determine how they will be packaged and marketed. A batch is classified as suitable for either premium, commercial, or non-brand distribution, depending on the number of undersized or broken nuts in the sample. The following table shows the total number of substandard nuts found in four 1-pound samples taken from each of 12 batches of nuts:

Batch	Substandard Nuts
1	42
2	18
3	86
4	34
5	10
6	95
7	12
8	55
9	23
10	47
11	101
12	29

The four batches with the fewest substandard nuts in their samples should be assigned the classification "premium," the four having the next fewest number are to be classified "commercial," and the remaining four should be classified "nonbrand."

Write a BASIC program to read the data (in the order given) into an array and then print a table showing the batch numbers classified as "premium," those classified as "commercial," and those classified as "nonbrand." The program should also print the average number of substandard nuts per sample pound in each classification ($4 \times 4 = 16$ sample pounds are taken from each batch).

EXERCISES FOR MECHANICAL TECHNOLOGY

6.22 A steel manufacturer produces 12 kinds of iron alloys, identified by numbers 1 through 12. The common name and important physical properties of each alloy are shown in the following table:

Identification Number	Common Name	Specific Gravity	Thermal Expansion Coefficient	Melting Point (°C)
1	Sterling steel	7.92	16.99	1425
2	Duraloy 18–8	7.86	14.99	1475
3	Sweetaloy 19	7.86	11.00	1495
4	Sweetaloy 17	7.86	15.98	1450
5	Ascoloy	7.90	16.2	1410
6	Ferronickel	8.1	18.0	1569
7	Invar	8.0	0.8	1495
8	Pyrasteel	7.89	17.1	1450
9	Platinite	8.2	7.5×10^6	1470
10	Misco	7.97	13.5	1540
11	Chromax	7.81	12.19	1480
12	Duriron	7.00	15.59	1265

After assigning numeric codes, as required, write a BASIC program that reads this data into an array and then prompts the user to enter an alloy identification number, to obtain a display of all relevant data for that alloy. Run your program to obtain the display for each alloy.

6.23 The *angular velocity* ω of a shaft is the angle through which it turns per unit time. If the angle is expressed in radians, angular velocity can then be expressed in radians per second. *Angular acceleration* α is the rate of change of angular velocity; it can be expressed as the *change* in ω that occurs per unit time and is therefore in units of (radians/sec) per second, or radians/sec^2. If α is positive, this means that ω is increasing, and if α is negative, ω is decreasing. If ω decreases with time up to an instant and then begins to increase, it follows that α changes from negative to positive. The only way that α can change from negative to positive is to become zero at the instant of time that ω ceases decreasing and begins increasing, that is, when ω reaches a minimum value.

The angular velocity and acceleration of a certain shaft changes with time, according to the following equations:

$$\omega = 2t^2 - 16t + 42 \quad \text{rad/sec}$$
$$\alpha = 4t - 16 \quad \text{rad/sec}^2$$

where t is in seconds. Write a BASIC program to compute ω and α at instants of time from $t = 0$ through $t = 10$ seconds, in 1-second intervals, and then print the time at which the angular velocity is a minimum, the value of the angular velocity at that time, and the angular acceleration at that time.

6.24 The efficiency η of a heat engine can be calculated from

$$\eta = 2554 \frac{\text{output horsepower}}{\text{rate of heat supplied, BTU/hr}}$$

Twelve engines of various types and manufacture were tested to determine their efficiencies. Based on measurements made by the investigators, the data in the following table was gathered:

Engine Type	Total Heat Supplied (BTU)	Total Time of Heat Supply (hrs)	Output Hp
steam	266,382	0.38	98.7
steam	283,096	0.50	106.3
gasoline	107,521	0.13	88.0
diesel	312,255	0.66	90.6
gasoline	201,130	0.28	78.1
gasoline	254,500	0.31	95.5
diesel	175,240	0.45	87.2
steam	440,680	0.83	100.7
diesel	280,990	0.28	80.4
gasoline	336,107	0.19	73.5
diesel	325,330	0.35	96.6
steam	473,652	0.99	84.7

Write a BASIC program to read the data (in the order given) and then print a table arranged in ascending order of efficiency (lowest efficiency first and largest efficiency last). The table should also show the output horsepower and the type of engine corresponding to each entry.

CHAPTER 7

FUNCTIONS AND SUBROUTINES

1 LIBRARY FUNCTIONS

BASIC reserves certain three-letter keywords (actually, abbreviations) that can be used by a programmer to evaluate some functions that are encountered frequently in technology. Recall that a function is simply a rule applied to a numeric quantity to generate a new number. For example, the mathematical function (of x) $f(x) = x^2$ tells us to take the value of x and square it to produce a new number. Thus, $f(2) = 4$ and $f(3) = 9$. The quantity in parentheses is called the *argument* of the function.

An example of a BASIC function is the *absolute value* function, which is expressed by writing ABS(a), where the argument a is a BASIC constant, variable, or expression. ABS(a) is expressed in mathematical notation by $|a|$. For example, ABS(-7) = 7 and ABS(12.5) = 12.5. The argument of any BASIC function can be either a constant, a variable, *or* an expression, but it must always be enclosed within parentheses. To illustrate how the absolute value function might be used in a BASIC program, suppose we want to calculate the square root of the difference between two numbers entered by the user. We can only take the square root of a positive number, but we do not know which of the two numbers entered by the user will be larger. We can solve this problem by using the ABS function, as follows:

```
10 INPUT N1,N2
20 LET D = N1 − N2
30 LET R = (ABS(D)) ** .5
```

Since ABS(D) is always positive, we can be sure that we are always taking the square root of a positive number in statement 30. Since the argument of a function can be an expression, we could also solve this problem by writing

```
10 INPUT N1,N2
20 LET R=(ABS(N1-N2))**.5
```

As we have already noted, the argument of a function must be enclosed in parentheses, so we must be very careful with the placement of parentheses in expressions involving arguments or in arguments involving expressions. For example,

$$(ABS(N1-N2)) ** .5$$

is the same as

$$ABS(N1-N2) ** .5$$

but if we want to find the absolute value of the square root of (N1 − N2), we would have to write

$$ABS((N1-N2) ** .5)$$

Note carefully how this expression differs from the previous two.

Consider how much clumsier the solution to the preceding example would have been without the ABS function:

```
10 INPUT N1,N2
20 IF N1>N2 THEN 50
30 LET D=N2-N1
40 GOTO 60
50 LET D=N1-N2
60 LET R=D**.5
```

Functions such as ABS are called *library* functions (also called "built-in" or *intrinsic* functions) because they are available to the programmer whenever needed. These functions can be selected and used at will, just as a book can be selected at will from those available in a library. The number and kinds of library functions available to a programmer depend on the version of BASIC being used. We will describe many of the most commonly available functions in this chapter, and then we will summarize them in a table, for reference purposes.

Another library function available in most versions of BASIC is the *square-root* function SQR(a), where the argument a must be positive (or zero). SQR(X) is equivalent to X**.5. Be careful not to confuse SQR(a) with the "square" of a, which it most definitely is not. To illustrate the use of the SQR function, we shall repeat the example in which we computed the square root of the difference between two user-supplied numbers:

```
10 INPUT N1,N2
20 LET D=ABS(N1-N2)
30 LET R=SQR(D)
```

Most versions of BASIC include some trigonometric functions, usually at least the sine and cosine functions. Other trigonometric functions can be obtained by algebraic manipulation of the sine and cosine, in accordance with standard trigonometric identities. It is most often true that the argument of a trigonometric function must be expressed in *radians* (rather than degrees), although some versions permit the user to select either radians or degrees through a system command. In this book we will assume that all trigonometric arguments must be in radians. Recall that

$$2\pi \text{ radians} = 360 \text{ degrees}$$

To convert θ degrees to radians, we calculate

$$\theta(2\pi/360), \quad \text{or} \quad 0.0174532\theta \text{ radians}$$

To convert ϕ radians to degrees, we calculate

$$\phi(360/2\pi), \quad \text{or} \quad 57.29578\phi \text{ degrees}$$

For example, 45 degrees = $(45)(2\pi/360)$ = 0.7853981 radians, and 0.2617993 radians = $0.2617993(360/2\pi)$ = 30 degrees.

The keywords for the sine and cosine functions are SIN(a) and COS(a). Thus, if we want to find the sine of 45° ($\pi/4$ radians), we would write SIN(.7853981). If we want to find the cosine of $x - 2y$, where x and y are given in degrees, we would then write COS((X−2*Y)*.0174532).

Example 7.1 Write a BASIC program that prompts the user to enter the magnitude and angle of a force vector and then prints the horizontal and vertical components of that force.

Recall that a force having a magnitude of F and an angle of θ (measured in a counterclockwise direction from the positive x axis) has a horizontal component of $F_x = F\cos\theta$ and a vertical component of $F_y = F\sin\theta$. (See Figure 7.1.)

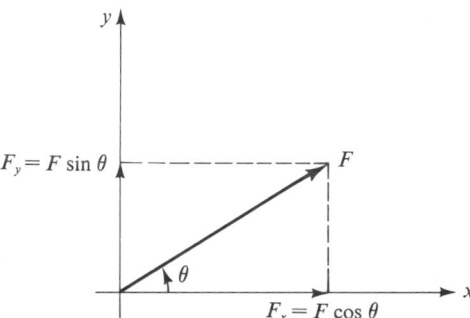

FIGURE 7.1 Horizontal and vertical components of a force vector (Example 7.1)

Figure 7.2 shows a BASIC program that computes F_x and F_y.

```
10 PRINT "ENTER FORCE MAGNITUDE AND ANGLE (DEG)"
20 INPUT F,A
30 PRINT "HORIZONTAL COMPONENT: "F*COS(.0174532*A)
40 PRINT "VERTICAL COMPONENT: "F*SIN(.0174532*A)
50 END
```

FIGURE 7.2 Program for Example 7.1

The results of three runs of this program are shown in Figure 7.3 on page 156. Note that a negative angle was entered in the second run, and the vertical component F_y is therefore negative (directed downward). Negative angles are angles that are measured in a *clockwise* direction from the positive x axis. In the third run, the angle entered is in the second quadrant, and the horizontal component is therefore negative (directed towards the left). Make sketches of these two cases to confirm that the algebraic signs of F_x and F_y are correct as computed.

Most versions of BASIC also include the tangent function, TAN(a). This function, and other trigonometric functions, can also be computed using the SIN and COS functions, in accordance with the following identities:

$$\tan\theta = \sin\theta/\cos\theta$$
$$\cot\theta = \cos\theta/\sin\theta = 1/\tan\theta$$
$$\sec\theta = 1/\cos\theta$$
$$\csc\theta = 1/\sin\theta$$

The following program segment computes the cotangent of an angle A entered by the user (in degrees):

```
10 INPUT A
20 LET B=.0174532*A
30 LET C=COS(B)/SIN(B)
```

If the tangent function (TAN) is available but the cotangent (COT) is not, statement 30 could also have been written LET C = 1/TAN(A). When we write BASIC expressions containing trigonometric functions, as in the preceding example, we must be careful to avoid the possibility of division by zero. For example, in the preceding computation, if A = 0, then B = 0 and SIN(B) = 0, so an error message is generated when the computer attempts to execute statement 30. A conditional branch instruction that checks the value of A should be used to bypass the computation at statement 30, if it is possible for A to have a value of zero.

The argument of a function can be another function. For example, we are permitted to write

$$\text{SIN(ABS(X))}$$

```
ENTER FORCE MAGNITUDE AND ANGLE (DEG)
?275,30
HORIZONTAL COMPONENT: 238.157
VERTICAL COMPONENT: 137.499

>

ENTER FORCE MAGNITUDE AND ANGLE (DEG)
?275,-30
HORIZONTAL COMPONENT: 238.157
VERTICAL COMPONENT: -137.499

>

ENTER FORCE MAGNITUDE AND ANGLE (DEG)
?275,120
HORIZONTAL COMPONENT: -137.497
VERTICAL COMPONENT: 238.159

>
```

FIGURE 7.3 The results of three runs of the program in Figure 7.2

which computes the sine of the absolute value of X, or

$$\text{ABS(SIN(X))}$$

which computes the absolute value of the sine of X. Note carefully how parentheses are used in these expressions. The argument of any function must always be enclosed within parentheses, even if that argument is another function. Functions can also appear in any expression where BASIC allows them as, for example, in

$$\text{IF (SIN(X)} - \text{ABS(X))/2} > .5 \text{ THEN } 100$$

Another function widely used in technology is the *exponential* function, e^x, where e is the base of the natural logarithm (e is a constant, approximately equal to 2.71828183). The exponential function is expressed in BASIC by

$$\text{EXP}(a)$$

where a is the argument. The value of the argument can be either a positive or negative real number. Recall that

$$e^{-x} = 1/e^x$$

so large values of x result in small values of e^{-x}.

Example 7.2 A *damped sine wave* has the general form

$$f(t) = Ae^{-\alpha t}\sin\omega t \tag{1}$$

where t is in seconds, A and α are positive constants, and ω is an angular frequency, in radians per second. A graph of Equation 1 plotted versus time t has the general appearance shown in Figure 7.4. Note that the graph is a sine wave whose peaks gradually diminish in amplitude. The function $Ae^{-\alpha t}$, which determines the values of the peaks, is called the *envelope* of f(t).

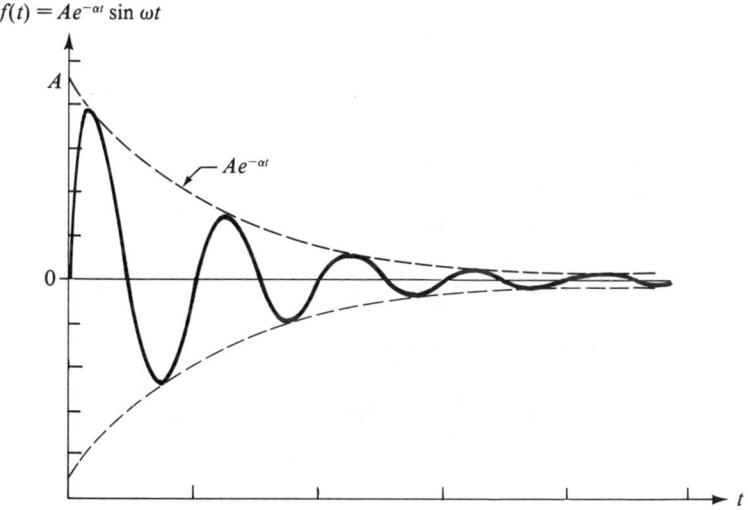

FIGURE 7.4 Graph of a damped sine wave (Example 7.2)

Write a BASIC program to print a table of the values of $f_1(t) = 10e^{-0.5t}\sin(6.28t)$ and $f_2(t) = 10e^{-t}\sin(6.28t)$ at 21 different values of t, ranging from $t = 0$ to $t = 2$ seconds. Use these values to construct graphs of $f_1(t)$ and $f_2(t)$ versus t.

The program and results are shown in Figure 7.5.

The oscillatory nature of $f_1(t)$ and $f_2(t)$ is clearly apparent in the data: Note the periodic sign reversals. Figure 7.6 shows graphs of $f_1(t)$ and $f_2(t)$ plotted on the same set of axes. Also note that $f_2(t)$ diminishes more rapidly toward zero ("damps out" faster) because the envelope e^{-t} approaches zero faster than $e^{-0.5t}$.

It is customary to say that a function "returns" a certain result when it is evaluated. Thus the ABS function returns the absolute value of its argument, and the COS function returns the cosine of its argument.

The SGN(a) function returns one of the values $-1, 0$, or $+1$, depending on whether its argument a is negative, zero, or positive, respectively. (For example, SGN(Y−X) = −1 for X>Y, SGN(A−2*B) = 0 for B = A/2, and SGN((M−N)**2) = 1 for M ≠ N). Note that ABS(X) = X*SGN(X) for any X.

As an example of how the SGN function might be used in a BASIC program, suppose that we want to determine whether the roots of a quadratic equation are real and unequal, real and equal, or imaginary. Recall that we wrote a program to perform this task, in a previous example illustrating conditional branching. The roots of the

```
10 PRINT "     T","     F1(T)","    F2(T)"
20 FOR T=0 TO 2 STEP .1
30 LET F1=10*EXP(-.5*T)*SIN(6.28*T)
40 LET F2=10*EXP(-T)*SIN(6.28*T)
50 PRINT T,F1,F2
60 NEXT T
70 END
```

T	F1(T)	F2(T)
0	0	0
.100000	5.58873	5.31617
.200000	8.60373	7.78498
.300000	8.18836	7.04778
.400000	4.82081	3.94695
.500000	1.24036E-02	9.65993E-03
.600000	-4.34296	-3.21734
.700000	-6.69711	-4.71937
.800000	-6.38038	-4.27690
.900000	-3.76266	-2.39918
1	-1.93198E-02	-1.17181E-02
1.10000	3.37485	1.94712
1.20000	5.21299	2.86095
1.30000	4.97159	2.59540
1.40000	2.93674	1.45834
1.50000	2.25694E-02	1.06610E-02
1.60000	-2.62253	-1.17838
1.70000	-4.05775	-1.73434
1.80000	-3.87385	-1.57499
1.90000	-2.29210	-.886449
2.00000	-2.34360E-02	-8.62163E-03

>

FIGURE 7.5 Program for Example 7.2 and the results of one program run

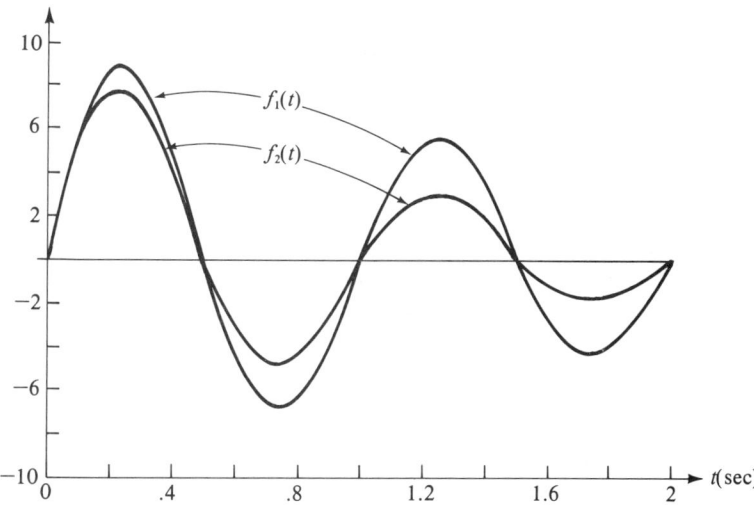

FIGURE 7.6 Graphs of two damped sine waves—Note that $f_2(t)$ diminishes more rapidly than $f_1(t)$

quadratic equation $Ax^2 + Bx + C$ are real and unequal if $B^2 - 4AC > 0$, real and equal if $B^2 - 4AC = 0$, and imaginary if $B^2 - 4AC < 0$. The following program uses the SGN function and the ON-GOTO statement to determine the nature of the roots:

```
10 INPUT A,B,C
20 ON SGN(B*B-4*A*C)+2 GOTO 30,50,70
30 PRINT "IMAGINARY"
40 GOTO 80
50 PRINT "REAL AND EQUAL"
60 GOTO 80
70 PRINT "REAL AND UNEQUAL"
80 END
```

Note that SGN(B*B − 4*A*C) + 2 in statement 20 has a value of 1, 2, or 3, depending on whether $B^2 - 4AC$ is less than, equal to, or greater than zero, respectively.

The *inverse* (arc) trigonometric functions have arguments that are the values of trigonometric functions; they return values of angles (in radians, in most versions of BASIC). For example, the arc tangent function ATN(a) returns the value of the angle whose tangent equals a. Recall that the equation

$$\tan \theta = a$$

conveys the same information as

$$\arctan(a) = \theta$$

The latter equation is also frequently written using the notation

$$\tan^{-1}(a) = \theta$$

The argument of the ATN function can be any positive or negative number. For example, since $\tan(\pi/4) = 1$ and $\tan(-\pi/3) = -1.73205$, it follows that arctan(1) = $\pi/4$ radians and arctan(−1.73205) = −$\pi/3$ radians.

Similarly, the inverse sine and inverse cosine functions are defined by their relations to the sine and cosine functions, as follows:

$$\sin \theta = a \Rightarrow \arcsin(a) = \theta$$
$$\cos \theta = a \Rightarrow \arccos(a) = \theta$$

The arguments of the arc sine and arc cosine functions are limited to the range of numbers between -1 and $+1$, inclusive, since all values of the sine and cosine functions are in that range. In versions of BASIC that have the inverse sine and cosine functions (many microcomputer versions do not), the arc sine and arc cosine are expressed by ASN(a) and ACS(a), respectively. The inverse trigonometric functions can be computed from the ATN function, using the following identities:

$$\arcsin(x) = \arctan\{x/(1 - x^2)^{1/2}\}$$
$$\arccos(x) = \pi/2 - \arctan\{x/(1 - x^2)^{1/2}\}$$
$$\text{arccot}(x) = \pi/2 - \arctan(x)$$

It is important to note that the inverse trigonometric functions return angles in the first and fourth quadrants only—that is, the angles lie between $+\pi/2$ and $-\pi/2$ radians. Thus, for example, ATN($-1/1$) and ATN($1/-1$) will both return the angle $-\pi/4$ radians ($-45°$), even though ATN($1/-1$) = $3\pi/4$ radians ($135°$). (See Figure 7.7.) This computation is ambiguous, since there is no way to distinguish that an argument of the form (Y/X) is negative because Y is negative or because X is negative. We encounter the same problem when evaluating inverse trigonometric functions on a pocket calculator. The following example shows how we can write a BASIC program to correct the arc tangent computation and eliminate this problem.

Example 7.3 Write a BASIC program that prompts the user to enter values of x and y and then computes the correct value of $\arctan(y/x)$, where x and y can have any positive or negative value or be equal to zero.

The first step in solving this problem is to determine if x is equal to zero. If so, we will not compute y/x, since this would cause division by zero and thus generate an error message. If $x = 0$, we know that arc tan(y/x) is either $+\pi/2$ radians or $-\pi/2$ radians ($\pm 90°$), depending on the sign of y. (We assume that x and y are not *both* zero, since this produces the undefined form 0/0.) If $x = 0$ and $y > 0$, then $\arctan(y/x) = \pi/2$ radians, and if $x = 0$ while $y < 0$, then $\arctan(y/x) = -\pi/2$ radians.

If x is not zero, we will proceed to compute the arctangent of the absolute value of y/x; $\arctan|y/x|$. We can then correct this computation for the algebraic signs of x

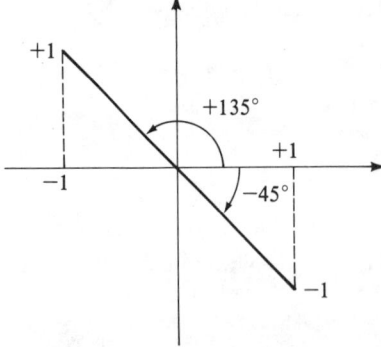

FIGURE 7.7 The arctangent of ($-1/1$) is 45° and the arctangent of ($1/-1$) is 135°. The BASIC ATN function returns $-45°$ (or $-\pi/4$ radians) in both cases.

and y. Let $\arctan|y/x| = \theta'$, and let the true value of $\arctan(y/x) = \theta$. We proceed as follows:

(1) If $x > 0$ and $y > 0$, then $\theta = \theta'$.
(2) If $x > 0$ and $y < 0$, then $\theta = -\theta'$.
(3) If $x < 0$ and $y > 0$, then $\theta = \pi - \theta'$.
(4) If $x < 0$ and $y < 0$, then $\theta = \theta' - \pi$.

Figure 7.8 illustrates these facts.

Figure 7.9 on page 162 shows a flowchart diagraming the logic used to correct the computation of $\arctan(y/x)$.

A BASIC program implementing the logic of the flowchart in Figure 7.9 is shown in Figure 7.10 (page 162).

Figure 7.11 on page 163 shows the results of several runs of the program in Figure 7.10.

Table 7.1 summarizes the library functions that we have described, along with others found in some versions of BASIC. Each argument a can be a constant, variable, or expression. Consult your programming manual to determine which of the functions are available in your version of BASIC.

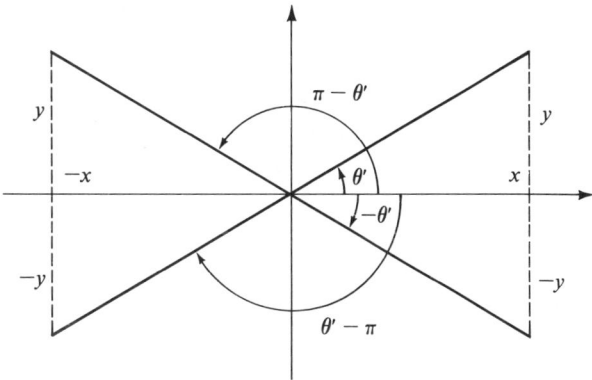

FIGURE 7.8 Relations among angles in four quadrants (Example 7.3)

EXERCISES

7.1 Show the display that would result if each of the following programs were run:

(a)
```
10 LET X=4
20 PRINT ABS(X**2-2*X**2)
30 END
```

(b)
```
10 LET T1=30
20 PRINT SIN(.0174532*T1)-COS(.0174532*T1*2)
30 END
```

(c)
```
10 INPUT A
20 PRINT SIN(A)**2+COS(A)**2
30 END
```

(d)
```
10 FOR I=1 TO 5
20 PRINT ((SGN(2-I))**3)*SGN(3-I)
30 NEXT I
40 END
```

(continued on page 163)

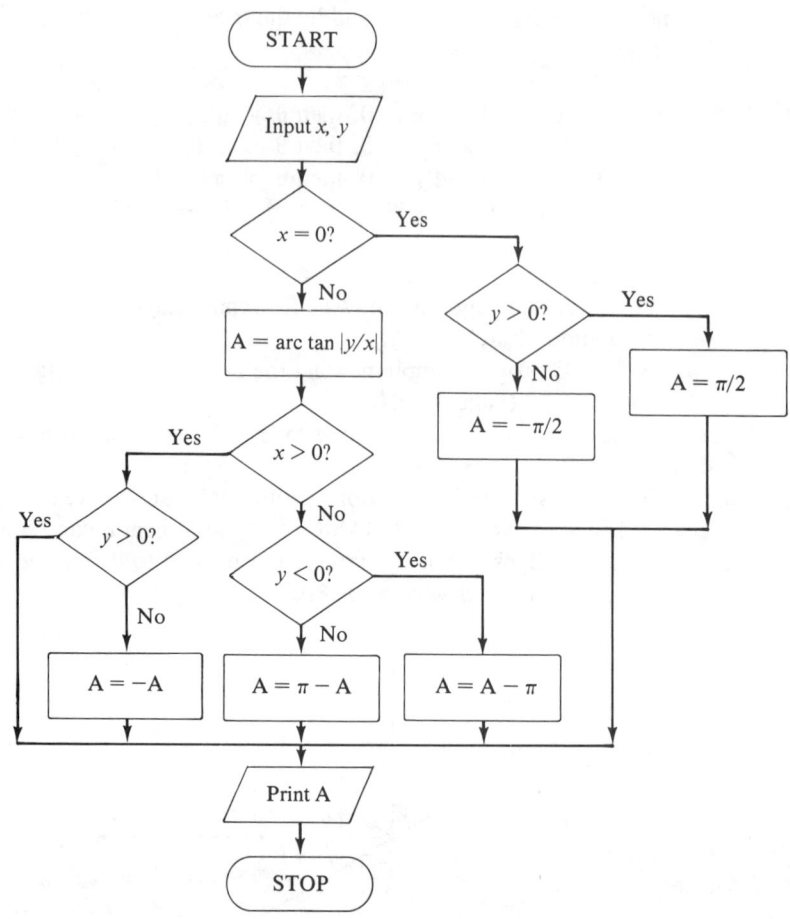

FIGURE 7.9 Flowchart for correcting the computation of ATN(y/x), according to the values of x and y (Example 7.3)

```
10 REM   A PROGRAM THAT COMPUTES ARCTANGENT (Y/X)
RADIANS
20 REM   USING THE BASIC ATN FUNCTION AND CORRECTS FOR
30 REM   THE SIGN OF X AND Y.
40 PRINT "ENTER X,Y. (X AND Y CANNOT BOTH BE ZERO.)"
50 INPUT X,Y
60 IF X=0 THEN 170
70 LET A=ATN(ABS(Y/X))
80 IF X>0 THEN 140
90 IF Y<0 THEN 120
100 LET A=3.1416-A
110 GOTO 210
120 LET A=A-3.1416
130 GOTO 210
140 IF Y>0 THEN 210
150 LET A= -A
160 GOTO 210
170 IF Y>0 THEN 200
180 LET A= -1.5708
190 GOTO 210
200 LET A=1.5708
210 PRINT "ARCTAN(Y/X)="A" RADIANS"
220 END
```

FIGURE 7.10 Program for Example 7.3

```
ENTER X,Y. (X AND Y CANNOT BOTH BE ZERO.)
?1,1
ARCTAN(Y/X)= .785398 RADIANS
>

ENTER X,Y. (X AND Y CANNOT BOTH BE ZERO.)
?1,-1
ARCTAN(Y/X)= -.785398 RADIANS
>

ENTER X,Y. (X AND Y CANNOT BOTH BE ZERO.)
?-1,1
ARCTAN(Y/X)=2.35620 RADIANS
>

ENTER X,Y. (X AND Y CANNOT BOTH BE ZERO.)
?-1,-1
ARCTAN(Y/X)= -2.35620 RADIANS
>

ENTER X,Y. (X AND Y CANNOT BOTH BE ZERO.)
?1,0
ARCTAN(Y/X)= 0 RADIANS
>

ENTER X,Y. (X AND Y CANNOT BOTH BE ZERO.)
?0,1
ARCTAN(Y/X)= 1.57080 RADIANS
>

ENTER X,Y, (X AND Y CANNOT BOTH BE ZERO.)
?0,-1
ARCTAN(Y/X)= -1.57080 RADIANS
>
```

FIGURE 7.11 The results of seven runs of the program in Figure 7.10

(e)
```
10 LET W=12
20 FOR I=1 TO 12
30 LET W=W-2*I
40 IF W>0 THEN 60
50 GOTO 70
60 NEXT I
70 PRINT EXP(W)*COS(W)
80 END
```

(f)
```
10 INPUT X
20 PRINT ATN(X/-X)*57.29578
30 PRINT ATN(-X/X)*57.29578
40 PRINT-ATN(X/X)*57.29578
50 PRINT-ATN(-X/-X)*57.29578
60 END
```

(continued on page 165)

Table 7.1

FUNCTION	DESCRIPTION
ABS(*a*)	Returns the absolute value of *a*
ACS(*a*)	Returns the arc cosine of *a*
ASN(*a*)	Returns the arc sine of *a*
ATN(*a*)	Returns the arc tangent of *a*
COS(*a*)	Returns the cosine of *a*
COT(*a*)	Returns the cotangent of *a*
CSC(*a*)	Returns the cosecant of *a*
EXP(*a*)	Returns e^a, where e is the base of the natural logarithm
FIX(*a*)	Returns the truncated value of *a*—that is, the value of *a* with any fractional part deleted (for example, FIX(3.7) = 3 and FIX(−2.5) = −2)
INT(*a*)	Returns the greatest integer less than or equal to *a* (for example, INT(3.7) = 3 and INT(−2.5) = −3)
LGT(*a*) *or* LOG10(*a*) *or* CLOG(*a*) (depending on the BASIC version used)	Returns the logarithm to the base 10 of *a* (the common logarithm)
LOG(*a*)	Returns the logarithm to the base e of *a* (the natural logarithm)
MAX(*list*)	Returns the maximum of the *list* of arguments, with each argument separated by a comma (for example, MAX(2,−3,5,1) = 5)
MIN(*list*)	Returns the minimum of the *list* of arguments, with each argument separated by a comma (for example, MIN(2,−3,5,1) = −3)
RND(*a*)	Returns a pseudorandom integer from the interval 1 through *a*
RND(0)	Returns a pseudorandom number between 0 and 1
SGN(*a*)	Returns −1 if *a* is negative, 0 if *a* equals 0, and +1 if *a* is positive
SIN(*a*)	Returns the sine of *a*
SQR(*a*)	Returns the square root of *a*
TAN(*a*)	Returns the tangent of *a*

(g) 10 LET M=2
    20 PRINT EXP(SIN(M-2))
    30 END

(h) 10 FOR J=1 TO 3
    20 PRINT SGN(SGN(2-J))
    30 NEXT J
    40 END

(i) 10 INPUT A
    20 PRINT SGN(ABS(SGN(A**2+1)))
    30 END

(j) 10 LET P=7.5
    20 PRINT EXP(-SQR(ABS(6.5-P)))
    30 END

7.2 Using library functions from Table 7.1, write BASIC expressions for each of the following:

(a) $|x|\sin x^2$

(b) $|x|^2 \sin^2 x$

(c) $\sin\left(\dfrac{x+y}{2}\right) \cos\left(\dfrac{x-y}{2}\right)$

(d) $1 + \tan^2(A - B^2)$

(e) an expression that is equal to A if $A < 0$, equal to 0 if $A = 0$, and equal to $-A$ if $A > 0$

(f) $e^{-\sin(3J)}$

(g) $|e^{|x|}\tan x|$

(h) $\sqrt{\arctan|a|}$

(i) $e^{e^x} - e^{-e^{-x}}$

(j) $\sqrt{1 - \sin^2 \sqrt{t}}$

7.3 Write a BASIC program that prompts the user to enter the horizontal and vertical components (F_x and F_y) of a force vector and then prints the magnitude F of that force ($F = \sqrt{F_x^2 + F_y^2}$) and its angle θ in degrees. $\theta = \arctan(F_y/F_x)$. Assume that F_x is always positive. Run your program for $F_x = F_y = 10$ and again for $F_x = 300$ and $F_y = -400$.

7.4 Write a BASIC program to print a table of values of the function $f(t) = 60e^{-2}\sin(\phi - \theta)$, where $\phi = 0.5$ radians, as θ varies from 0 through 360 degrees, in 15-degree increments.

7.5 The *hyperbolic sine* (sinh) and *hyperbolic cosine* (cosh) functions are defined as follows:

$$\sinh(x) = \frac{e^x - e^{-x}}{2}$$

$$\cosh(x) = \frac{e^x + e^{-x}}{2}$$

Write a BASIC program to print a table of values of x, e^x, $\sinh(x)$, $\cosh(x)$, and $\sinh(x) + \cosh(x)$. Run your program for values of x from $x = 0$ through $x = 2$, using increments of 0.1. Based on your results, write an apparent identity involving the sinh and cosh functions.

7.6 For small values of θ, sinθ is approximately equal to θ, when θ is expressed in radians. The percent error caused by using this approximation is as follows:

$$\text{percent error} = \frac{|\sin\theta - \theta|}{\sin\theta} \times 100\%$$

Write a BASIC program to print a table showing θ, sin θ, and the percent error in the approximation for θ = 0.05 radians through 1 radian, in increments of 0.05 radians.

7.7 For small values of θ, the approximation

$$\ln(\tan\sqrt{\theta}) \simeq 0.5(\ln\theta) + \frac{\theta}{3}$$

is valid where θ is in radians and ln is the natural logarithm. The percent error in this approximation is as follows:

$$\text{percent error} = \frac{\left|\ln(\tan\sqrt{\theta}) - 0.5(\ln\theta) - \frac{\theta}{3}\right|}{\ln(\tan\sqrt{\theta})} \times 100\%$$

Write a BASIC program to print the percent error in the approximation, after the user enters a value of θ in radians. Run your program for θ = 0.5, 0.1, and 0.01 radians.

2 USER-DEFINED FUNCTIONS

In most versions of BASIC, functions can be created by a programmer and then used in the same way that library functions are used. Functions are created using the "define function" (DEF) statement, whose format is

DEF FN*y*(*x*) = *expr*(*x*)

where *y* is any valid variable name, used to name (i.e., identify) the particular function being defined; *x* is a *dummy variable* which we will discuss in a moment; and *expr*(*x*) is an expression involving the variable *x*. The expression on the right-hand side of the equation in the DEF statement represents the function (of *x*) being defined by the programmer. The variable *x* is called a dummy variable because it is used only to define the function. Instead of *x*, any valid variable name can be used for this purpose. When the function FN*y*(*x*) is used in a program as a library function is used, we can substitute any BASIC constant, variable, or expression for the argument *x*.

This method of defining and using a function is best understood by example. Suppose we want to define a function that produces a value equal to 4.5 times its argument, plus 2—that is, we want $f(x) = 4.5x + 2$. We decide to name the function FNA. Then the function is defined by the statement

DEF FNA(X)=4.5*X+2

After we define the function FNA, we can use it like a library function anywhere in a BASIC program. The computer will evaluate FNA(*a*), where *a* is any argument, by multiplying the value of *a* times 4.5 and adding 2. For example,

FNA(2) = (4.5)(2) + 2 = 11

Similarly, if the variable K has previously been assigned the value -3, then

$$\text{FNA(K)} = (4.5)(-3) + 2 = -11.5$$

The following program segment shows how the function could be used in a conditional branch statement:

```
10 DEF FNA(X)=4.5*X+2
20 INPUT T
30 IF FNA(T)/2>1 THEN 60
```

A branch to statement 60 occurs in statement 30 if the value of $\dfrac{(4.5)(T) + 2}{2}$ is greater than 1. Note that the value assigned to T by the INPUT statement is substituted for X when FNA is evaluated.

The dummy variable X used in the DEF statement of the preceding example could have been any other variable. If we wish, the dummy variable can have the same name as the actual variable used in the program when the function is evaluated. For example, the preceding program segment could have been written as follows:

```
10 DEF FNA(T)=4.5*T+2
20 INPUT T
30 IF FNA(T)/2>1 THEN 60
```

We now present a few notes, precautions, and variations on the use of the DEF statement:

(1) Most (but not all) versions of BASIC require the DEF statement to appear in a program *before* the function it defines is used. This procedure is good practice in any case, for program legibility, and we will follow it in this book.
(2) The function name y in FNy can be any valid variable name, depending on the version of BASIC being used. The same is true for the dummy variable x. In this book, we will always use a single letter for each of these.
(3) Some versions of BASIC permit the definition of a function of several variables. In these versions, the dummy variable x can be replaced by a list of dummy variables. For example, DEF FNA(X,Y,Z) = X + Y − Z.
(4) The expression used to define a function can itself contain a function (for example, DEF FNB(M) = 3 * EXP(−M)). When a function name is used in a program, it is customary to say that we "call" that function. A function cannot call itself—that is, the function name cannot appear in its own definition or in its argument. Thus, BASIC permits LET Y = FNA(X + FNB(X)) but *not* LET Y = FNA(X + FNA(X)).

Example 7.4 The equation $f(t) = Ke^{-t/\tau}$, where K and τ are constants and t is time, describes a form known as a *decaying exponential*. At $t = 0, f(t) = K$; as t becomes very large, $f(t)$ approaches zero. The constant τ is called the *time-constant*, and its value determines how rapidly $f(t)$ approaches zero as t increases. Figure 7.12 shows graphs of $10e^{-t/1}$ and $10e^{-t/4}$ to illustrate this point. The time-constants are $\tau = 1$ and $\tau = 4$, and you can see that the smaller the value of the time-constant, the more quickly $f(t)$ approaches zero. Whatever the time-constant, $f(t)$ is for all practical purposes equal to zero at $t = 5\tau$.

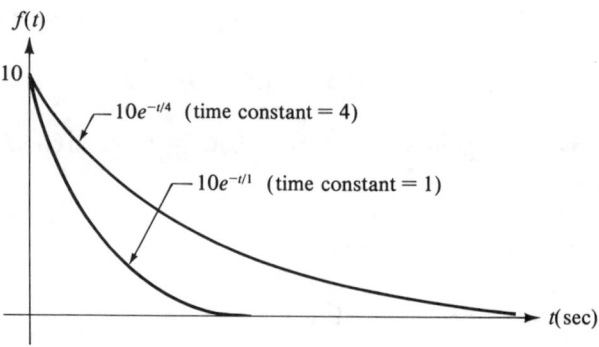

FIGURE 7.12 Graphs of two decaying exponential functions (Example 7.4)

In technology problems, we frequently encounter equations that are the sum or difference of decaying exponentials, such as

$$f(t) = K_1 e^{-t/\tau_1} + K_2 e^{-t/\tau_2}$$

It is of interest to determine the value of t required to make f(t) very small—that is, determine how large t must be so that $f(t)$ will decay to a value less than a certain small percentage of its value at $t = 0$. Note that $f(0) = K_1 + K_2$ in this case.

Write a BASIC program that prompts the user to enter values for K_1, K_2, τ_1, and τ_2 and then prints the value of t beyond which $f(t) = K_1 e^{-t/\tau_1} + K_2 e^{-t/\tau_2}$ is less than 1.0% of $K_1 + K_2$. Assume that K_1, K_2, τ_1, and τ_2 must all be positive values. Incorporate user-defined functions and print the values of $K_1 e^{-t/\tau_1}$, $K_2 e^{-t/\tau_2}$, and $f(t)$ for the value of t that satisfies the criterion.

We know that f(t) should be very close to zero if t is greater than, say, 10 times the *larger* of τ_1 and τ_2. We therefore decide to compute f(t) for 41 values of t between $t = 0$ and $t = 10\tau$, where τ is the larger of τ_1 and τ_2. If, for example, $\tau_1 = 0.5$ and $\tau_2 = 8$, we will calculate $f(t)$ at $t = 0$, $t = 2$, $t = 4$, . . . , $t = 80$ (at intervals of $\frac{1}{4}$ of the larger time-constant). After each computation of $f(t)$, we will check to determine whether we have reached a value of $f(t)$ that satisfies the criterion $f(t) < 0.01(K_1 + K_2)$.

Note that we are not attempting to find the *smallest* value of t for which $f(t) < 0.01(K_1 + K_2)$: That would be a much more difficult programming task. We are content to find a value of t within $\frac{1}{4}\tau$ of the smallest value of t that satisfies the criterion. A flowchart outlining the logic for this program is shown in Figure 7.13.

Figure 7.14 on page 170 is a BASIC program that implements the logic of the flowchart in Figure 7.13. Also shown are the results of a program run in which $K_1 = 10$, $\tau_1 = 2$ and $K_2 = 5$, $\tau_2 = 4$. You can see that f(t) < 0.01(5 + 10) = 0.15 for t > 15 seconds.

A few versions of BASIC permit the definition of *multiline* functions. In a multiline function, more than one statement is used to define the function. Special keywords denote the beginning and the end of the definition. The statements between these keywords define the function itself, and at least one of these statements must be in the form LET FNy = *expr*. Typically, the DEF statement is used to mark the beginning of the multiline function, and the keyword FNEND is used to end it. Following is an example:

```
10 DEF FNW(X)
20 IF X>20 THEN 50
```

FUNCTIONS AND SUBROUTINES

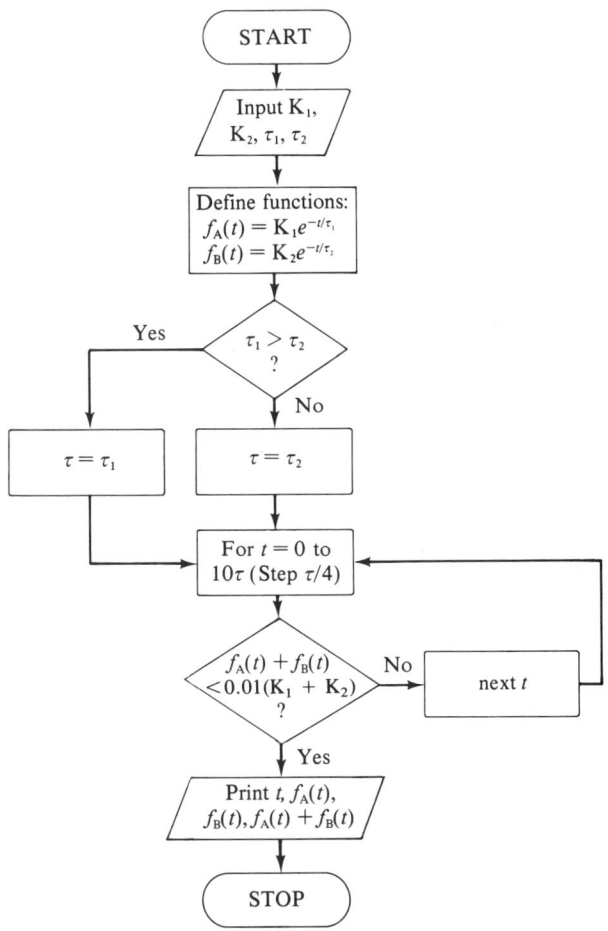

FIGURE 7.13 Flowchart for computing a value of t that satisfies $f(t) \leq .01(K_1 + K_2)$ (Example 7.4)

```
30 LET FNW=100
40 GOTO 60
50 LET FNW=35
60 FNEND
```

The formats for defining multiline functions vary widely among those versions of BASIC that permit them, so you should consult your programming manual for specifics about your particular version.

EXERCISES

7.8 Show the display that would result if each of the following programs were run:

(a)
```
10 DEF FNI(I)=I**2-2
20 DEF FNJ(J)=10-J/2
30 FOR I=1 TO 5
40 PRINT FNI(I)-FNJ(I)
50 NEXT I
60 END
```

```
10 REM   PROGRAM THAT FINDS A VALUE OF T SUCH THAT THE
20 REM   MAGNITUDE OF F(T)=K1*EXP(-T/T1)+K2*EXP(-T/T2)
30 REM   IS < MAGNITUDE OF 0.1(K1+K2).
40 PRINT "F(T)=K1*EXP(-T/T1)+K2*EXP(-T/T2)"
50 PRINT "ENTER K1 AND T1"
60 INPUT K1,T1
70 PRINT "ENTER K2 AND T2"
80 INPUT K2,T2
90 DEF FNA(T)=K1*EXP(-T/T1)
100 DEF FNB(T)=K2*EXP(-T/T2)
110 REM   FIND LARGER OF T1 AND T2
120 IF T1>T2 THEN 150
130 LET T3=T2
140 GOTO 160
150 LET T3=T1
160 FOR T=0 TO 10*T3 STEP T3/4
170 IF FNA(T)+FNB(T)< .01*(K1+K2) THEN 190
180 NEXT T
190 PRINT "AT T=";T;":"
200 PRINT "FNA(T)=";FNA(T)
210 PRINT "FNB(T)=";FNB(T)
220 PRINT "F(T)=";FNA(T)+FNB(T)
230 END


F(T)=K1*EXP(-T/T1)+K2*EXP(-T/T2)
ENTER K1 AND T1
?10,2
ENTER K2 AND T2
?5,4
AT T= 15:
FNA(T)= 5.53084E-03
FNB(T)= .117589
F(T)= .123120
>
```

FIGURE 7.14 Program for Example 7.4 and the results of one program run

(b)
```
10 DEF FNA(X)=4*SQR(X+1)
20 DEF FNB(X)=ABS(X-10)
30 LET M=3
40 PRINT FNB(FNA(M))+FNB(1)
50 END
```

(c)
```
10 DIM A(2,4)
20 DEF FNX(A)=2*(A+I)
30 DEF FNY(B)=B+J
40 FOR I=1 TO 2
50 FOR J=1 TO 4
60 LET A(I,J)=FNX(I)+FNY(2*J)
70 NEXT J
80 NEXT I
90 FOR J=1 TO 4
100 PRINT A(2,FNY(0))
```

FUNCTIONS AND SUBROUTINES

```
   110 NEXT J
   120 END
(d) 10 DEF FNK(X)=(X**2-1)/(X+1)
   20 DEF FNW(Y)=Y-1
   30 PRINT "ENTER Z, Z CANNOT = 0"
   40 INPUT Z
   50 PRINT SIN(FNK(Z)-FNW(Z))
   60 PRINT COS(FNW(Z)-FNK(Z))
   70 PRINT EXP(FNK(Z)-FNW(Z))
   80 END
```

7.9 Use the DEF statement to define the following BASIC functions:

(a) $f(x) = \sqrt{(x^2 + 1)^2}$

(b) $f(a) = e^{-\sin(a)} - e^{-\sin(b)}$

(c) $f(k) = |\tan^2(k - \pi/2) - |j||$

(d) $$f(n) = \begin{cases} m - n & \text{if } n > 0 \\ m & \text{if } n = 0 \\ m + n & \text{if } n < 0 \end{cases}$$

7.10 Write a BASIC program to print 11 values of the function $Ke^{-t/\tau}$, where $\tau = (x + 1)^2$, and the user enters values for K and x. The 11 values should be calculated at equal intervals between $t = 0$ and $t = 5\tau$, inclusive. Incorporate user-defined function(s) in your program, and run it for the following combinations of inputs:

(a) $K = 1, x = 0$
(b) $K = 10, x = 4$
(c) $K = 10, x = 0.5$

3 SUBROUTINES

In many programming situations, it is necessary to repeat the same set of computations several times, each time using a new set of data. Suppose, for example, that we want to find the averages of three sets of numbers, each consisting of 10 numbers. The statements required to find the average of one set of 10 numbers consist of a FOR-TO, NEXT loop that computes their sum, followed by division of the sum by 10. This same set of statements must be repeated for each of the three sets of 10 numbers.

A *subroutine* is a set of statements that is set aside from the main program, often for facilitating its repetition. In the averaging example that we just discussed, a subroutine could be the set of statements that calculates the average of any 10 numbers supplied to it. Every time this computation is required, the computer branches to the subroutine, the average is found, and the computer then branches back to the main program. Subroutines are also useful for isolating particularly complex or long sets of computations from the main program and thus improving legibility. The logical flow of a program is easier to understand when an involved computation is given its own set of statement numbers outside the range of the main program.

The GOSUB statement is used to cause a branch to a subroutine. The format for this statement is

GOSUB s*n*

where *sn* is the statement number of the first statement in the subroutine. When the GOSUB statement is executed, the computer immediately branches to the specified statement number. The subroutine can consist of any set of BASIC statements—including remarks, loops, input/output statements, and so forth. The last statement in the subroutine must be RETURN. The RETURN statement marks the end of a subroutine and, when executed, causes the computer to branch back to the main program. *The next statement executed after the RETURN statement is the statement following GOSUB in the main program.* Note that the computer *automatically* branches back to the correct location in the main program. In other words, the programmer does not need to specify a statement number with the RETURN statement. This is an exceptionally important aspect of subroutines, because it allows us to branch to the same subroutine from any number of locations in a main program and always return to the correct place: the statement number following the GOSUB statement that took us to the subroutine. Figure 7.15 illustrates this concept.

In Figure 7.15, the circled numbers indicate the sequence of branches that occurs when two GOSUB statements "call" (cause a branch to) the same subroutine. In this example, the main program consists of statements 10 through 450, and the subroutine consists of statements 1000 through 1100. When the first GOSUB 1000 statement is executed, the program branches to statement 1000, and the subroutine statements are then executed. The RETURN statement at 1100 then causes a branch back to statement 60, and execution of the main program resumes. When the second GOSUB statement is executed, this process is repeated, but this time the RETURN statement causes a branch back to statement 190. In each case, the RETURN statement causes a branch back to the point in the main program where execution of that program left off.

As we have already indicated, a subroutine has its own set of statement numbers outside the range of the main program. As a general rule, the first statement number of a subroutine should be greater than the largest statement number in the main program. The END statement for the entire program, including any subroutines, should then appear after the RETURN statement in the last subroutine—that is, the END statement should have the largest statement number in the program, as usual. But this raises a new problem. Remember that the computer executes statements in the order that the statement numbers appear, until it encounters an END statement. What then is to prevent the computer from entering a subroutine once the statements in the main program have been executed? That is, how can we prevent the computer from "falling into" a subroutine following the main program, when no GOSUB

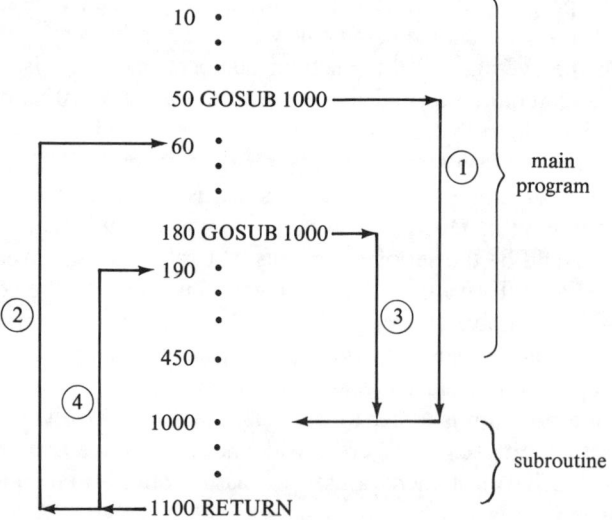

FIGURE 7.15 Sequence of branches that occurs when a subroutine is called twice in a main program

statement has instructed it to perform the subroutine? If this situation is allowed to occur, the computer will generate an error message.

The dilemma we have just described can be resolved through the use of a STOP statement. A STOP statement has the same effect on program execution as an END statement does—namely, the computer ceases all statement execution. Therefore, we can (and must) place a STOP statement between the last statement of a main program and the first statement of a subroutine, to prevent the computer from falling into the subroutine. In Figure 7.15, for example, we must insert a STOP statement with a statement number ranging somewhere from 451 through 999.

The difference between a STOP statement and an END statement is that an END statement tells the computer's compiler or interpreter (see Chapter 1) how far to proceed before ceasing conversion of BASIC statements into machine-language instructions. The STOP statement merely stops program execution, after a run has been initiated. However, in many versions of BASIC, the STOP and END statements can be used interchangeably, since conversion of BASIC statements to machine language continues as long as statement numbers are present. In most versions of BASIC, program execution can be resumed after a STOP or END statement, by using a system command entered by the user. (See Chapter 4.) STOP statements are useful in interactive programming, when the user needs to know the outcome of certain intermediate computations before proceeding, as, for example, when the data to be entered depends upon the outcome of such computations.

Example 7.5 The product of two complex numbers, $a + jb$ and $c + jd$, where $j = \sqrt{-1}$, can be found by converting each to their respective polar forms, $A \angle \theta$ and $B \angle \phi$, where

$$A = \sqrt{a^2 + b^2}, \quad \theta = \arctan(b/a)$$

and

$$B = \sqrt{c^2 + d^2}, \quad \phi = \arctan(d/c)$$

A and B are the *magnitudes* of the two complex numbers, and θ and ϕ are their *angles*, measured clockwise from the positive real axis. (See Figure 7.16. Note that c is negative in the complex number $c + jd$ illustrated by the figure.) The polar form of the product of the complex numbers is then $AB \angle \theta + \phi$.

Write a BASIC program that prompts the user to enter the real and imaginary parts of each number (a,b and c,d) and then prints the magnitude AB and the angle $\theta + \phi$ of their product. Use a subroutine to perform the magnitude and angle computations required to convert the numbers to their individual polar forms.

Since the numbers a, b, c, and d can be either positive or negative, the angles θ and ϕ can fall in any one of the four quadrants. We must therefore incorporate the

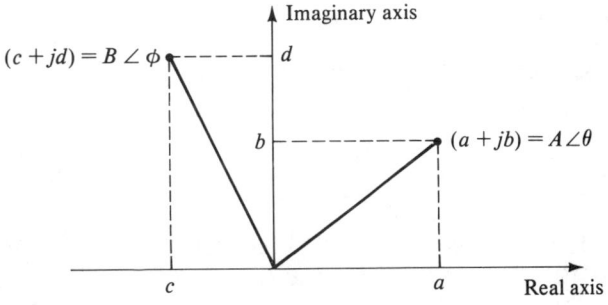

FIGURE 7.16 Polar forms of two complex numbers (Example 7.5)

correction that we discussed in Example 7.3 when using the ATN function. We will use the correction algorithm as part of the subroutine that performs the conversions to polar form. Figure 7.17 is a BASIC program that includes this subroutine.
Note the following points pertaining to the program shown in Figure 7.17:

(1) The subroutine computes the magnitude M and angle R of the complex number $x + jy$. Therefore, we set X = A and Y = B before calling the subroutine in statement 90.
(2) On returning from the subroutine (at statement 100), we must *save* the values of M and R computed by the subroutine. We can accomplish this by letting M1 = M and R1 = R.

```
10  REM   A PROGRAM THAT FINDS THE POLAR FORM OF
20  REM   THE PRODUCT OF TWO COMPLEX NUMBERS.
30  PRINT "ENTER A AND B (FOR A+JB)"
40  INPUT A,B
50  PRINT "ENTER C AND D (FOR C+JD)"
60  INPUT C,D
70  LET X=A
80  LET Y=B
90  GOSUB 500
100 LET M1=M
110 LET R1=R
120 LET X=C
130 LET Y=D
140 GOSUB 500
150 LET M2=M
160 LET R2=R
170 PRINT "MAGNITUDE=:"M1*M2
180 PRINT "ANGLE=:"(R1+R2)*57.29578" DEGREES"
190 STOP
500 REM   SUBROUTINE THAT CONVERTS X+JY TO POLAR FORM
510 REM   M=MAGNITUDE;R=ANGLE,IN RADIANS,CORRECTED FOR SIGN
520 IF X=0 THEN 630
530 LET R=ATN(ABS(Y/X))
540 IF X>0 THEN 600
550 IF Y<0 THEN 580
560 LET R=3.1416-R
570 GOTO 670
580 LET R=R-3.1416
590 GOTO 670
600 IF Y>0 THEN 670
610 LET R=-R
620 GOTO 670
630 IF Y>0 THEN 660
640 LET R=-1.5708
650 GOTO 670
660 LET R=1.5708
670 LET M=SQR(X**2+Y**2)
680 RETURN
690 END
```

FIGURE 7.17 Program for Example 7.5

FUNCTIONS AND SUBROUTINES   **175**

(3) Before calling the subroutine a second time to find the polar form of $c + jd$, we must set $X = C$ and $Y = D$.
(4) On returning from the subroutine the second time, we set $M2 = M$ and $R2 = R$. These are the computed values of the magnitude and angle of $c + jd$.
(5) A STOP statement (190) is placed between the main program and the subroutine to prevent the subroutine from being erroneously executed a third time.

The results of two runs of the program in Figure 7.17 are shown in Figure 7.18. In the first run, we see that $(3 + j4)(1 + j2) = 11.1803 \underline{/116.565°}$ and the second run shows that $(0 - j50)(-100 + j100) = 7071.07 \underline{/45°}$.

Example 7.6 Write a BASIC program to read numeric data into a 4 × 4 array and then find the maximum number in each row and the maximum number in each column. Use one subroutine to find the maximum entry in a row and another subroutine to find the maximum entry in a column.

We will use the algorithm described in Chapter 6 for finding the maximum of a set of numbers. Figure 7.19 on page 176 shows an abbreviated flowchart for the program. Figure 7.20 (page 177) shows the program and the results of a run made for the array:

$$\begin{bmatrix} 3 & 7 & -2 & 10 \\ 1 & -5 & 4 & 0 \\ -6 & 11 & 0 & 6 \\ 8 & -1 & 5 & 12 \end{bmatrix}$$

Subroutines can be *nested*, in the sense that a GOSUB statement can appear in one subroutine and cause a branch out of that subroutine to a second subroutine. When the RETURN instruction in the second subroutine is executed, the computer branches back to the correct location in the first subroutine. We say that the second subroutine is nested in the first. Figure 7.21 on page 178 illustrates the branching involved when one subroutine is nested in another; the circled numbers indicate the sequence in which the branches occur.

```
ENTER A AND B (FOR A+JB)
?3,4
ENTER C AND D (FOR C+JD)
?1,2
MAGNITUDE=: 11.1803
ANGLE=: 116.565 DEGREES
>

ENTER A AND B (FOR A+JB)
?0,-50
ENTER C AND D (FOR C+JD)
?-100,100
MAGNITUDE=: 7071.07
ANGLE=: 45.0002 DEGREES
>
```

FIGURE 7.18 The results of two runs of the program in Figure 7.17

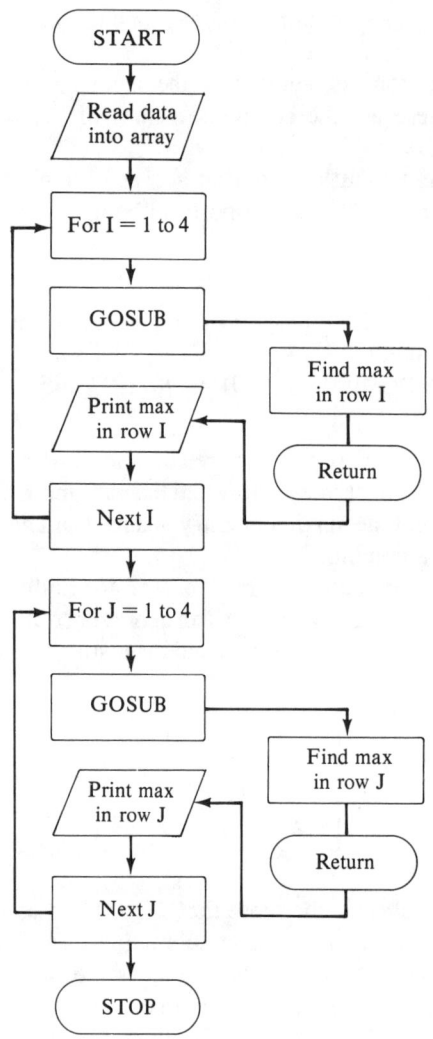

FIGURE 7.19 Abbreviated flowchart for Example 7.6

A nested subroutine can have another subroutine nested in it, and this nesting can continue indefinitely, limited only by the capacity of the computer.

Example 7.7 Each checking account record maintained by a certain bank consists of the following data, in the order listed: customer name; previous balance; total withdrawals, in dollars; total deposits, in dollars; number of checks written; lowest balance during the record period; and type of account (1 or 2). In type 1 accounts, no interest is paid on deposits and checks are free, through 50. A 10¢ charge is assessed for each check written over 50. In type 2 accounts, 5 percent interest is paid on the lowest balance during the record period and a 10¢ charge is assessed on all checks written, through 50. A 15¢ charge is assessed for each check written over 50.

Write a BASIC program to read N records (N is entered by the user) and then print the following information: all data in the recod, the new balance, the interest credited (if any), and the total charges (if any) assessed for checks.

We will use a subroutine to compute the new balance, by adding deposits and subtracting withdrawals from the old balance. If the account is a type 2 account, we will then branch to another subroutine to compute the interest credited and the check

```
10 REM  A PROGRAM THAT FINDS THE MAXIMUM NUMBER IN
   EACH ROW
20 REM  AND IN EACH COLUMN OF A 4x4 ARRAY
30 DIM A(4,4)
40 FOR I=1 TO 4
50 FOR J=1 TO 4
60 READ A(I,J)
70 NEXT J
80 NEXT I
90 FOR I=1 TO 4
100 GOSUB 200
110 PRINT "THE MAXIMUM IN ROW"I" IS:"R
120 NEXT I
130 FOR J=1 TO 4
140 GOSUB 400
150 PRINT "THE MAXIMUM IN COLUMN"J" IS:"C
160 NEXT J
170 DATA 3,7,-2,10,1,-5,4,0
180 DATA -6,11,0,6,8,-1,5,12
190 STOP
200 REM  SUBROUTINE THAT FINDS MAX IN ROW I
210 LET R=A(I,1)
220 FOR J=2 TO 4
230 IF A(I,J)<R THEN 250
240 LET R=A(I,J)
250 NEXT J
260 RETURN
400 REM  SUBROUTINE THAT FINDS MAX IN ROW J
410 LET C=A(1,J)
420 FOR I=2 TO 4
430 IF A(I,J)<C THEN 450
440 LET C=A(I,J)
450 NEXT I
460 RETURN
470 END


THE MAXIMUM IN ROW 1 IS: 10
THE MAXIMUM IN ROW 2 IS: 4
THE MAXIMUM IN ROW 3 IS: 11
THE MAXIMUM IN ROW 4 IS: 12
THE MAXIMUM IN COLUMN 1 IS: 8
THE MAXIMUM IN COLUMN 2 IS: 11
THE MAXIMUM IN COLUMN 3 IS: 5
THE MAXIMUM IN COLUMN 4 IS: 12
>
```

FIGURE 7.20 Program for Example 7.6 and the results of a program run

charges. On returning to the first subroutine, we will modify the previously computed new balance and then return to the main program. On the other hand, if the account is a type 1 account, we will compute the check charges, modify the new balance, and return to the main program. The program is shown in Figure 7.22 on page 179. Note that data must be supplied by the user before this program is run. (See statement 190.)

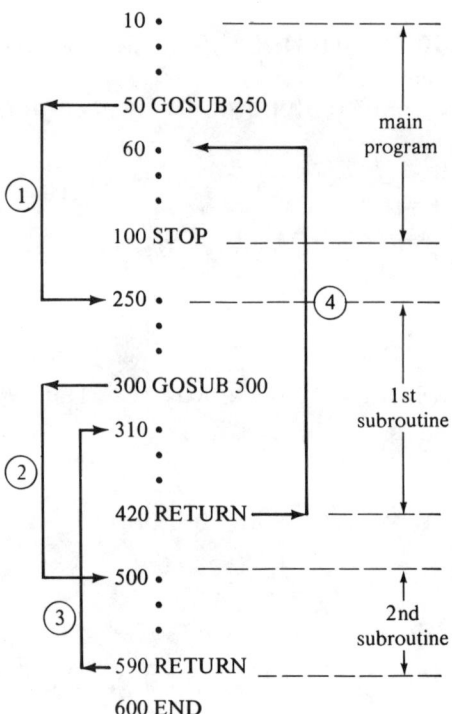

FIGURE 7.21 Sequence of branches that occurs when one subroutine calls another—In this case, the second subroutine is said to be *nested* in the first.

EXERCISES

7.11 Show the display that would result if each of the following programs were run:

(a)
```
10 FOR I=1 TO 5
20 GOSUB 100
30 NEXT I
40 STOP
100 PRINT I-3
110 RETURN
120 END
```

(b)
```
10 LET X=100
20 LET Y=SQR(X)
30 GOSUB 200
40 GOSUB 200
50 PRINT X
60 STOP
200 LET X=X+Y
210 RETURN
220 END
```

(c)
```
10 LET A=0
20 GOSUB 50
30 PRINT B
40 STOP
```

(program continued on page 179)

```
10 REM   PROGRAM FOR COMPUTING CHECKING ACCOUNT
   BALANCES
20 PRINT "ENTER TOTAL NUMBER OF RECORDS"
30 INPUT N
40 FOR J=1 TO N
50 READ N$,P,W,D,C,L,T
60 PRINT N$
70 PRINT "PREVIOUS BALANCE= $"P
80 PRINT "TOTAL WITHDRAWALS= $"W
90 PRINT "TOTAL DEPOSITS= $"D
100 PRINT "CHECKS WRITTEN:"C
110 PRINT "LOWEST BALANCE= $"L
120 PRINT "TYPE OF ACCOUNT:"T
130 GOSUB 300
140 PRINT "CHECK CHARGES= $"M
150 PRINT "INTEREST CREDITED= $"I
160 PRINT "NEW BALANCE= $"B
170 PRINT
180 NEXT J
190 REM   USER SUPPLIES DATA IN 200 THROUGH 298
299 STOP
300 REM   SUBROUTINE THAT COMPUTES NEW BALANCE B
310 LET B=P+D-W
320 LET M=0
330 LET I=0
340 IF T=2 THEN 390
350 IF C<51 THEN 380
360 LET M=0.1*(C-50)
370 LET B=B-M
380 RETURN
390 GOSUB 500
400 LET B=B-M+I
410 RETURN
500 REM   SUBROUTINE THAT COMPUTES CHARGES AND INTEREST
    ON TYPE 2 ACCT.
510 LET I=.05*L
520 IF C>50 THEN 550
530 LET M=0.1*C
540 RETURN
550 LET M=5+0.15*(C-50)
560 RETURN
570 END
```

FIGURE 7.22 Program for Example 7.7—Note that the user must supply data in statements 200 through 298.

```
50 LET B=EXP(EXP(A))
60 IF B>2 THEN 80
70 RETURN
80 GOSUB 110
90 PRINT A-B
100 RETURN
110 PRINT SIN(SIN(A))
120 RETURN
```

(d) ```
    10 LET T=0
    20 GOSUB 330
    30 GOTO 20
    100 GOSUB 500
    110 STOP
    330 LET T=T+1
    340 PRINT T
    350 IF T<6 THEN 370
    360 GOTO 100
    370 RETURN
    500 PRINT T*T
    510 RETURN
    520 END
```

**7.12** Write a BASIC program to perform the following:

(1) Read and print five integers.
(2) If the sum of these integers is less than 10, branch to a subroutine that computes and prints the product of the integers.
(3) If the sum of the integers is equal to or greater than 10, branch to a subroutine that finds and prints the minimum of the integers.

Run your program for the following sets of integers:

(a) $-4, 2, 9, -1, 3$
(b) $6, 0, -5, 11, 12$

## 4  A SUBROUTINE FOR GRAPHIC OUTPUT

The specialty field in computer science known as *graphics* deals with the use of a computer to generate pictorial displays of geometric figures, diagrams, charts, and graphs. Some versions of BASIC have special statements that can be used to create displays of simple figures such as circles, straight lines, and rectangles. These statements can then be used in a program to generate more complex figures for display on the screen of a video terminal. Printed copy of such displays requires the use of a graphics *plotter*.

Graphics has become an important tool in many areas of technology. For example, in the field called computer-aided design (CAD), engineering drawings are created by using computer techniques. Such diverse objects as machine drawings, schematic electrical diagrams, maps, and architectural drawings can be created using these techniques. Dynamic graphic displays (those that change their shape, form, or color with time) are used to study the effects of changes of one or more physical variables on another variable. Applications include aerodynamic flow patterns, space vehicle trajectories, weather formations, magnetic and electric field patterns, and studies of strain in structures.

Since there is a wide variation in the number and types of BASIC statements used to produce graphic output among the versions of BASIC having this capability, we will not discuss them any further here. We can, however, use standard BASIC statements to create one very useful form of graphic output, the two-dimensional graph, and we will discuss this form in some detail. Of course, graphs are important in all fields of technology, because they give us visual insight into the effect of changes in one variable (the independent variable) on another (dependent) variable, whenever there is some functional relationship between these two.

We will devote the remainder of this chapter to the development of a subroutine that can be used to create a plot of the values of a dependent variable, which we will
```

call y, versus an independent variable x. Since we will not use any special BASIC graphics statements, the plot can either be displayed on a video terminal or printed on a conventional printer, or both. A graphics plotter is not necessary. We will require any program using this subroutine to compute the values of x and y to be plotted, and we will assume that these values are stored by that main program in arrays called X and Y. Thus, when the program calls the subroutine, the values of x and y will be immediately accessible to the subroutine through references to arrays X and Y.

Our plot of y versus x will be created with a vertical x axis (contrary to conventional x-y plots). This means that values of x will occur, in the same order as in array X, downward from the top of the printed page. Corresponding values of y will be plotted horizontally. For example, a plot of the function $y = \sin x$ will have the same general form as that shown in Figure 7.23(a).

Instead of a continuous line, such as that shown in Figure 7.23(a), we will plot a succession of asterisks, as shown in Figure 7.23(b). Each asterisk is displaced to the right a distance proportional to the magnitude of the single y value it represents. We will use the TAB function to displace each asterisk on each new line by the correct distance.

Now consider in detail how the TAB function can be used for the intended purpose. If the y value to be plotted is the *maximum* of all y values, we must then tab right the entire width of the printer (or screen). If it is the *minimum* of all y values, we tab only to the first (leftmost) column of the display. For any other y value, the number of spaces we tab must have a ratio to the total width of the display equal to the ratio of the difference between that y value and the minimum y value to the total range between y values. For example, suppose the minimum y value is 10 and the maximum y value is 30. Then a y value of, say, 20 should cause an asterisk to be positioned in the middle column of the display, since $y = 20$ is halfway between $y = 10$ and $y = 30$. Thus, we should TAB one-half the total number of columns (the total width) available for the display. We obtain the ratio of one-half by calculating $(y - 10)/(30 - 10) = (20 - 10)/(30 - 10) = 10/20$.

Let M1 = the minimum of all y values to be plotted and M2 = the maximum of those values. Let C = the number of the rightmost column in the display. If the first (leftmost) column is number 1, note that there are then C − 1 total columns available for the display. Let J = the column to which we must tab, for a given value of y. Then, as previously described, the following equality of ratios must hold:

$$\frac{J - 1}{C - 1} = \frac{y - M1}{M2 - M1} \tag{1}$$

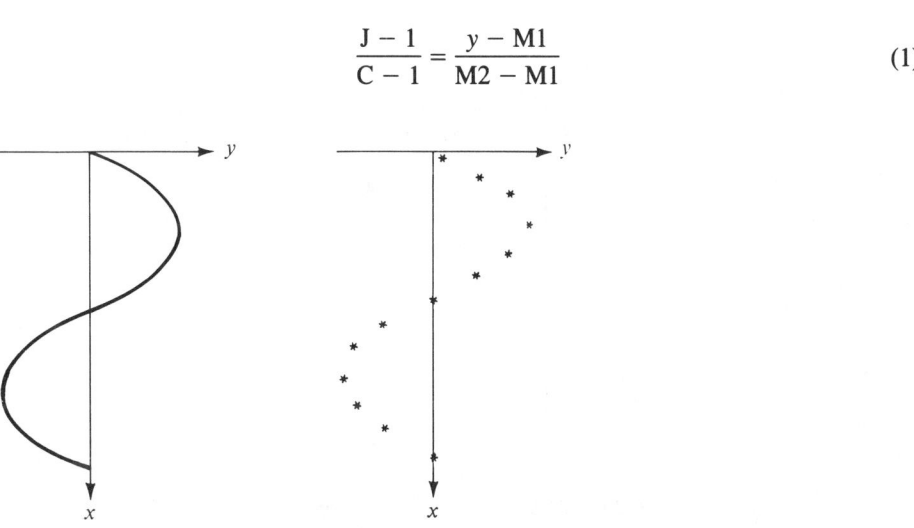

FIGURE 7.23 Graph of a sine wave plotted with a vertical x axis and horizontal y axis (b) Sine wave graph as it appears when asterisks are printed to represent y values

Figure 7.24 shows the y values, column numbers, and distances involved in Equation 1. Numeric values of the variable y are shown at the top of the figure, and corresponding column numbers are shown at the bottom.

Using Equation 1, we solve for J and find

$$J = \frac{(C-1)(y-M1)}{(M2-M1)} + 1 \tag{2}$$

Equation 2 tells us how to calculate the column J to which we must tab for a given value of y. Note, for example, if $y = M2$, the maximum y value, then Equation 2 tells us to tab to column

$$J = \frac{(C-1)(M2-M1)}{(M2-M1)} + 1 = C - 1 + 1 = C$$

which, as required, is the rightmost column.

Now consider that every tab must be to an *integer*-numbered column. That is, J in Equation 2 must always be an integer (whole) number. Since the quantity $(C-1)(y-M1)/(M2-M1)$ in Equation 2 will not necessarily equal an integer number, we must modify the computation of J so that only integer values will result. The best solution is to *round off* the value computed in Equation 2 to the nearest integer. We can use the INT function for that purpose. Recall (see Table 7.1) that INT(a) returns the greatest integer that is less than or equal to a. For example, INT(4.6) = 4 and INT(-2.1) = -3. As shown by these last two examples, the INT function does *not* round off its argument. However, it is an important fact that INT($a + 0.5$) returns a number rounded to the nearest integer value of a. In the two previous examples, INT(4.6 + 0.5) = INT(5.1) = 5 and INT($-2.1 + 0.5$) = INT(-1.6) = -2.* Each result is now the correct rounded value of the original argument. We must therefore modify the computation of J in Equation 2, as follows:

$$J = \text{INT}\left[0.5 + \frac{(C-1)(y-M1)}{M2-M1} \right] + 1 \tag{3}$$

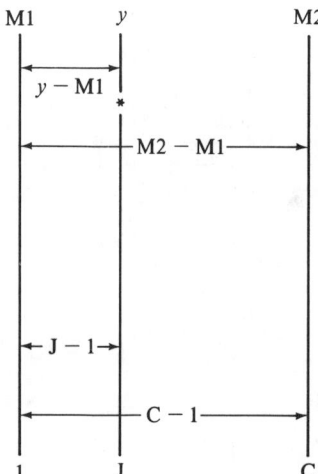

FIGURE 7.24 y values and their range (top) and the corresponding column numbers of the display (bottom)

---

*In a few versions of BASIC, the INT function *truncates* its argument, i.e., is the same as the FIX function in Table 7.1. Provided a is positive, INT($a + 0.5$) still rounds a. However, in those versions, if a is negative, you must use INT($a - 0.5$) to round.

or, as a BASIC statement:

```
LET J=INT(.5+(C-1)*(Y-M1)/(M2-M1))+1
```

To illustrate the application of the ideas developed so far, suppose we wish to plot the function $y = x^2$ for the values $x = -1, 0, 1, 2, 3$, over a display ranging between columns 1 and 51. Then $y(-1) = 1$, $y(0) = 0$, $y(1) = 1$, $y(2) = 4$, and $y(3) = 9$. We assume that the main program has created the arrays X and Y, as follows:

$$X = \begin{bmatrix} -1 \\ 0 \\ 1 \\ 2 \\ 3 \end{bmatrix} \quad Y = \begin{bmatrix} 1 \\ 0 \\ 1 \\ 4 \\ 9 \end{bmatrix}$$

Now $M1 = 0$, $M2 = 9$, and $C = 51$. The tabs $J(1), J(2), \ldots, J(5)$ required for each of the y values $Y(1), Y(2), \ldots, Y(5)$ are computed using Equation 3, as follows:

$$J(1) = \text{INT}\left[0.5 + \frac{50(1-0)}{9-0}\right] + 1 = \text{INT}(6.055) + 1 = 6 + 1 = 7$$

$$J(2) = \text{INT}\left[0.5 + \frac{50(0-0)}{9-0}\right] + 1 = \text{INT}(0.5) + 1 = 0 + 1 = 1$$

$$J(3) = \text{INT}\left[0.5 + \frac{50(1-0)}{9-0}\right] + 1 = \text{INT}(6.055) + 1 = 6 + 1 = 7$$

$$J(4) = \text{INT}\left[0.5 + \frac{50(4-0)}{9-0}\right] + 1 = \text{INT}(22.72) + 1 = 22 + 1 = 23$$

$$J(5) = \text{INT}\left[0.5 + \frac{50(9-0)}{9-0}\right] + 1 = \text{INT}(50.5) + 1 = 50 + 1 = 51$$

The display would then have the appearance shown in Figure 7.25, where column numbers corresponding to the calculated tabs have been added for clarity.

In the subroutine, we will print an asterisk on every *other* line, to improve visual continuity or "smoothness" of the plotted curve. On the lines between asterisks we will print dots in the columns corresponding to M1, 0, and M2. Let K be the column to which we must tab to print a dot for the zero axis. Assuming that $M1 < 0 < M2$, we have a situation as illustrated in Figure 7.26 (page 184).

```
1       7           23                                              51
        *
*
        *
                    *
                                                                    *
```

FIGURE 7.25 Asterisk locations for five values of the function $y = x^2$, from $x = -1$ through $x = 3$

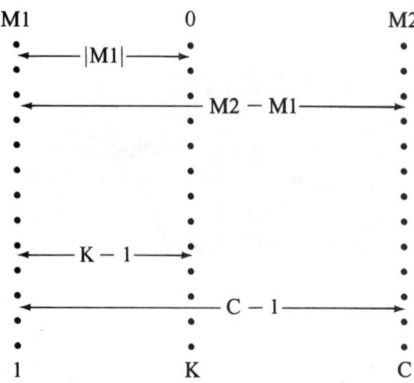

FIGURE 7.26 y values (top) and column numbers (bottom) when a zero axis is located between a negative minimum value (M1) and a positive maximum value (M2)

Applying Equation 3, we find that

$$K = INT[0.5 + (C - 1)|M1|/(M2 - M1)] + 1 \qquad (4)$$

Note that we are using $|M1|$ in Equation 4 instead of M1, since M1 is assumed to be a negative number.

On the other hand, if the values of both M1 and M2 are positive (if $0 < M1 < M2$), our plot will not include the value zero. That is, asterisks will be printed between column 1 (corresponding to $y = M1$), and column C (corresponding to $y = M2$), so that the full variation of y is displayed over the full width available for plotting. Whenever $0 < M1$, our subroutine will not print a column of dots for the zero axis. The same is true if $M1 < M2 < 0$—that is, when both the minimum and maximum values of y are negative.

In some situations, it is desirable to obtain a plot displaying the location of the zero axis, even if none of the y values reach zero. In our subroutine, we will provide the user with the option of obtaining an additional such plot whenever $0 < M1 < M2$ or $M1 < M2 < 0$. The zero axis will always be displayed in the only other possible case: $M1 < 0 < M2$.

Figure 7.27 (pages 186 and 187) is a flowchart for the required subroutine. Note the following points in connection with this flowchart:

(1) The first computation to be performed is the determination of the minimum and maximum y values, M1 and M2. This computation uses the values Y(I) that the main program is assumed to have stored in array Y.

(2) The subroutine then determines whether the range of y values includes zero. If M1 is positive or M2 is negative, then zero is not included in the range, and the variable Z is set equal to 1. Otherwise, Z is set equal to 0. (Z is tested later in the subroutine to ensure correct branching, according to the case at hand.)

(3) The user supplies the greatest column number (C) that is available for the plot. This is a user-supplied value, because different types of printers (and video screens) have different widths. Of course, C can be any number less than the maximum available, but for best accuracy and resolution, the greatest possible value should be used.

(4) For reference purposes, the values of X(I) and Y(I) are printed in a table in the same order as they were stored in arrays X and Y by the main program. The subroutine does not use the values of X(I), so it is up to the user to interpret each line on which an asterisk appears as corresponding to each successive value of X(I).

(5) If Z = 1, then zero is not in the range and only the "MIN" and "MAX" column headers (for the minimum and maximum axes) are printed. If Z = 0, a "0-axis" column header is also printed.
(6) Whether Z = 0 or 1, the tab values J(I) are computed for each *y* value, and asterisks are printed at those tab positions. If Z = 0, dots are printed between asterisks to form the minimum, zero, and maximum axes. If Z = 1, dots are printed only for the minimum and maximum axes.
(7) After printing the plot, the value of Z is checked again. If Z = 0, a zero axis has already been displayed, and execution returns to the main program. If Z = 1, then zero was not in the range, and the user is given the option of obtaining another plot that includes the zero axis. If this option is declined, a RETURN to the main program is executed.
(8) If the user elects to obtain another plot, the subroutine first checks to determine whether M2 is negative. If it is, then M1 < M2 < 0, and the zero axis will be column C. If M2 is not negative, then 0 < M1 < M2, and the zero axis will be the first column of the plot.
(9) If M2 < 0, then the tab value K is computed for printing dots down the maximum axis, and tab values J(I) are computed for printing asterisks. If M2 > 0, the tab value K is computed for printing dots down the minimum axis, and tab values J(I) are computed for printing asterisks. Note that the computations for J(I) will be different, depending on whether M2 < 0 or M2 > 0. These are also different from the original computations for J(I) that were made to display the plot over the entire range from M1 to M2. These differences exist because the numeric range is computed differently in each case.

Following is a listing of the subroutine, which we will hereafter call the "PLOTTER" subroutine. Note that it begins with statement 1000, so a program calling it will use the statement GOSUB 1000.

```
1000 REM   A SUBROUTINE THAT PRINTS AND PLOTS Y(I) VS.
1010 REM   X(I) ON A DISPLAY THAT EXTENDS TO COLUMN C.
1020 PRINT "ENTER NUMBER OF VALUES TO BE PLOTTED."
1030 INPUT N
1040 LET M1=Y(1)
1050 LET M2=Y(1)
1060 FOR I=1 TO N
1070 IF Y(I)>M1 THEN 1090
1080 LET M1=Y(I)
1090 IF Y(I)<M2 THEN 1110
1100 LET M2=Y(I)
1110 NEXT I
1120 IF M1>0 THEN 1160
1130 IF M2<0 THEN 1160
1140 LET Z=0
1150 GOTO 1170
1160 LET Z=1
1170 PRINT "ENTER MAXIMUM COLUMN NO. AVAILABLE FOR PLOT"
1180 INPUT C
1190 PRINT "     X"," Y"
1200 PRINT "---------","---------"
1210 FOR I=1 TO N
1220 PRINT X(I), Y(I)
1230 NEXT I
```

(continued on page 188)

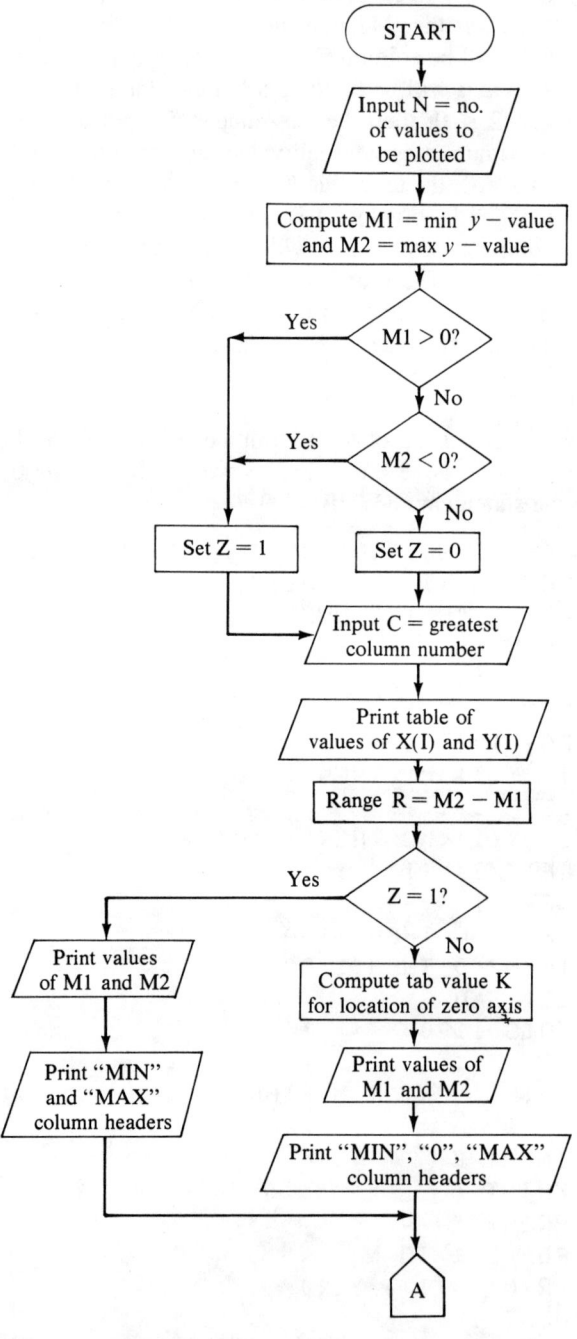

FIGURE 7.27 Flowchart for the graph-plotting subroutine (continued on page 187)

FUNCTIONS AND SUBROUTINES 187

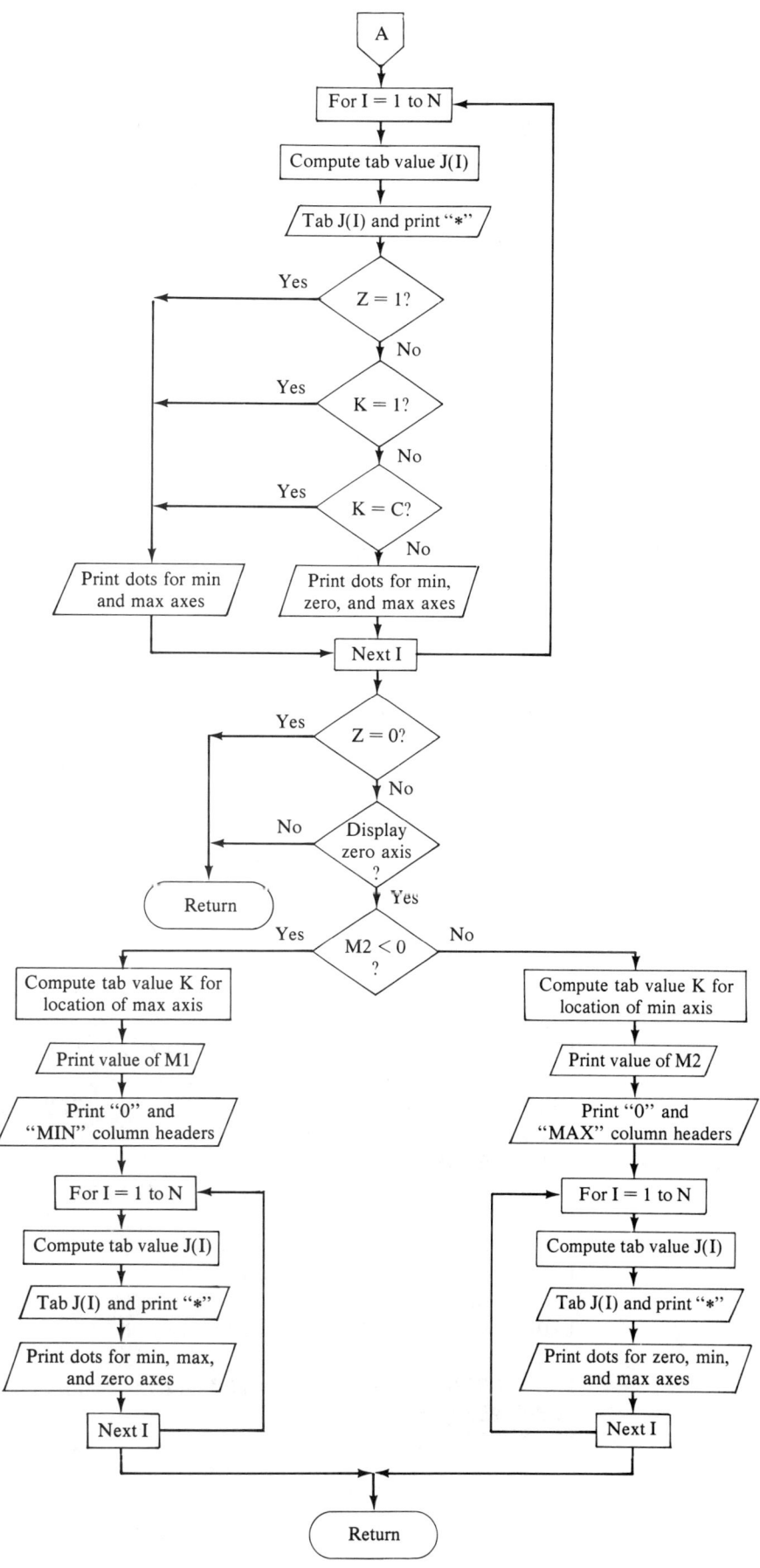

FIGURE 7.27 (cont.)

```
1240 PRINT
1250 PRINT
1260 LET R=M2-M1
1270 IF Z=1 THEN 1350
1280 LET K=INT(.5+(C-1)*ABS(M1)/R)+1
1290 PRINT "MIN=" M1
1300 PRINT "MAX=" M2
1310 PRINT
1320 PRINT "MIN" TAB(C-3) "MAX"
1330 PRINT TAB(K) "0"
1340 GOTO 1390
1350 PRINT "MIN=" M1
1360 PRINT "MAX=" M2
1370 PRINT
1380 PRINT "MIN" TAB(C-3) "MAX"
1390 FOR I=1 TO N
1400 LET J=INT(.5+(C-1)*(Y(I)-M1)/R)+1
1410 PRINT TAB(J);"*"
1420 IF Z=1 THEN 1470
1430 IF K=1 THEN 1470
1440 IF K=C THEN 1470
1450 PRINT TAB(1) "." TAB(K) "." TAB(C) "."
1460 GOTO 1480
1470 PRINT TAB(1) "." TAB(C) "."
1480 NEXT I
1490 IF Z=0 THEN 1740
1500 PRINT "DO YOU WANT ZERO-AXIS DISPLAYED? ENTER YES OR NO."
1510 INPUT A$
1520 IF A$="NO" THEN 1740
1530 PRINT
1540 IF M2<0 THEN 1650
1550 LET K=INT(.5+(C-1)*M1/M2)+1
1560 PRINT "MAX=" M2
1570 PRINT
1580 PRINT TAB(1) "0" TAB(C-3) "MAX"
1590 FOR I=1 TO N
1600 LET J=INT(.5+(C-1)*Y(I)/M2)+1
1610 PRINT TAB(J) ; "*"
1620 PRINT TAB(1) "." TAB(K) "." TAB(C) "."
1630 NEXT I
1640 RETURN
1650 LET K=INT(.5+(C-1)*R/(-M1))+1
1660 PRINT "MIN=" M1
1670 PRINT
1680 PRINT "MIN" TAB(C) "0"
1690 FOR I=1 TO N
1700 LET J=INT(.5+(C-1)*(Y(I)-M1)/(-M1))+1
1710 PRINT TAB(J) "*"
1720 PRINT TAB(1) "." TAB(K) "." TAB(C) "."
1730 NEXT I
1740 RETURN
```

FUNCTIONS AND SUBROUTINES

We should note that some versions of BASIC treat the first column of a display as column number zero rather than as column number one. The PLOTTER subroutine is written so that it can be used with either case. Several PRINT statements contain TAB(1); therefore, the leftmost axis of the plot is always column one, although that column might actually be the second column in some versions of BASIC.

Example 7.8 Write a BASIC program using the PLOTTER subroutine to print graphs of the following functions:

(a) $y = 10\sin 2\pi x$; (b) $y = 20 + 10\sin 2\pi x$; (c) $y = -30 - 10\sin 2\pi x$.

Each graph should cover a range of values from $x = 0$ through $x = 1$ in 20 intervals. (This range covers exactly one cycle of the sine wave.) Obtain additional plots showing the location of the zero axis for any function whose range of values does not include zero.

The required program is shown in Figure 7.28. We begin by assigning a dimension of 21 to arrays X and Y, since we must compute 21 values when using 20 intervals. The constant W equals 2π. As I ranges from 1 through 21, X(I) ranges from 0 through 1.0. In the first FOR-TO, NEXT loop, Y(I) is assigned the corresponding values of function (a). The PLOTTER subroutine is then called in statement 70. (The subroutine listing is not shown in the program.) Upon returning to the main program at statement 80, the y values of function (b) are computed in another FOR-TO, NEXT loop. Note that the values of function (b) are equal to 20, plus the corresponding values of function (a), which were already stored in array Y. Also note that it is not necessary to recompute values of X(I). The program then calls PLOTTER again, and on returning to statement 120, enters another FOR-TO, NEXT loop. This loop computes values of function (c). PLOTTER is then called a third and last time.

The results of a run of the program in Figure 7.28 are shown in Figure 7.29. Figure 7.29(a) shows the plotted graph of function (a), which is symmetrical about the zero axis. Since the zero axis is displayed, the question "DO YOU WANT ZERO-AXIS DISPLAYED?" is not printed. Figure 7.29(b) shows the plotted graph of function (b). As you can see, function (b) varies between +10 and +30, so the zero axis is not displayed. The user responds "YES" to the question, and Figure 7.29(c) shows the same function with the zero axis displayed. Figure 7.29(d) shows the plot of function (c), which varies between −20 and −40. Figure 7.29(e) shows that function regraphed with its zero axis.

```
10  DIM X(21),Y(21)
20  LET W=6.2831852
30  FOR I=1 TO 21
40  LET X(I)=(I-1)/20
50  LET Y(I)=10*SIN(W*X(I))
60  NEXT I
70  GOSUB 1000
80  FOR I=1 TO 21
90  LET Y(I)=20+Y(I)
100 NEXT I
110 GOSUB 1000
120 FOR I=1 TO 21
130 LET Y(I)=-30-10*SIN(X(I))
140 NEXT I
150 GOSUB 1000
160 STOP
```

FIGURE 7.28 Program that calls the PLOTTER subroutine to graph several sine wave functions (Example 7.8)

```
ENTER NUMBER OF VALUES TO BE PLOTTED.
? 21
ENTER MAXIMUM COLUMN NO. AVAILABLE FOR PLOT
? 50
     X              Y
 ---------      ---------
 0              0
 .05            3.09017
 .1             5.87785
 .15            8.09017
 .2             9.51057
 .25            10
 .3             9.51057
 .35            8.09017
 .4             5.87785
 .45            3.09017
 .5             1.87254E-06
 .55            -3.09017
 .6             -5.87785
 .65            -8.09017
 .7             -9.51057
 .75            -10
 .8             -9.51057
 .85            -8.09017
 .9             -5.87786
 .95            -3.09017
 1              -3.74507E-06

MIN=-10
MAX= 10

MIN                                          MAX
                       0
```

FIGURE 7.29(a) Graph created by the PLOTTER subroutine for $y = 10\sin(2\pi x)$

```
ENTER NUMBER OF VALUES TO BE PLOTTED.
? 21
ENTER MAXIMUM COLUMN NO. AVAILABLE FOR PLOT
? 50
     X              Y
 ---------      ---------
 0              20
 .314159        23.0902
 .628319        25.8779
 .942478        28.0902
 1.25664        29.5106
 1.5708         30
 1.88496        29.5106
 2.19911        28.0902
 2.51327        25.8779
 2.82743        23.0902
 3.14159        20
 3.45575        16.9098
 3.76991        14.1222
 4.08407        11.9098
 4.39823        10.4894
 4.71239        10
 5.02655        10.4894
 5.34071        11.9098
 5.65487        14.1221
 5.96903        16.9098
 6.28319        20

MIN= 10
MAX= 30

MIN                                          MAX


DO YOU WANT ZERO-AXIS DISPLAYED? ENTER YES OR NO.
? YES
```

FIGURE 7.29(b) Graph created by the PLOTTER subroutine for $y = 20 + 10\sin(2\pi x)$

MAX= 30

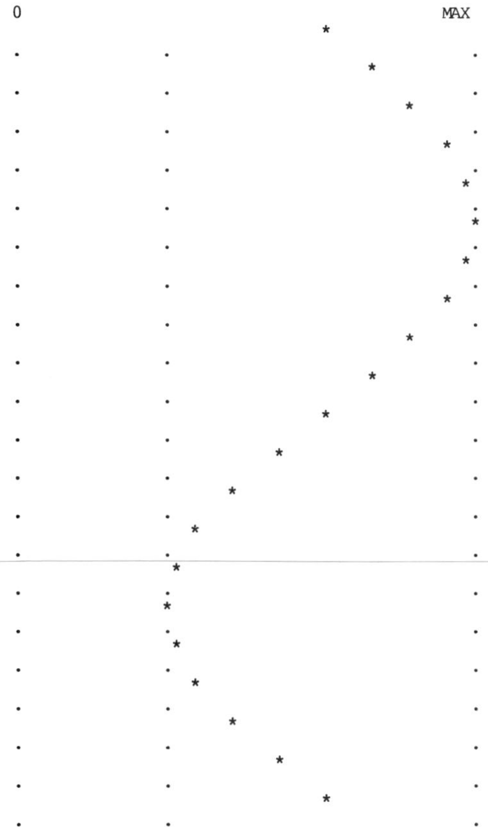

0   MAX

```
ENTER NUMBER OF VALUES TO BE PLOTTED.
? 21
ENTER MAXIMUM COLUMN NO. AVAILABLE FOR PLOT
? 50
     X              Y
  ---------      ---------
  0              -30
   .314159       -33.0902
   .628319       -35.8779
   .942478       -38.0902
  1.25664        -39.5106
  1.5708         -40
  1.88496        -39.5106
  2.19911        -38.0902
  2.51327        -35.8779
  2.82743        -33.0902
  3.14159        -30
  3.45575        -26.9098
  3.76991        -24.1222
  4.08407        -21.9098
  4.39823        -20.4894
  4.71239        -20
  5.02655        -20.4894
  5.34071        -21.9098
  5.65487        -24.1221
  5.96903        -26.9098
  6.28319        -30

MIN=-40
MAX=-20
```

MIN   MAX

DO YOU WANT ZERO-AXIS DISPLAYED? ENTER YES OR NO.
? YES

FIGURE 7.29(c) Graph of $y = 20 + 10\sin(2\pi x)$ with zero axis displayed

FIGURE 7.29(d) Graph created by the PLOTTER subroutine for $y = -30 - 10\sin(2\pi x)$

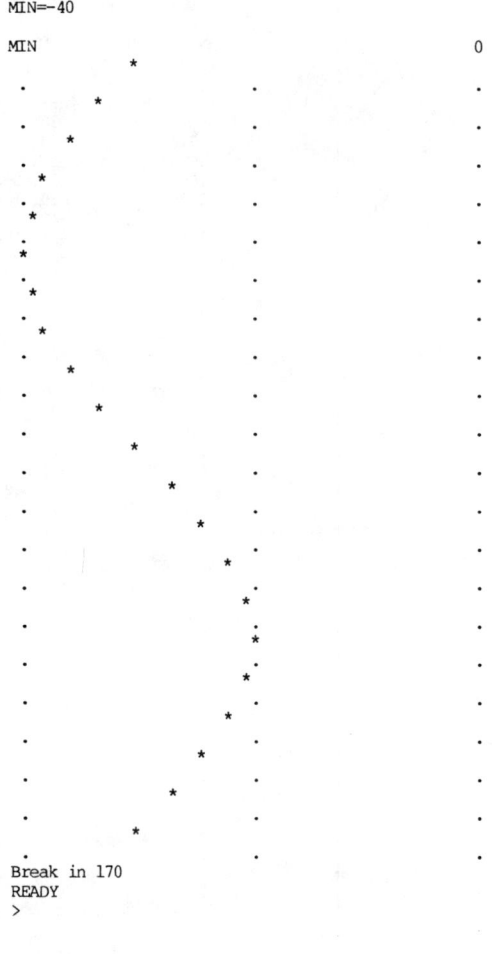

FIGURE 7.29(e) Graph of $y = -30 - 10\sin(2\pi x)$ with zero axis displayed

---

EXERCISES

7.13 Write a BASIC statement to set A equal to the value of B rounded to the nearest one-tenth. For example, if B = 46.27, then A = 46.3. Use the INT function.

7.14 Explain the purpose of statements 1420 and 1430 in the PLOTTER subroutine.

7.15 Write a BASIC program using the PLOTTER subroutine to obtain printed graphs for each of the following functions:

(a) $y = 12e^{-x/0.5}$
(b) $y = 20 - 12e^{-x/0.5}$
(c) $y = 6 - 12e^{-x/0.5}$

The graphs should be plotted for 20 equal intervals of x from $x = 0$ through $x = 2$. Obtain an additional graph showing the zero axis for function (b). Use the maximum column width available on your printer.

7.16 Write a BASIC program using the PLOTTER subroutine to obtain a printed graph of the following function:

$$y = 5e^{-2t}\sin(2\pi t + \pi/4)$$

Your graph should cover two cycles of the sine wave and should contain enough plotted points to provide a clear picture of the variation of y with t over the required range, beginning at $t = 0$.

7.17 Use the PLOTTER subroutine to obtain a graph of any function of the form $y = A(1 - e^{-t/\tau})$ where A and τ are constants. Then sketch a smooth curve through the points of your graph and use it to determine the values of y, expressed as percents of A, at each of the following values of t: $t = 0.2\tau, 0.6\tau, 1.1\tau, 1.8\tau, 2.5\tau,$ and 3τ.

GENERAL EXERCISES

7.18 The kinetic energy E of a body having a rest mass of M_0 and velocity v is given by

$$E = \left[\left(1 - \frac{v^2}{c^2}\right)^{-1/2} - 1\right] M_0 c^2$$

where c is the velocity of light (3×10^8 m/sec). If the relativistic effects are neglected, then

$$E \simeq \frac{1}{2} M_0 v^2$$

Write a program that incorporates user-defined BASIC function(s) to print a table showing the kinetic energy of a body having a rest mass of 1 kg when its velocity is changed from 0.08c through 0.99c in increments of 0.01c. The table should list (1) the kinetic energy, computed using the exact equation; (2) the approximate kinetic energy, computed when the relativistic effects are neglected; and (3) the percent error caused by using the approximation instead of the exact equation.

$$\text{Percent error} = \frac{|\text{exact} - \text{approximate}|}{\text{exact}} \times 100\%.$$

The units of kinetic energy are newton-meters. In this system of units, mass is in kilograms, so the magnitude of 1 kg is 1.0 (not 10^3).

7.19 Write a BASIC program to read the elements of a 5 × 5 array and then call one of two subroutines, depending on user-supplied input. If the user inputs "ROW", the computer should branch to a subroutine that rearranges the entries in the array so that they appear in ascending order in each *row*. If the user inputs "COLUMN", the computer should branch to a subroutine that rearranges the entries in the array so that they appear in ascending order in each *column*. In either case, the main program should print the rearranged array. For example, suppose the array is as follows:

$$\begin{bmatrix} 5 & 6 & 2 & 3 & 1 \\ 2 & 5 & 9 & 0 & -4 \\ 1 & 0 & -9 & 2 & 8 \\ 6 & 4 & 5 & -1 & 3 \\ -4 & 2 & 3 & 4 & 5 \end{bmatrix}$$

If the user inputs "ROW", the program should print:

$$\begin{array}{rrrrr} 1 & 2 & 3 & 5 & 6 \\ -4 & 0 & 2 & 5 & 9 \\ -9 & 0 & 1 & 2 & 8 \\ -1 & 3 & 4 & 5 & 6 \\ -4 & 2 & 3 & 4 & 5 \end{array}$$

Run your program using the preceding example array once for the "ROW" input and again for the "COLUMN" input.

7.20 The rectangular form of a complex number is $a + jb$, where a is the real part, and $j = \sqrt{-1}$. The polar form is M $\angle\theta$, where M is the magnitude and θ is the angle, measured counterclockwise from the positive real axis. To convert M $\angle\theta$ to rectangular form, we write $M\cos\theta + jM\sin\theta$—that is, the real part is $M\cos\theta$ and the imaginary part is $M\sin\theta$. The sum of the two complex numbers $(a + jb)$ and $(c + jd)$ is $(a + c) + j(b + d)$ and their difference is $(a - c) + j(b - d)$.

Write a BASIC program to perform the following:

(1) Prompt the user to enter the magnitudes and angles of two complex numbers.
(2) Call a subroutine that converts these to rectangular form and prints their sum in rectangular form.
(3) If the magnitude of the sum $\sqrt{(a + c)^2 + (b + d)^2}$ is greater than 1, call another (nested) subroutine to compute and print their difference in rectangular form.

Run your program using the following pairs of complex numbers:

(a) $5\angle 30°$, $8\angle 60°$
(b) $2\angle 95°$, $2\angle -95°$

EXERCISES FOR ARCHITECTURAL/CIVIL/CONSTRUCTION TECHNOLOGY

7.21 In stadia surveying, the horizontal and vertical distances between the instrument and the rod, in terms of the rod intercept i, are given by

$$H = Ki(\cos a)^2 + (f + c)\cos a \quad \text{feet and}$$
$$V = \tfrac{1}{2}Ki(\sin 2a) + (f + c)\sin a \quad \text{feet}$$

where i = rod intercept in feet,
 K = stadia factor,
 a = vertical inclination of line of sight (in degrees) measured from the horizontal, and
 $f + c$ = the distance constant.

Assuming $K = 100$ and $f + c = 1$ foot, write a BASIC program that incorporates library and user-defined functions to print a table of values of H and V, after the user enters the rod intercept i. The table should list values of H and V for values of vertical inclination ranging from 1 degree to 30 degrees in 30-minute (0.5 degree) increments.
Run your program for $i = 1.3$ ft and for $i = 0.62$ ft.

7.22 The cantilevered beam shown in Figure 7.30 has a concentrated load P applied at distance a from the unsupported end.

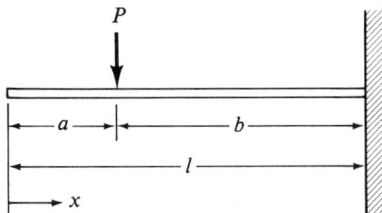

FIGURE 7.30 Exercise 7.22

The equation for the deflection Δ of the beam, as a function of the distance x measured from the unsupported end, depends on whether x is less than, equal to, or greater than a:

$$\text{If } x < a, \text{ then} \quad \Delta(x) = \frac{Pb^2}{6EI}(3l - 3x - b) \text{ inches}$$

$$\text{If } x = a, \text{ then} \quad \Delta(x) = \frac{Pb^3}{3EI} \text{ inches}$$

$$\text{If } x > a, \text{ then} \quad \Delta(x) = \frac{P(l-x)^2}{6EI}(3b - l + x) \text{ inches}$$

where E is the modulus of elasticity in pounds per square inch; I is the moment of inertia of the beam cross-section, in in^4; P is the concentrated load, in pounds; and l, a, b, and x are measured in inches. Note that $l = a + b$.

Assuming that $E = 30 \times 10^6$ lb/in^2, write a BASIC program to perform the following:

(1) Prompt the user to enter values for P, I, a, b, and the value of x at which the deflection is desired.
(2) Branch to one of three subroutines that computes and prints $\Delta(x)$, (according to whether x is less than, equal to, or greater than a).
(3) Return to the main program, where the percentage of the maximum deflection represented by $\Delta(x)$ is computed and printed. The maximum deflection occurs at the unsupported end, and is equal to

$$\frac{Pb^2}{6EI}(3l - b) \text{ inches.}$$

Run your program for the following combinations of inputs:
(a) $P = 10^4$ lbs, $I = 76$in^4, $a = 48$in, $b = 48$in, $x = 12$in
(b) $P = 10^4$ lbs, $I = 76$in^4, $a = 48$in, $b = 48$in, $x = 72$in
(c) $P = 10^4$ lbs, $I = 76$in^4, $a = 48$in, $b = 48$in, $x = 48$in

7.23 Based on the results of a survey made in one metropolitan area, a construction cost estimator derived the following formula for estimating the cost of erecting a certain type of residential building:

$$C = 1483\sqrt{S} + 1040B + 706.9\ln U + 9750$$

where C is the construction cost, in dollars,
S is the square feet of living area,
B is the number of bathrooms,

U is the square feet of unheated, exterior area (carport, etc.), and
$\ln U$ is the natural logarithm (base e) of U.

If the structure consists of two stories, the formula is modified by addition of the following quantity:

$$207B - 0.181S + 4060$$

Write a BASIC program to perform the following:

(1) Prompt the user to enter values for S, B, and U and either "1" or "2", corresponding to the number of stories.
(2) Call a subroutine to compute the cost C of a one-story dwelling.
(3) If the number of stories is 2, call a nested subroutine to modify the cost C by the appropriate quantity.
(4) Return to the main program, where the total cost C and the cost per square foot (C/S) are both computed and printed.

Run your program for the following combinations of inputs:

(a) A 1654-square-foot, one-story dwelling having 2 and one-half baths and a 155-square-foot carport
(b) A 2040-square-foot, two-story dwelling having 3 and one-half baths and a 240-square-foot carport

---

EXERCISES FOR ELECTRICAL/ELECTRONICS/COMPUTER TECHNOLOGY

7.24 When the switch in the RC network of Figure 7.31 is placed in position 1, the current flowing in the circuit as a function of time t is $i(t) = (E/R)e^{-t/RC}$ amperes, where e is the base of the natural logarithm. The voltage across the capacitor as a function of time t is $v_c(t) = E(1 - e^{-t/RC})$ volts. If the switch is placed in position 2 after the capacitor has fully charged, the current $i(t)$ and voltage $v_c(t)$ are

$$i(t) = \frac{-E}{R} e^{-t/RC} \text{ amperes}$$

and

$$v_c(t) = E e^{-t/RC} \text{ volts}$$

Write a BASIC program incorporating library and user-defined functions to print a table of values of $i(t)$ and $v_c(t)$, after the switch is placed in position 1, and another table of values of $i(t)$ and $v_c(t)$, after the switch is placed in position 2. Values of $i(t)$ should be listed in *milliamps*. The values of E, R, and C are entered by the user (E

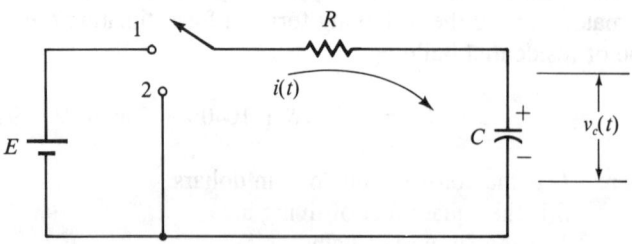

FIGURE 7.31 Exercise 7.24

in volts, R in ohms, and C in farads). The tables should list values of $i(t)$ and $v_c(t)$ for t ranging from $t = 0$ through $t = 5RC$ seconds, in intervals of $0.25RC$ seconds. (Note that the time-constant τ of the circuit equals RC seconds.)

Run your program for the following parameter values:

(a) $E = 20V$, $R = 100K$, $C = .015\mu F$
(b) $E = 12V$, $R = 47K$, $C = 100\mu F$

7.25 Figure 7.32 shows an RLC circuit in which a current i flows, after the dc voltage E is applied to it by closing the switch.

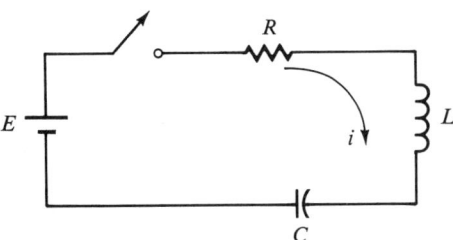

FIGURE 7.32 Exercise 7.25

The equation for i as a function of the time t that has elapsed after the switch is closed, $i(t)$, depends on the value of R in the circuit, according to the following:

(1) If $R > 2\sqrt{L/C}$, then

$$i(t) = \frac{E}{2\alpha L}\left\{\exp\left[-\left(\frac{R}{2L} - \alpha\right)t\right] - \exp\left[-\left(\frac{R}{2L} + \alpha\right)t\right]\right\} \text{ amps}$$

where $\alpha = \sqrt{\left(\frac{R}{2L}\right)^2 - \frac{1}{LC}}$ and $\exp[\ ]$ means $e^{[\ ]}$.

This is called the *overdamped* case.

(2) If $R = 2\sqrt{L/C}$, then

$$i(t) = (E/L)\, t \exp\left[-\left(\frac{R}{2L}\right)t\right] \text{ amps}$$

This is called the *critically damped* case.

(3) If $R < 2\sqrt{L/C}$, then

$$i(t) = \frac{E}{L\beta}\exp\left[-\left(\frac{R}{2L}\right)t\right]\sin\beta t \text{ amps}$$

where

$$\beta = \sqrt{\frac{1}{LC} - \left(\frac{R}{2L}\right)^2}$$

This is called the *underdamped* case.

Write a BASIC program that prompts the user to enter values for E, R, L, and C and then branches to one of three subroutines to compute values of $i(t)$, depending on which of these three cases is applicable. The subroutine should compute and print a table of values of $i(t)$ for t ranging from $t = 0$ through $t = 10L/R$ seconds in intervals of $0.2L/R$ seconds. The main program should print "OVERDAMPED", "UNDERDAMPED", or "CRITICALLY DAMPED", as appropriate.

Run your program for the following parameter values:

(a) $E = 10V, R = 40\Omega, L = 20mH, C = 50\mu F$
(b) $E = 10V, R = 120\Omega, L = 4mH, C = 10\mu F$
(c) $E = 10V, R = 10\Omega, L = 40mH, C = 100\mu F$

7.26 According to the *maximum power transfer theorem* for ac circuits, an active source delivers maximum power to a load when the load impedance Z_L is the complex conjugate of the source impedance Z_S. Figure 7.33 shows a source containing inductive and resistive elements and a load impedance Z_L connected across its terminals. The source voltage is E VRMS.

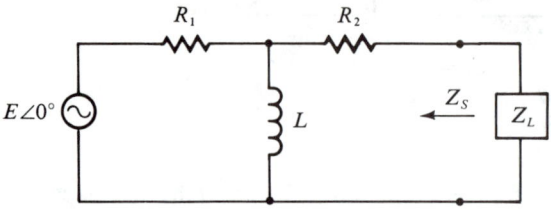

FIGURE 7.33 Exercise 7.26

The source impedance for the network in Figure 7.33 is

$$Z_S = R_2 + \frac{j\omega R_1 L}{R_1 + j\omega L} = \frac{R_1^2 R_2 + (\omega L)^2(R_1 + R_2)}{R_1^2 + (\omega L)^2} + j\frac{\omega L R_1^2}{R_1^2 + (\omega L)^2} \quad (1)$$

where ω is the angular frequency of the source in radians per second. Thus the real (resistive) part of the source impedance is

$$\frac{R_1^2 R_2 + (\omega L)^2 (R_1 + R_2)}{R_1^2 + (\omega L)^2}$$

and the imaginary (reactive) part of the source impedance is

$$\frac{\omega L R_1^2}{R_1^2 + (\omega L)^2}.$$

The complex conjugate of Z_S is given by Equation 1, with the $+$ sign between the real and imaginary parts changed to a $-$ sign. When the load impedance Z_L equals the conjugate of Z_S, the power delivered to the load is

$$P = \frac{E^2\{R_1^2 + (\omega L)^2\}}{4\{R_1^2 R_2 + (\omega L)^2(R_1 + R_2)\}} \text{ watts}$$

Write a BASIC program to perform the following:

(1) Prompt the user to enter values for E in VRMS, ω in radians per second, L in henries, and R_1 and R_2 in ohms;

(2) Call a subroutine to compute and print the real and imaginary parts of the load impedance required for maximum power transfer; and
(3) Call a (nested) subroutine to compute the maximum power delivered to the load.

Run your program for the following combinations of inputs:

(a) $E = 120$ VRMS, $\omega = 377$ rad/sec, $L = 0.03$ H, and $R_1 = R_2 = 10$ ohms
(b) $E = 24$ VRMS, $\omega = 10^3$ rad/sec, $L = 2.5$ H, $R_1 = 2.2$K, $R_2 = 1$K

EXERCISES FOR INDUSTRIAL/MANUFACTURING/PRODUCTION TECHNOLOGY

7.27 The total length L of a flat belt required to join two pulleys depends on the size of the pulleys and the separation S between their centers. (See Figure 7.34.)

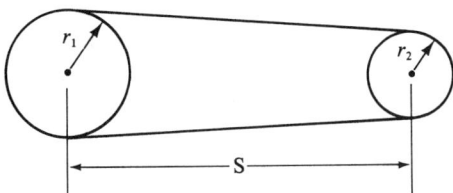

FIGURE 7.34 Exercise 7.27

Assuming that $r_1 \geq r_2$, the length L is given by

$$L = 2S\cos\theta + \pi\{(r_1 + r_2) + (r_1 - r_2)\theta/90\} \text{ inches}$$

where

$$\theta = \arcsin\left(\frac{r_1 - r_2}{S}\right) \text{ degrees,}$$

and r_1, r_2, and S are in inches.

Write a BASIC program that incorporates library and user-defined functions to print a table of values of L versus S, after the user enters r_1 and r_2. The table should list L for values of S from 48 inches through 240 inches, in increments of 6 inches. If your version of BASIC does not have the arc sine library function, use the identity

$$\arcsin(x) = \arctan\{x/(1 - x^2)^{1/2}\}$$

Run your program for the following combinations of inputs:

(a) $r_1 = 18$ inches, $r_2 = 6$ inches
(b) $r_1 = r_2 = 9$ inches

7.28 In a machine shaft, the ratio R of the maximum intensity of the stress resulting from an axial load to the average axial stress is given by

$$R = \begin{cases} \dfrac{1}{1 - 0.0044(L/k)} & \text{for } \dfrac{L}{k} < 115 \\ \dfrac{S_y}{n\pi^2 E}\left(\dfrac{L}{k}\right)^2 & \text{for } \dfrac{L}{k} \geq 115 \end{cases}$$

where L = the length between bearings, in inches;
 k = the radius of the gyration of the shaft, in inches;
 s_y = compression yield stress, in pounds per square inch;
 E = modulus of elasticity, in pounds per square inch; and
 n = 2.5 to 3 for fixed ends (rigid bearings).

Write a BASIC program that prompts the user to enter values for s_y, E, L, and k and then calls one of two subroutines to compute and print R, depending on the value of L/k. If L/k ≥ 115, print values of R for n = 2.5 through 3 in increments of 0.05.

Run your program for the following combinations of inputs:

(a) $s_y = 38 \times 10^3$ lbs/in², $E = 30 \times 10^6$ lbs/in², k = 0.25 in, L = 30 in
(b) $s_y = 38 \times 10^3$ lbs/in², $E = 30 \times 10^6$ lbs/in², k = 0.25 in, L = 12.5 in

7.29 The quality-control department of a manufacturing plant performs *acceptance sampling* to determine whether manufacturing lots should be accepted or rejected. The plant manufactures two products, A and B. Twenty samples of each product are taken from each production lot, and if no more than one defective unit is found in a sample of product A, that production lot is accepted. If no more than two defective units are found in a sample of product B, that production lot is accepted.

The probability P(r) of finding r defective units in a sample of 20, when p is the true proportion of defective units in the lot, can be approximated for large lot sizes by

$$P(r) = \frac{(20p)^r}{r!} e^{-20p}$$

where e is the base of the natural logarithm. Consequently, the probability that no more than one defective unit will be found in a sample of 20 is the probability P(0) that zero defectives will be found plus the probability P(1) that one defective unit will be found. Similarly, the probability that no more than two defective units will be found is P(0) + P(1) + P(2).

The quality control director wants to know the probability that a given lot of either product A or product B will be accepted, when the proportion p of defectives in the lot has an assumed value (between 0 and 1).

Write a BASIC program to perform the following:

(1) Prompt the user to enter an (assumed) proportion of defectives p in the lot, and the type of product ("A" or "B").
(2) Call a subroutine to compute the probability of accepting a type A lot, that is, P(0) + P(1).
(3) If the product is type B, call a nested subroutine to compute P(3) and then return to the first subroutine to print the probability of acceptance and the probability of rejection (1.0 − probability of acceptance).

Run your program for the following combinations of inputs:

(a) type A, p = 0.08
(b) type B, p = 0.035

EXERCISES FOR MECHANICAL TECHNOLOGY

7.30 If a constant force F is suddenly applied to a spring-damper combination, as shown in Figure 7.35, the displacement x of the system as a function of time t is then given by

$$x(t) = \frac{F}{K}(1 - e^{-Kt/B}) \text{ meters}$$

and the velocity is

$$v(t) = \frac{F}{B} e^{-Kt/B} \text{ meters/sec}$$

where F = the magnitude of the force in newtons,
K = the spring-constant in newtons per meter, and
B = the damping, in newtons per meter per second.

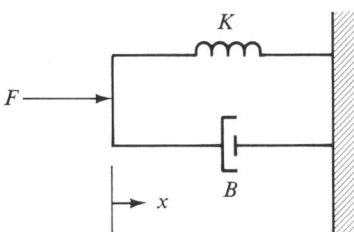

FIGURE 7.35 Exercise 7.30

If the force is suddenly removed after the system has reached its maximum displacement (F/K meters), then

$$x(t) = \frac{F}{K} e^{-Kt/B} \text{ meters}$$

and

$$v(t) = \frac{-F}{B} e^{-Kt/B} \text{ meters/sec}$$

where t is the time elapsed after removal of the force.

Write a BASIC program that incorporates library and user-defined functions to print a table of values of $x(t)$ and $v(t)$, after the force is applied, and another table of values of $x(t)$ and $v(t)$, after the force is removed. The values of F, K, and B are entered by the user in newtons, newtons per meter, and newtons per meter per second, respectively. The tables should list values of $x(t)$ and $v(t)$ for t ranging from $t = 0$ through $t = 5B/K$ seconds in intervals of $0.25B/K$ seconds.

Run your program for the following combinations of inputs:

(a) $F = 20\text{N}$, $K = 5 \times 10^3$ N/m, $B = 200$ N/m/sec
(b) $F = 450\text{N}$, $K = 9 \times 10^6$ N/m, $B = 1.8 \times 10^4$ N/m/sec

7.31 Figure 7.36 shows a mass M resting on a frictionless surface and attached to a spring-damper combination that is fixed at one end. A force F is suddenly applied to the mass, causing it to move to the right. $x(t)$ is the displacement of the mass as a function of the time t that has elapsed after application of the force.
In the international (SI) system of units, F is in newtons, M is in kilograms (kg), K is in newtons per meter, B is in newtons per meter per second, and x is in meters. The equation for $x(t)$ depends upon the magnitude of the quantity B/M:

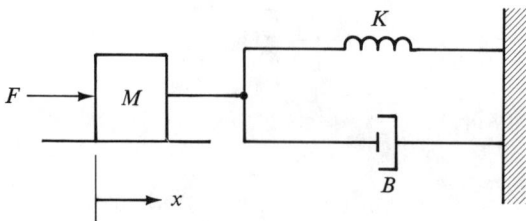

FIGURE 7.36 Exercise 7.31

(1) If $(B/M)^2 > 4K/M$, then

$$x(t) = \frac{F}{K}\left[1 - \left(\frac{b}{b-a}\right)e^{-at} + \left(\frac{a}{b-a}\right)e^{-bt}\right]$$

where

$$a = \frac{B/M + \sqrt{(B/M)^2 - 4K/M}}{2}$$

and

$$b = \frac{B/M - \sqrt{(B/M)^2 - 4K/M}}{2}$$

(2) If $(B/M)^2 = 4K/M$, then

$$x(t) = \frac{F/M}{a^2}\left[1 - e^{-at} - ate^{-at}\right]$$

where

$$a = \frac{B}{2M}$$

(3) If $(B/M)^2 < 4K/M$, then

$$x(t) = \frac{F}{K} + \frac{Fe^{-\zeta\omega t}}{K\sqrt{1-\zeta^2}}\sin(\omega\sqrt{1-\zeta^2}\,t - \theta)$$

where

$$\zeta = B/2\sqrt{KM}$$
$$\omega = \sqrt{K/M}$$
$$\theta = \arctan\left(\frac{\sqrt{1-\zeta^2}}{-\zeta}\right)$$

Write a BASIC program that prompts the user to enter values for F, M, K, and B and then branches to one of three subroutines to compute values of $x(t)$, depending on which of the three preceding cases is applicable. The subroutine should compute and print a table of values of $x(t)$ for t ranging from $t = 0$ through $t = 10\ M/B$ seconds, in intervals of $0.2\ M/B$ seconds. Take special notice of the following points:

(1) In the SI system of units, 1 kilogram of mass is entered as 1.0 (not 1×10^3).
(2) The angle computation in the third case must take into account the signs of the numerator and denominator of the argument of the arctan function; that is, arctan $(\sqrt{1-\zeta^2}/-\zeta^2)$ is not the same as arctan$(-\sqrt{1-\zeta^2}/\zeta^2)$.

Run your program for the following combinations of inputs:

(a) $F = 200\text{N}, M = 2 \text{ kg}, K = 5 \times 10^3 \text{ N/m}, B = 200 \text{ N/m/sec}$
(b) $F = 200\text{N}, M = 2 \text{ kg}, K = 5 \times 10^3 \text{ N/m}, B = 100 \text{ N/m/sec}$
(c) $F = 200\text{N}, M = 2 \text{ kg}, K = 5 \times 10^3 \text{ N/m}, B = 500 \text{ N/m/sec}$

7.32 The friction loss in a circular pipe through which water is flowing can be calculated from

$$H_f = fLv^2/64.4D$$

where H_f = loss in the head due to friction, in feet of water;
f = the friction factor (dimensionless),
L = the length of the pipe in feet;
v = the velocity of water flow, in feet per second; and
D = the diameter of the pipe in feet.

If an abrupt expansion occurs in the diameter of the pipe, there is then an additional loss in head given by

$$H_f = \frac{v_1^2}{64.4}\left[1-\left(\frac{D_1}{D_2}\right)^2\right]^2$$

where v_1 is the velocity, in feet per second, in the smaller section; and
D_1 and D_2 are the diameters (in feet) of the smaller and larger sections of the pipe, respectively.

Figure 7.37 shows two kinds of water pipe installations, one having a diameter of D_1 throughout its total length ($L_1 + L_2$) and another with a diameter of D_1 for length L_1, followed by a section having a diameter D_2 for length L_2.

Write a BASIC program to perform the following:

(1) Prompt the user to enter values for L_1, L_2, D_1 and D_2 in feet, v_1 in feet per second, and the friction factor f.

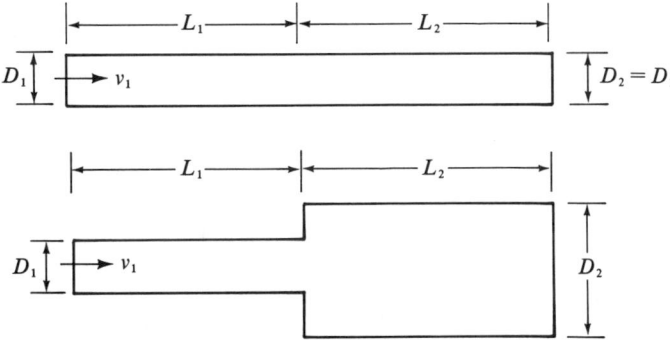

FIGURE 7.37 Exercise 7.32

(2) Call a subroutine to compute the loss in head through a pipe of uniform diameter D_1 and length L_1.
(3) Call a nested subroutine to compute the loss in head in a pipe of diameter D_2 and length L_2, and if $D_2 > D_1$, add the loss due to expansion.
(4) Return to the first subroutine, where the loss computed in the second subroutine is added to that computed in the first, and then print the total loss.

Run your program for the following combinations of inputs:

(a) $L_1 = 800\,\text{ft}, L_2 = 1200\,\text{ft}, D_1 = 1.75\,\text{ft}, D_2 = 1.75\,\text{ft}, v_1 = 6\,\text{ft/sec},$
$f = 0.022$
(b) $L_1 = 800\,\text{ft}, L_2 = 1200\,\text{ft}, D_1 = 1.75\,\text{ft}, D_2 = 3.5\,\text{ft}, v_1 = 6\,\text{ft/sec},$
$f = 0.022$

CHAPTER 8

SOLVING SIMULTANEOUS EQUATIONS

1 SIMULTANEOUS LINEAR EQUATIONS

To obtain complete solutions to many applied problems in technology, we are required to find the values of two or more variables (unknowns). Recall that we must have as many independent equations as unknowns, to find all solutions to a given problem. As an example, the two equations

$$3x + 2y = 16$$
$$2x - y = 6$$

involving the two unknowns x and y, can be solved using standard algebraic techniques to find $x = 4$ and $y = 2$. Both equations are satisfied when 4 is substituted for x and 2 is substituted for y. A set of n equations containing n unknowns is said to be a set of *simultaneous* equations, and the process of finding the n values that satisfy all equations is called solving the equations simultaneously.

In this chapter, we are only concerned with sets of simultaneous *linear* equations, those in which all variables are raised to the unity power. Thus, we will exclude equations containing quantities such as x^2 or $y^{1/2}$. The two equations in the previous paragraph are examples of linear equations. The graph of a linear equation in an $x - y$ coordinate plane is a straight line.

Example 8.1 The initial displacement of particle A in relation to a fixed reference point is 4 meters; it is traveling with a fixed velocity of 8 m/sec. The initial displacement of particle B is 2 meters, and it travels along the same line as particle A with a velocity of 10 m/sec. At what time t and at what displacement does particle B overtake A?

This problem can be solved by formulating two simultaneous linear equations involving displacement s and time t. Recall that the distance traveled by a body moving at a constant velocity equals velocity multiplied by time. Therefore, the total displacement of particle A equals its initial displacement plus its velocity multiplied by the time t of travel:

$$s = 4 + 8t \text{ meters}$$

where t is in seconds. Similarly, the equation for the displacement of particle B is

$$s = 2 + 10t \text{ meters}$$

The graphs of these two equations are shown in Figure 8.1, where it can be seen that the simultaneous solution of the two equations corresponds to the point (s_0, t_0) where the two lines intersect. At time t_0, the displacement of each particle equals s_0, and B therefore overtakes A at time t_0.

One way to solve the equations simultaneously is to "eliminate" s by substituting the equation for the displacement s of A into the equation for s of B:

$$4 + 8t = 2 + 10t$$

From this we find that $t_0 = 1$ second and $s_0 = 12$ meters.

We have mentioned that n *independent* equations are required to solve for n unknowns. Two equations are independent if one equation is not simply a multiple of the other—that is, if one cannot be obtained from the other by multiplying it through by a constant. Thus, the equations

$$3x - 12y = 8$$
$$1.5x - 6y = 4$$

are not independent, since the second can be obtained from the first by multiplying the first through by 0.5. Therefore, these equations cannot be solved to obtain single (unique) values for x and y that simultaneously satisfy both equations. Equations that are *not* independent are said to be dependent. If any two of n simultaneous linear equations are dependent, a unique set of solution values cannot be found.

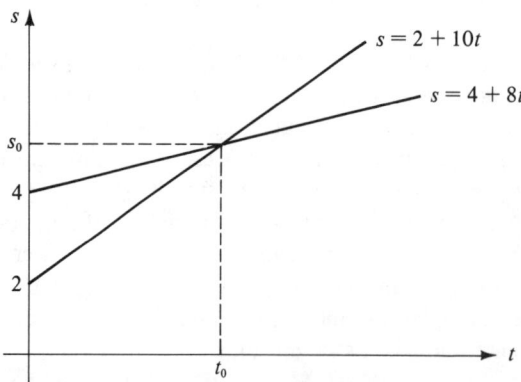

FIGURE 8.1 The simultaneous solution of two displacement equations is the point where their graphs intersect (Example 8.1).

SOLVING SIMULTANEOUS EQUATIONS

EXERCISES

8.1 Determine which of the following sets of simultaneous equations are linear sets for which a unique solution can be found:

(a) $2x - 3y = 27$
$8x - 12y = 108$

(b) $x = 2y - 11$
$y = 2x - 11$

(c) $-a + 7b = 0$
$2a - c = 7$
$a + b + c = 5$

(d) $2x - 2y = 10$
$x + y^2 = 4$

(e) $3u - 2w + 4 = 0$
$u + 6v - 8 = 0$

(f) $14P - 12Q + R = 18$
$3P - 7Q - R = 11$
$-7P + 6Q - 0.5R = -9$

8.2 In a certain consulting firm, the chief engineer is twice as old as the head draftsman and three times as old as the receptionist. The sum of their ages is 110. Write and solve three simultaneous linear equations to determine the age of each.

2 CRAMER'S RULE

To use BASIC for solving simultaneous equations, we need a systematic approach that can be programmed and used with any set of n independent linear equations. In short, we need an algorithm. *Cramer's Rule* is an algebraic technique that provides such an algorithm. We must first review the concept of the *determinant* of a matrix.

Consider the 2×2 matrix

$$A = \begin{bmatrix} a & b \\ c & d \end{bmatrix}$$

where a, b, c, and d are numbers. The determinant of A is defined by

$$\det A = ad - bc$$

In other words, the determinant is the number that results when the product of the numbers b and c on one diagonal is subtracted from the product of the numbers a and d on the other diagonal. For example,

$$\det \begin{bmatrix} 3 & 4 \\ -2 & -1 \end{bmatrix} = (3)(-1) - (4)(-2) = 5$$

Cramer's rule involves finding the quotient of two determinants. When applied to two simultaneous equations with two unknowns, the first step is to arrange the equations so that all the x terms line up, all the y terms line up, and all the constants line up, each in a vertical column. For example, suppose we want to solve the two equations

$$y = -2x - 1$$
$$2y - 10 = -8x$$

Rearranging as we described, we obtain

$$2x + y = -1$$
$$8x + 2y = 10$$

Notice that $2x$ and $8x$ are in the same column (as are y and $2y$) and that the constants -1 and 10 are written in a column on the opposite side of the equations.

The next step is to find the determinant of the 2×2 matrix formed by writing the coefficients of the variables in exactly the same rows and columns as they appear in the rearranged equations. In the current example, we find

$$\det \begin{bmatrix} 2 & 1 \\ 8 & 2 \end{bmatrix} = 2(2) - (1)(8) = -4$$

This determinant is called delta (Δ). According to Cramer's rule, the value of x is found by performing the following computations:

(1) Substitute the constants (-1 and 10 in this example) into the *first* column of the matrix for which the delta determinant was found. In this example, we obtain the modified matrix

$$\begin{bmatrix} -1 & 1 \\ 10 & 2 \end{bmatrix}$$

(2) Find the determinant of the modified matrix. In this case,

$$\det \begin{bmatrix} -1 & 1 \\ 10 & 2 \end{bmatrix} = (-1)(2) - (10)(1) = -12$$

(3) The solution for x is the value of this determinant divided by Δ. In this example,

$$x = \frac{-12}{\Delta} = \frac{-12}{-4} = 3$$

To solve for y using Cramer's rule, we repeat the above computations, except that the constants are substituted into the *second* column of the original matrix instead of the first. In our example,

$$y = \frac{\det \begin{bmatrix} 2 & -1 \\ 8 & 10 \end{bmatrix}}{\Delta} = \frac{28}{-4} = -7$$

Note that we substituted the column of constants into the *first* column of the matrix when solving for x, because the x coefficients were originally in the first column, and we substituted the constants into the *second* column when solving for y, because the y coefficients were originally in the second column. If we had rearranged the two original equations so that the y terms were in the first column and the x terms in the second column, the substitution procedure just described would be reversed. In other words, if we had written

$$y + 2x = -1$$
$$2y + 8x = 10$$

then

$$\Delta = \det \begin{bmatrix} 1 & 2 \\ 2 & 8 \end{bmatrix} = 4$$

$$y = \frac{\det \begin{bmatrix} -1 & 2 \\ 10 & 8 \end{bmatrix}}{\Delta} = \frac{-28}{4} = -7$$

$$x = \frac{\det \begin{bmatrix} 1 & -1 \\ 2 & 10 \end{bmatrix}}{\Delta} = \frac{12}{4} = 3$$

Note that we obtain the same solutions as we obtained before. Substitute the results $x = 3$, $y = -7$ into the two original equations, to verify that both equations are satisfied by these solutions.

Using Cramer's rule, we can now easily write a BASIC program to solve any two linear, independent equations containing two unknowns. However, before running such a program, the user must arrange the two equations so that the x terms, y terms, and constants line up in individual columns, as we have described. This is a necessary preliminary step because the application of Cramer's rule requires the matrix of coefficients to conform to that alignment, as we have seen. If we write a BASIC program to perform the Cramer's rule computations, assuming that the first column of the matrix represents x coefficients and the second column represents y coefficients, then the coefficients supplied to the program (the data for the program) must conform to that assumption. To illustrate this, the following program prompts the user to enter x coefficients, y coefficients, and constants, and then computes x and y, assuming that the x coefficients are in the first column:

```
10 REM   PROGRAM THAT FINDS X AND Y IN
20 REM   AX+BY=K1; CX+DY=K2.
30 PRINT "AX+BY=K1; CX+DY=K2."
40 PRINT "ENTER A,B,K1"
50 INPUT A,B,K1
60 PRINT "ENTER C,D,K2"
70 INPUT C,D,K2
80 LET M=A*D-B*C
90 PRINT "X=" (K1*D-K2*B)/M
100 PRINT "Y=" (A*K2-C*K1)/M
110 END
```

Of course, the preceding program can be used to solve any two linear, independent equations simultaneously, regardless of the variable names used. The inputs A and C, though, must be the coefficients of whatever variable appears in the first column of the equations; B and D must be the coefficients of whatever variable appears in the second column. Also, the constants K1 and K2 must be the values obtained when the two equations are arranged, so that all constants appear on opposite sides of the equations from the variables.

When we are solving three equations simultaneously for three unknowns, we must begin by writing a 3×3 matrix. As in the case of two equations, this matrix represents the coefficients in the same order as they appear when the equations are arranged so that all variables line up in columns and the constants appear on the opposite sides. For example, the three equations

$$y + z = 4x - 1$$
$$z = 5 - 3y - 6x$$
$$6x = 2y + 2z$$

are rearranged to give

$$4x - y - z = 1$$
$$6x + 3y + z = 5$$
$$6x - 2y - 2z = 0$$

so the 3×3 matrix is:

$$\begin{bmatrix} 4 & -1 & -1 \\ 6 & 3 & 1 \\ 6 & -2 & -2 \end{bmatrix}$$

To apply Cramer's rule for three equations, we must once again find the determinant (Δ) of the matrix of coefficients. This computation is not quite so simple as that for the 2×2 matrix, and there are a number of algorithms that can be used to perform it. We will demonstrate one of these, called *expansion by the first column*. Consider the general 3×3 matrix A shown below:

$$A = \begin{bmatrix} a & b & c \\ d & e & f \\ g & h & i \end{bmatrix}$$

The determinant of A is found through expansion by the first column, as follows:

$$\det A = (a)\det \begin{bmatrix} e & f \\ h & i \end{bmatrix} - (d)\det \begin{bmatrix} b & c \\ h & i \end{bmatrix} + (g)\det \begin{bmatrix} b & c \\ e & f \end{bmatrix}$$

Note that each term in this equation is found by multiplying an entry in the first column by the determinant of the 2×2 matrix that results when the row and column of the 3×3 matrix in which that entry appears are deleted. For example, the first term is the first entry (a) in the first column of the 3×3 matrix, multiplied by the determinant of the 2×2 matrix that results when the first row and first column of A are deleted. We now express the computation of each of the 2×2 determinants in the preceding equation to obtain the general form

$$\det A = a(ei - fh) - d(bi - ch) + g(bf - ce) \tag{1}$$

In the three-equation example that we began earlier, we find

$$\Delta = \det A$$
$$= 4\{(3)(-2) - (1)(-2)\} - 6\{(-1)(-2) - (-1)(-2)\} + 6\{(-1)(1) - (-1)(3)\} = -4$$

The determinant of a 3×3 matrix can also be found by expanding down any other column or across any row of the matrix, using a similar procedure. The only variation

SOLVING SIMULTANEOUS EQUATIONS

in the procedure is the algebraic sign (+ or −) inserted in front of each product obtained when an entry is multiplied by a 2×2 determinant. The sign in each case is $(-1)^{r+c}$, where r is the number of the row and c is the number of the column where the entry appears. For example, suppose we expanded across the second row of the general matrix A. Then,

$$\Delta = \det A$$

$$= (-1)^{2+1}(d)\det\begin{bmatrix} b & c \\ h & i \end{bmatrix} + (-1)^{2+2}(e)\det\begin{bmatrix} a & c \\ g & i \end{bmatrix} + (-1)^{2+3}(f)\det\begin{bmatrix} a & b \\ g & h \end{bmatrix}$$

$$= -d(bi - ch) + e(ai - cg) - f(ah - bg)$$

Verify that the signs shown in the expression for det(A), derived by expansion down the first column, agree with the procedure just described.

Cramer's rule is used to solve three equations simultaneously in the same way that it is used to solve two equations. The solution for x is the determinant of the matrix obtained by substituting the constants in the first column of A, divided by Δ. In our example,

$$x = \frac{\det\begin{bmatrix} 1 & -1 & -1 \\ 5 & 3 & 1 \\ 0 & -2 & -2 \end{bmatrix}}{\Delta}$$

$$= \frac{1\det\begin{bmatrix} 3 & 1 \\ -2 & -2 \end{bmatrix} - 5\det\begin{bmatrix} -1 & -1 \\ -2 & -2 \end{bmatrix} + 0\det\begin{bmatrix} -1 & -1 \\ 3 & 1 \end{bmatrix}}{-4}$$

$$= \frac{1\{(3)(-2) - (1)(-2)\} - 5\{(-1)(-2) - (-1)(-2)\} + 0}{-4}$$

$$= \frac{-4}{-4} = 1$$

Similarly, substituting the constants in the second column, we find

$$y = \frac{\det\begin{bmatrix} 4 & 1 & -1 \\ 6 & 5 & 1 \\ 6 & 0 & -2 \end{bmatrix}}{\Delta} = -2$$

and, finally,

$$z = \frac{\det\begin{bmatrix} 4 & -1 & 1 \\ 6 & 3 & 5 \\ 6 & -2 & 0 \end{bmatrix}}{\Delta} = 5$$

Example 8.2 Write a BASIC program to solve the following three equations simultaneously for p_1, p_2 and p_3.

$$p_1 + 4p_2 = 12 - p_3$$
$$9 + 2p_1 = 3p_3$$
$$p_2 = 7p_1 - 4p_3$$

The first step is to rearrange the equations so that the p_1, p_2 and p_3 terms are aligned in columns and so that the constants appear on the opposite sides of the equations:

$$p_1 + 4p_2 + p_3 = 12$$
$$2p_1 \quad\quad - 3p_3 = -9$$
$$7p_1 - p_2 - 4p_3 = 0$$

Note that the coefficient of p_2 in the second equation is 0. The matrix A is therefore:

$$\begin{bmatrix} 1 & 4 & 1 \\ 2 & 0 & -3 \\ 7 & -1 & -4 \end{bmatrix}$$

We will write a BASIC program to read these values, and the constants, using READ and DATA statements. The program can be used, however, to solve any set of three simultaneous, linear, independent equations, by simply changing the numbers in the DATA statements. The following program is based on computation of the determinant of a 3×3 matrix by expansion down the first column (Equation 1):

```
10 READ A,B,C,D,E,F,G,H,I
20 READ K1,K2,K3
30 LET D1 = A*(E*I-F*H)-D*(B*I-C*H)+G*(B*F-C*E)
40 LET D2=K1*(E*I-F*H)-K2*(B*I-C*H)+K3*(B*F-C*E)
50 PRINT "P1=" D2/D1
60 LET D3 = A*(K2*I-F*K3)-D*(K1*I-C*K3)+G*(K1*F-C*K2)
70 PRINT "P2=" D3/D1
80 LET D4 = A*(E*K3-K2*H)-D*(B*K3-K1*H)+G*(B*K2-K1*E)
90 PRINT "P3=" D4/D1
100 DATA 1,4,1,2,0,-3,7,-1,-4
110 DATA 12,-9,0
120 END
```

A run of this program reveals that $p_1 = 3$, $p_2 = 1$, and $p_3 = 5$.

EXERCISES

8.3 One way to determine whether a set of linear equations can be solved is to compute the determinant (delta) of the matrix of coefficients. If this determinant equals zero, the equations cannot be solved. Using this test, determine which of the following sets of equations can be solved.

(a) $x + 7 = -y$
$2x = 2y + 4$

(b) $35.7u = 58.1v + 7.7$
$8.3u = 5.1v$

SOLVING SIMULTANEOUS EQUATIONS

(c) $s = t$
$s = -t$

8.4 Using Cramer's rule, find solutions, wherever possible, to the sets of equations in Exercise 8.3.

8.5 The total cost of manufacturing machine part A is $\$D_1$ per part, plus a setup cost of $\$S_1$. The cost of manufacturing part B is $\$D_2$ per part, plus a setup cost of $\$S_2$. Write a BASIC program that prompts the user to enter values for D_1, D_2, S_1, and S_2 and then computes and prints the number of parts of each kind that can be produced for equal cost and the value of that cost. Run your program for $D_1 = \$20$, $D_2 = \$30$, $S_1 = \$500$, and $S_2 = \$250$.

8.6 Write a BASIC program that prompts the user to enter values for A_1, A_2, A_3, B_1, B_2, and B_3 and then solves the equations

$$A_1\tan\theta_1 + A_2\tan\theta_2 = A_3$$
$$B_1\tan\theta_1 + B_2\tan\theta_2 = B_3$$

for θ_1 and θ_2, in degrees. (Hint: Use Cramer's rule to solve for $\tan\theta_1$ and $\tan\theta_2$ and then use the ATN function.) Your program should print an error message if the equations cannot be solved for the combination of values entered by the user. (See Exercise 8.3.)

Run your program for the following combinations of input:

(a) $A_1 = 10$, $A_2 = 4.2265$, $A_3 = 10$, $B_1 = 5$, $B_2 = 3$, $B_3 = 5.886751$
(b) $A_1 = 3.75$, $A_2 = -9.25$, $A_3 = 12$, $B_1 = 2.8125$, $B_2 = -6.9375$, $B_3 = 9$

8.7 Given the matrix

$$A = \begin{bmatrix} a & b & c \\ d & e & f \\ g & h & i \end{bmatrix}$$

(a) derive an expression for det(A) by expansion down the third column
(b) derive an expression for det(A) by expansion across the first row

8.8 The efficiency E of a certain engine is directly proportional to its speed S in RPM, its temperature T in degrees Fahrenheit, and the air pressure P at its intake in pounds per square inch. The relationship is $E = k_1S + k_2T + k_3P$, where k_1, k_2, and k_3 are constants of proportionality. The following table summarizes the results of a series of tests in which the efficiency was determined at different speeds, temperatures, and air pressures.

E(%)	S(RPM)	T(°F)	P(lbs/in²)
49.7	5000	200	14.7
61.9	8000	160	13.9
42.5	4000	170	14.0

Write a BASIC program to perform the following:

(1) compute and print the constants of proportionality
(2) compute and print the efficiency when $S = 6000$ RPM, $T = 150°F$, and $P = 14.1$ lbs/in^2.

8.9 Write a BASIC program to solve the three following equations simultaneously for a, b, and c, where c is in degrees:

$$A_1e^{-a} + B_1\ln(b) + C_1\tan(c) = K_1$$
$$A_2e^{-a} + B_2\ln(b) + C_2\tan(c) = K_2$$
$$A_3e^{-a} + B_3\ln(b) + C_3\tan(c) = K_3$$

The user enters values for $A_1, B_1, C_1, K_1, A_2, B_2, C_2, K_2, A_3, B_3, C_3$, and K_3. Run your program for

$$A_1 = 5, B_1 = 6, C_1 = 3, K_1 = 8$$
$$A_2 = 2, B_2 = -9, C_2 = -2, K_2 = 0$$
$$A_3 = -1, B_3 = 4, C_3 = 6, K_3 = 5$$

(Recall that $e^x = y$ implies $x = \ln(y)$: Use this fact to solve for a, once e^{-a} is found and to solve for b, once $\ln(b)$ is found.)

8.10 Write an expression for the determinant of a 4×4 matrix by expansion down the first column. Use the rule previously described for determining the algebraic sign of each term: $(-1)^{r+c}$. Then write a BASIC program to find the determinant of

$$\begin{bmatrix} 1 & 2 & 3 & 4 \\ 5 & 6 & 7 & 8 \\ 9 & 10 & 11 & 12 \\ 13 & 14 & 15 & 16 \end{bmatrix}$$

3 MATRIX ALGEBRA

Some versions of BASIC (although not many microcomputer versions) are capable of performing matrix manipulations leading *directly* to the solutions of simultaneous equations. If you have access to one of these versions and also have a rudimentary knowledge of matrix algebra, then you will not need to implement Cramer's rule in your programs. Programs using Cramer's rule can be very long and tedious, when the number of equations is four or more (for example, see Exercise 8.10). Later in this chapter, we will discuss BASIC statements used to perform this timesaving matrix algebra, but we will now review some fundamental concepts in that algebra.

We have already discussed matrices in the context of a two-dimensional array, and we have, in particular, been using matrices of coefficients in our study of Cramer's rule. These coefficient matrices are always *square matrices*—that is, they always have the same number of rows and columns. A matrix of the coefficients in a solvable set of n linear equations must be an $n \times n$ square matrix, since there must be as many equations as unknowns. We will also deal with *column matrices,* or $n \times 1$ arrays, which are often called vectors.

The product of matrix A and matrix B (written AB) can be defined if the number of columns in A equals the number of rows in B. In other words, the number of columns in the *first* matrix (the one on the *left* side of the product expression) must equal the number of rows in the *second* matrix (the one on the *right* side of the product expression). Thus, if A is an $n \times n$ matrix and B is an $n \times 1$ matrix, the product AB can be defined, but the product BA cannot. In the latter case, the number of columns in the first matrix (B) is 1, and the number of rows in the second matrix (A) is n.

You do not need to know how the product of two matrices is actually computed. Let it suffice to say that the matrix representing the product AB has the same number of rows as A and the same number of columns as B. To check two matrices to determine whether they can be multiplied and to determine the dimension of the product matrix (if the product is defined), write the dimensions of each matrix in the same order as they are to be multiplied, as follows:

$$(m \times n)(p \times q)$$

Here we are testing to see whether an $m \times n$ matrix on the left can multiply a $p \times q$ matrix on the right. The multiplication is permitted if $n = p$ and, in that case, the product matrix has a dimension of $(m \times q)$. For example, a (4×4) matrix on the left can multiply a (4×1) matrix on the right, and the result is a (4×1) matrix:

$$\overbrace{(4 \times 4)(4 \times \underbrace{1)}_{\text{OK}}}^{4 \times 1 \text{ product}}$$

However, this order cannot be reversed:

$$(4 \times \underbrace{1)(4}_{\text{not OK}} \times 4)$$

Two matrices are *equal* if and only if the corresponding entries in each matrix are equal. That is, the entry in the first row and first column of one matrix must equal the entry in the first row and first column of the equal matrix; the entry in the second row, first column of one must equal the entry in the second row, first column of the other; and so forth. For example, if

$$\begin{bmatrix} a \\ b \end{bmatrix} = \begin{bmatrix} -2 \\ 3 \end{bmatrix}$$

then it must be true that $a = -2$ and $b = 3$. Clearly, two equal matrices must have the same dimensions: Each must have the same number of rows and columns as the other.

An $n \times n$ *identity matrix,* designated by I, is a square matrix having ones for every entry on its diagonal and zeros everywhere else. For example, the 4×4 identity matrix is

$$I = \begin{bmatrix} 1 & 0 & 0 & 0 \\ 0 & 1 & 0 & 0 \\ 0 & 0 & 1 & 0 \\ 0 & 0 & 0 & 1 \end{bmatrix}$$

The $n \times n$ identity matrix I has an important property: When it is multiplied by any other $n \times n$ matrix A, on the right or left, the result is always A.

$$AI = IA = A$$

Similarly, if an $n \times n$ identity matrix multiplies an $n \times 1$ column matrix C, the result is C.

$$\underset{(n \times n)}{\text{I}} \quad \underset{(n \times 1)}{\text{C}} \quad = \quad \underset{(n \times 1)}{\text{C}}$$

Note that CI is not defined, in this case.

The *inverse* A^{-1} of an $n \times n$ matrix A is defined to be the matrix that, when multiplied by A on the left or right, results in the $n \times n$ identity matrix I:

$$AA^{-1} = I = A^{-1}A$$

The inverse is defined only for $n \times n$ square matrices and exists only if the determinant of the matrix is *nonzero*. Thus, A^{-1} exists if and only if $\det(A) \neq 0$. A matrix whose determinant is zero is called a *singular* matrix, and one whose determinant is nonzero is called *nonsingular*. The procedure for finding the inverse of a given nonsingular, square matrix can be very involved, and for large matrices it is an exceptionally time-consuming task. Versions of BASIC capable of computing matrix inverses require a large amount of memory, so small computers do not generally have this capability. It is the ability to invert matrices that makes versions of BASIC having that capability so extremely powerful for solving simultaneous equations and eliminates the need for long programs implementing Cramer's rule.

Simultaneous equations can be expressed in matrix form, as we will illustrate using the following set of two equations with two unknowns:

$$ax + by = k_1$$
$$cx + dy = k_2$$

Here a, b, c, and d are coefficients, and k_1 and k_2 are constants, all assumed to be known values. In matrix form, this set of equations is written as:

$$\begin{bmatrix} a & b \\ c & d \end{bmatrix} \begin{bmatrix} x \\ y \end{bmatrix} = \begin{bmatrix} k_1 \\ k_2 \end{bmatrix}$$

Note that the leftmost matrix is the (2×2) matrix of coefficients discussed earlier in connection with Cramer's rule. The entries in the matrix are in the same order as the coefficients appear in the equations, when the equations are arranged so that the x terms are aligned and the y terms are aligned. The (2×1) column matrix $\begin{bmatrix} x \\ y \end{bmatrix}$ contains the names of the unknowns (variables) in the same order that the columns in the coefficient matrix correspond to the variables. In other words, if the first column in the coefficient matrix represents x coefficients, x is then the first entry in the column matrix. Note that the (2×2) coefficient matrix can multiply the (2×1) column matrix but not vice versa. Indeed, the preceding matrix equation expresses this product as the (2×1) column matrix containing the constants k_1 and k_2: $(2 \times 2)(2 \times 1) = (2 \times 1)$.

As a specific example, we will write the matrix equation corresponding to the following simultaneous equations:

$$3x = 5 - y$$
$$2y = x + 3$$

SOLVING SIMULTANEOUS EQUATIONS

Rearranging as usual, we obtain

$$3x + y = 5$$
$$-x + 2y = 3$$

Then, in matrix form:

$$\begin{bmatrix} 3 & 1 \\ -1 & 2 \end{bmatrix} \begin{bmatrix} x \\ y \end{bmatrix} = \begin{bmatrix} 5 \\ 3 \end{bmatrix}$$

If the procedure for computing the product of two matrices (which we have not described) were applied to the left side of the preceding matrix equation, the result would equal the left sides of the two original equations:

$$3x + y$$
$$-x + 2y$$

Let us now return to the general form of the matrix equation representing two simultaneous equations:

$$\begin{bmatrix} a & b \\ c & d \end{bmatrix} \begin{bmatrix} x \\ y \end{bmatrix} = \begin{bmatrix} k_1 \\ k_2 \end{bmatrix}$$

Suppose that each side of this equation is multiplied by the inverse of the coefficient matrix, which we will designate as

$$\begin{bmatrix} a & b \\ c & d \end{bmatrix}^{-1}$$

Then we have

$$\begin{bmatrix} a & b \\ c & d \end{bmatrix}^{-1} \begin{bmatrix} a & b \\ c & d \end{bmatrix} \begin{bmatrix} x \\ y \end{bmatrix} = \begin{bmatrix} a & b \\ c & d \end{bmatrix}^{-1} \begin{bmatrix} k_1 \\ k_2 \end{bmatrix}$$

Note that the inverse matrix must multiply on the left of each side, so that the multiplication will be defined. Now the product of the coefficient matrix and its inverse yields the 4×4 identity matrix that, when multiplied by the column matrix $\begin{bmatrix} x \\ y \end{bmatrix}$, yields that column matrix again. So the equation reduces to

$$\begin{bmatrix} x \\ y \end{bmatrix} = \begin{bmatrix} a & b \\ c & d \end{bmatrix}^{-1} \begin{bmatrix} k_1 \\ k_2 \end{bmatrix}$$

This important result illustrates the main principle underlying the use of matrices for solving simultaneous equations: *The unknowns (x and y, in this case) are computed by forming the product of the inverse of the coefficient matrix and the column matrix containing the constants.*

You can extend the concepts we have just discussed to three, four, or any number of simultaneous equations. Let us make the notation more succinct by designating the coefficient matrix as A, the column of variables (unknowns) as V, and the column of constants as C. The set of simultaneous equations can then be expressed in matrix form as

$$AV = C$$

Multiplying both sides by the inverse of A, we have

$$A^{-1}AV = A^{-1}C$$
$$IV = A^{-1}C$$
$$V = A^{-1}C$$

This last equation again expresses the fact that the variable values are equal to the inverse of the coefficient matrix times the matrix of constants.

Recall that a set of linear, simultaneous equations can be solved only if the determinant of the coefficient matrix is nonzero (see Exercise 8.3). Also recall that an n × n square matrix has an inverse only if its determinant is nonzero. You can now see how these two facts are related: The determinant of the coefficient matrix must be nonzero if it is to have an inverse, and the inverse must exist to obtain the solution values from the product of the inverse and the constants.

As a final example, suppose we want to solve the following four equations simultaneously for p, q, r, and s:

$$2p + q - 5r + s = 6$$
$$p + 5q - 3r - 2s = 15$$
$$p - 4q + 2r + 2s = -6$$
$$3p + 3q + 7r + 5s = 0$$

The coefficient matrix A is

$$A = \begin{bmatrix} 2 & 1 & -5 & 1 \\ 1 & 5 & -3 & -2 \\ 1 & -4 & 2 & 2 \\ 3 & 3 & 7 & 5 \end{bmatrix}$$

The variable matrix is

$$V = \begin{bmatrix} p \\ q \\ r \\ s \end{bmatrix}$$

and the constant matrix is

$$C = \begin{bmatrix} 6 \\ 15 \\ -6 \\ 0 \end{bmatrix}$$

Thus, the four simultaneous equations are expressed by

$$\begin{bmatrix} 2 & 1 & -5 & 1 \\ 1 & 5 & -3 & -2 \\ 1 & -4 & 2 & 2 \\ 3 & 3 & 7 & 5 \end{bmatrix} \begin{bmatrix} p \\ q \\ r \\ s \end{bmatrix} = \begin{bmatrix} 6 \\ 15 \\ -6 \\ 0 \end{bmatrix}$$

or, in abbreviated form,

$$AV = C$$

It can be shown that the inverse of A is

$$A^{-1} = \begin{bmatrix} -.104167 & .58333 & .68750 & -.028333 \\ .031250 & -.041667 & -.239583 & .072917 \\ -.177083 & .1250 & .135417 & .031250 \\ .291667 & -.50 & -.458333 & .1250 \end{bmatrix}$$

Therefore, $V = A^{-1}C$, or

$$\begin{bmatrix} p \\ q \\ r \\ s \end{bmatrix} = \begin{bmatrix} -.104167 & .58333 & .68750 & -.028333 \\ .031250 & -.041667 & -.239583 & -.072917 \\ -.177083 & .1250 & .135417 & .031250 \\ .291667 & -.50 & -.458333 & .1250 \end{bmatrix} \begin{bmatrix} 6 \\ 15 \\ -6 \\ 0 \end{bmatrix}$$

Multiplication of the 4×4 matrix A^{-1} by the 4×1 matrix C on the right side of this equation leads to the result

$$\begin{bmatrix} p \\ q \\ r \\ s \end{bmatrix} = \begin{bmatrix} 4 \\ 1 \\ 0 \\ -3 \end{bmatrix}$$

From this we conclude that the solution is $p = 4$, $q = 1$, $r = 0$, and $s = -3$.

EXERCISES

8.11 (a) What is the inverse of a 4×4 identity matrix?
(b) Can a 3×3 identity matrix multiply a 4×4 identity matrix? Why (or why not)?
(c) What is the dimension of the matrix product AB if A has a dimension of (3×2) and B has a dimension of (2×3)? Is BA defined?
(d) Repeat the preceding exercise when A has a dimension of $1 \times n$ and B has a dimension of $n \times 1$.

8.12 Write the matrix form of each of the following sets of equations:

(a) $x - 9y = z$
$4z + 17 - 2x = 3y$
$y = x + 2z + 10$

(b) $3(a_1 + a_2) = 5$
$4a_2 = 6(2a_1 - 1)$

(c) $1.5R_3 = 2.5R_4$
$R_1 - 2R_3 = R_4 - 7.6$
$9R_2 = R_3 - 2R_4 + 6$
$R_1 + R_2 = R_3 + R_4$

8.13 For each part of Exercise 8.12, write the matrix equation that expresses the solution to the set of equations. Show all entries in each matrix. Inverses can be shown using the -1 notation in the upper right corner of the matrix; for example:

$$\begin{bmatrix} 1 & 2 \\ 3 & 4 \end{bmatrix}^{-1}$$

8.14 For each part of Exercise 8.12, use a single letter to designate each matrix. (Define the matrices that each letter represents.) Using these letter designations, write the matrix equation representing the original set of simultaneous equations and the matrix equation representing the solution to each set of equations.

4 MATRIX OPERATIONS IN BASIC

As we indicated earlier, some versions of BASIC are capable of performing matrix algebra and are therefore quite powerful for solving simultaneous equations. Instead of writing a long sequence of statements involving many intricate computations (as required to solve a large number of equations simultaneously using Cramer's rule), the programmer can use special matrix statements to accomplish these computations automatically.

BASIC statements used for matrix operations generally have the prefix MAT. For example, the statement

MAT PRINT A

causes the contents of matrix A to be displayed, one row per line. All matrices must be dimensioned before being used in any matrix statement. Matrices are dimensioned exactly as arrays are dimensioned (using the DIM statement); in this case, the MAT prefix is not required. For example, a 4×4 matrix A and a 4×1 column matrix B are dimensioned by the statement

DIM A(4,4),B(4)

It is not necessary to dimension B using B(4,1) since B(4) has the same meaning. Once a matrix has been dimensioned, we can refer to any single entry in that matrix the same way that we refer to an entry in an array. For example, the following sequence assigns the values 2, 4, and 6 to the 3×1 column matrix C.

```
10 DIM C(3)
20 FOR I=1 TO 3
30 LET C(I)=2*I
40 NEXT I
```

Similarly, we can use the LET and PRINT statements to assign a value to—or display the contents of—any single entry in a matrix. For example:

```
LET A(2,1)=10
PRINT M(5,3)
```

The MAT READ statement assigns values found in a DATA statement to the elements of a matrix. Values are assigned from left to right across each row, beginning with the first row and ending with the last. Following is an example:

```
DIM A(2,2),B(2)
MAT READ A,B
    .
    .
    .
DATA 5,7,0,3,1,6
```

The matrices created by this example are

$$A = \begin{bmatrix} 5 & 7 \\ 0 & 3 \end{bmatrix} \text{ and } B = \begin{bmatrix} 1 \\ 6 \end{bmatrix}$$

To multiply two matrices together, we must create a new matrix that is equal to the product. The standard multiplication symbol (*) is used to show multiplication of matrices. As an illustration, the following statement creates a matrix P that is equal to the product of the matrices A and B:

$$\text{MAT P} = A * B$$

Remember that the dimensions of A and B must be such that the product A*B is defined. It is also important to remember that the product matrix P must be dimensioned before it is created. For example, if A is (4×4) and B is (4×1), you must write

$$\text{DIM A(4,4),B(4),P(4)}$$

before defining P as the product of A and B.

The inverse of a matrix A is generated by INV(A). To create an inverse, you must set a new matrix equal to INV(A), as in

$$\text{MAT B} = \text{INV}(A)$$

Here, B becomes the inverse (A^{-1}) of A. Remember that A must be a square matrix, for the inverse to be defined, and that B must be dimensioned (with the same dimension as A) before it is set equal to INV(A).

Some versions of BASIC destroy the matrix A in the process used to create its inverse. Therefore, if the original matrix A must be used again, it should be preserved by setting another matrix equal to it, before creating INV(A). A matrix C can be set equal to a matrix A by the statement

$$\text{MAT C} = A$$

Again, C must be dimensioned before this statement is executed.

We now have all the tools necessary to solve a set of simultaneous equations using matrix operations. Recall that a set of n simultaneous equations can be represented in matrix form by

$$AV = C$$

where A is an $(n \times n)$ coefficient matrix, V is the $(n \times 1)$ matrix of unknowns, and C is the $(n \times 1)$ matrix of constants. The solution is then

$$V = A^{-1}C$$

Thus, the BASIC programmer need only create matrices equal to A and C, create INV(A), and multiply INV(A) by C.

Once the solution to a set of equations has been found by the computation $V = A^{-1}C$, it is a good idea to check the solution. Checking can be easily accomplished using the computed value of the matrix V to create the product AV. If V is correct, then AV equals C (the original matrix of constants). Remember, though, that you might have to preserve A before its inverse is created, so you can use it in this check.

Example 8.3 Write a BASIC program using matrix operations to solve the following set of equations simultaneously:

$$x + 2y = z - 9$$
$$4z - y = 3x$$
$$x = -y - z + 3$$

Rearranging the equations so that they appear in standard form

$$x + 2y - z = -9$$
$$-3x - y + 4z = 0$$
$$x + y + z = 3$$

You can see that we must create a coefficient matrix A equal to

$$A = \begin{bmatrix} 1 & 2 & -1 \\ -3 & -1 & 4 \\ 1 & 1 & 1 \end{bmatrix}$$

and a matrix C of constants equal to

$$C = \begin{bmatrix} -9 \\ 0 \\ 3 \end{bmatrix}$$

The following program performs the required matrix operations:

```
10 DIM A(3,3),C(3),D(4,4),E(4,4)V(4),F(4)
20 MAT READ A,C
30 MAT D=A
40 MAT E=INV(A)
50 MAT V=E*C
60 PRINT "X=";V(1), "Y=";V(2), "Z=";V(3)
70 MAT F=D*V
80 MAT PRINT F
90 DATA 1,2,-1,-3,-1,4,1,1,1,-9,0,3
100 END
```

Note that in statement 30 we preserve A by setting D equal to it. Statement 40 creates a matrix E equal to A^{-1}, so statement 50 is equivalent to $V = A^{-1}C$, the solution matrix. Since D=A, statement 70 is equivalent to F=AV, the check matrix, which should equal C. The results of a run of the program are as follows:

```
X=6.00000        Y=-6.00000       Z=3.00000
-9.00000         2.66454E-15      3.00000
```

The solution to the set of equations is seen to be $x = 6$, $y = -6$, $z = 3$. Also note that the check matrix F equals the original set of constants ($2.66454E - 15 \approx 0$).

EXERCISES

8.15 (a) Given the matrix

$$A = \begin{bmatrix} .20 & .35 & .40 & .30 \\ .15 & .30 & .50 & .10 \\ .50 & .35 & .10 & .25 \\ .40 & .20 & .30 & .20 \end{bmatrix}$$

Write a BASIC program to print A, A^{-1}, I*A, A*I, $A*A^{-1}$, $A^{-1}*A$, and $A*A^{-1}*A$, where I is a 4×4 identity matrix. Based on your results, what can you conclude about a matrix product involving the identity matrix and about the product of a matrix and its inverse?

(b) In the matrix

$$X = \begin{bmatrix} 2 & 8 & -4 \\ 6 & 0 & 3 \\ 4 & 16 & -8 \end{bmatrix}$$

the third row equals twice the first row. Write a BASIC program to print the inverse of any 3×3 matrix. Then run your program for the preceding matrix X and report the results.

8.16 Write a BASIC program using matrix operations to solve the following set of equations simultaneously:

$$6a - 4d + 10c = 8b$$
$$20b + 7c = 12d - 14$$
$$15c + 8d = -20a - 230$$
$$10a + 7b - d + c = 139$$

Include in your program a provision to check the solution.

8.17 Repeat Exercise 8.16 for the following set of equations:

$$3(T_1 - 2T_2) = T_1 + 4(T_3 + T_2) - 3.66$$
$$8T_1 + 4T_3 = 0.48 + 6(T_3 - 4T_1)$$
$$T_2 + 2T_1 = T_3 + 0.11$$

(Note: The remaining exercises can be solved by using either Cramer's rule or BASIC matrix operations, if available in your version of BASIC.)

EXERCISES FOR ARCHITECTURAL/CIVIL/CONSTRUCTION TECHNOLOGY

8.18 Figure 8.2 shows a beam supported at each end with a force F applied l_1 feet from the left end. The total length of the beam is l feet, and the reaction forces are R_1 and R_2.

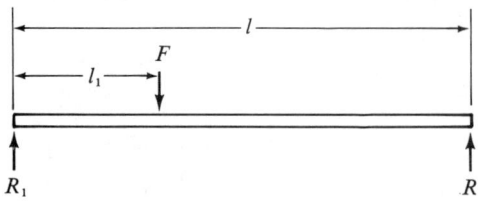

FIGURE 8.2 Exercise 8.18

Since the beam is static (motionless), the sum of the vertical forces must be zero:

$$F = R_1 + R_2$$

By summing moments about the point where the force F is applied, we obtain the equation:

$$l_1 R_1 + (l - l_1) R_2 = 0$$

Given F, l_1, and l, the two previous equations can be solved simultaneously for R_1 and R_2.

Write a BASIC program that prompts the user to enter values for l_1 and l in feet and F in pounds and then computes and prints R_1 and R_2 in pounds. Include a provision in your program to check the results (by substitution of R_1 and R_2 into the original equations *or* by forming the appropriate matrix product).

Run your program for the following input: $l_1 = 6$ ft, $l = 18$ ft, F = 1500 lbs.

8.19 Based on the results of a survey in a certain area, it was determined that the cost of constructing one type of commercial building could be estimated by the equation

$$C = k_1 A + k_2 W + k_3 L$$

where A is the area of the floor space in square feet,
 W is the total window area in square feet,
 L is the lot size in square feet, and
 k_1, k_2, and k_3 are constants of proportionality.

Based on this equation, the cost C of three such structures and the values of A, W, and L for each are summarized in the following table:

Cost ($)	A (ft^2)	W (ft^2)	L (ft^2)
283,800	10,500	470	55,000
248,800	8,400	520	60,000
594,200	25,670	880	110,200

Write a BASIC program that uses this data to estimate the cost of constructing a commercial building of this type, having an area of 15,000 square feet, a window area of 670 square feet, and a lot size of 85,000 square feet.

EXERCISES FOR ELECTRICAL/ELECTRONICS/COMPUTER TECHNOLOGY

8.20 The technique called *mesh analysis* can be used to solve for the loop currents in an electrical network containing more than one voltage source. Figure 8.3 shows a network containing three loops and two dc sources.

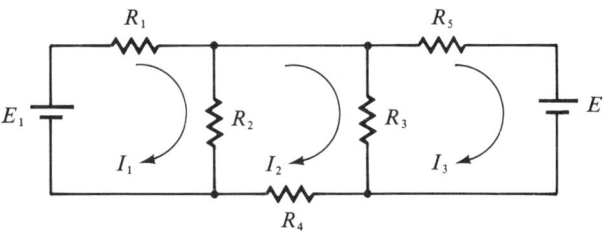

FIGURE 8.3 Exercise 8.20

Writing Kirchhoff's voltage law around each loop leads to the following equations:

$$E_1 = I_1 R_1 + (I_1 - I_2) R_2$$
$$0 = (I_2 - I_1) R_2 + I_2 R_4 + (I_2 - I_3) R_3$$
$$-E_2 = (I_3 - I_2) R_3 + I_3 R_5$$

Write a BASIC program that prompts the user to enter values for E_1, E_2, R_1, R_2, R_3, R_4, and R_5 and then solves for and prints the loop currents I_1, I_2, and I_3. Run your program for the following combination of input: $E_1 = 50$ V, $E_2 = 75$ V, $R_1 = 470\ \Omega$, $R_2 = 330\ \Omega$, $R_3 = 220\ \Omega$, $R_4 = 100\ \Omega$, and $R_5 = 220\ \Omega$.

8.21 The technique called *nodal analysis* can be used to solve for the node voltages in an electrical network containing more than one current source. Figure 8.4 shows a network containing two dc current sources and three nodes (including the reference or ground node V_0, assumed to be at 0V).

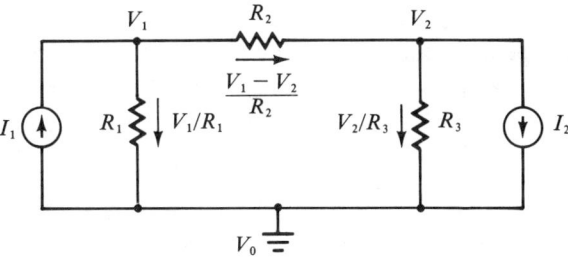

FIGURE 8.4 Exercise 8.21

The unknowns in Figure 8.4 are the node voltages V_1 and V_2. The arrows show currents in the circuit and are labeled with the current values written in terms of the node voltages. Writing Kirchhoff's current law at each node leads to the following equations:

$$I_1 = \frac{V_1}{R_1} + \frac{V_1 - V_2}{R_2}$$

$$\frac{V_1 - V_2}{R_2} = I_2 + \frac{V_2}{R_3}$$

Write a BASIC program that prompts the user to enter values for I_1, I_2, R_1, R_2, and R_3, solves the nodal equations for V_1 and V_2, and then computes and prints the current through each resistor. The output should consist of three statements similar to the following:

```
THE CURRENT IN R1=current in R₁ AMPS
```

where *current in* R_1 is the computed value of the current in R_1.

Run your program for the following combination of input: $I_1 = 25$ mA, $I_2 = 40$ mA, $R_1 = 3.3$K, $R_2 = 1.5$K, and $R_3 = 1$K.

EXERCISES FOR INDUSTRIAL/MANUFACTURING/PRODUCTION TECHNOLOGY

8.22 The total cost C of manufacturing N machine parts is given by

$$C = Nc_1 + Tc_2$$

where N = the number of parts manufactured,
c_1 = material cost in dollars per part,
T = the total time required to manufacture N parts (in hours), and
c_2 = the labor cost in dollars per hour.

It requires 7.5 hours to manufacture 100 parts at a total cost of $2,278.40 and 18 hours to manufacture 250 parts at a total cost of $5,671.40. Write a BASIC program to compute and print the total cost of manufacturing N parts in T hours, after the user enters N and T. Run your program for each of the following:

(a) N = 150 parts, T = 12 hours
(b) N = 100 parts, T = 7.5 hours

8.23 The results of an operations research study showed that the efficiency of a certain production operation could be maximized by controlling the total number of operating hours H per week, the average weekly inventory I in parts, and the weekly average number of employees E engaged in the operation. The study showed that, for maximum efficiency, these factors should be related by the following equations:

$$8.2E + 10.1H = I - 220.45$$
$$12E = 11.4H + 390.9$$
$$6(I - 9E) = 3(8E - H) + I + 658.5$$

The total weekly cost of the operation is $5780 + 18.2H + 0.15I + 9.75E$.

Write a BASIC program to compute the weekly operating hours, inventory, and number of employees required for maximum efficiency and then display the total weekly cost of operating based on those values.

EXERCISES FOR MECHANICAL TECHNOLOGY

8.24 Figure 8.5 shows a cylinder containing three springs having spring-constants K_1, K_2 and K_3 pounds per inch.

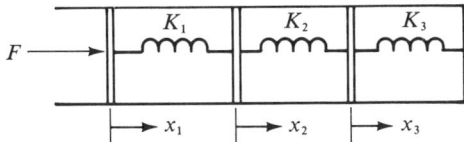

FIGURE 8.5 Exercise 8.24

Assuming that the system is free to move within the frictionless cylinder and that all masses are negligible, the force equations expressing equilibrium conditions are:

$$F = K_1(x_1 - x_2)$$
$$K_1(x_1 - x_2) = K_2(x_2 - x_3)$$
$$K_2(x_2 - x_3) = K_3 x_3$$

Write a BASIC program that prompts the user to enter values for F in pounds and K_1, K_2, K_3 in pounds per inch, and then computes and prints the values of x_1, x_2, and x_3. Run your program for $F = 20$ lbs, $K_1 = 30$ lbs/in, $K_2 = 10$ lbs/in, and $K_3 = 20$ lbs/in.

8.25 The connecting rod shown in Figure 8.6 is subjected to two downward forces, F_1 and F_2, and two upward forces, F_3 and F_4.

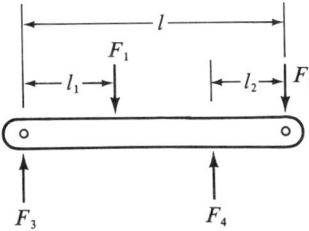

FIGURE 8.6 Exercise 8.25

Assuming the rod is in static equilibrium,

$$F_1 + F_2 = F_3 + F_4$$

and

$$F_3(l - l_2) + F_1(l - l_1 - l_2) = F_2 l_2$$

The last equation is derived by summing the moments about the point where F_4 is applied.

Write a BASIC program that prompts the user to enter values for l, l_1, l_2, F_4, and *either* F_2 or F_3. If the user enters a value for F_2, the program should compute and print the values of F_1 and F_3; if the user enters a value for F_3, the program should compute and print the values of F_1 and F_2.

Run your program for the following combinations of input:

(a) $l = 18$ in, $l_1 = 6$ in, $l_2 = 6$ in, $F_4 = 20$ lbs, and $F_2 = 10$ lbs
(b) $l = 24$ in, $l_1 = 8$ in, $l_2 = 6$ in, $F_4 = 30$ lbs, and $F_3 = 10$ lbs

CHAPTER 9

USING BASIC FOR PROBLEMS IN APPLIED CALCULUS

1 DIFFERENTIATION OF POLYNOMIALS

Recall the *power rule*, used to find the derivative of x^n with respect to x

$$\frac{dx^n}{dx} = nx^{n-1}$$

where n can be any real number—positive, negative, zero, or fractional. Following are examples:

$$\frac{dx^5}{dx} = 5x^4; \quad \frac{dy^{-2}}{dy} = -2y^{-3}; \quad \frac{dt^{1/2}}{dt} = \frac{1}{2}t^{-1/2}; \quad \frac{dz}{dz} = 1$$

The *linearity principle* for derivatives states that the derivative of a sum is the sum of the derivatives and that the derivative of a constant multiplied by a function equals the constant multiplied by the derivative of the function

$$\frac{d(u+v)}{dx} = \frac{du}{dx} + \frac{dv}{dx}; \quad \frac{dcu}{dx} = c\frac{du}{dx}$$

where u and v are functions of x, and c is a constant. For example,

$$\frac{d(x+x^3)}{dx} = 1 + 3x^2; \quad \frac{d5x^2}{dx} = 10x$$

Using these principles and knowing that the derivative of a constant is zero, we can write an expression for the derivative of a general n^{th} degree polynomial in x:

If
$$u(x) = a_n x^n + a_{n-1} x^{n-1} + \cdots + a_1 x + a_0$$

then
$$u'(x) = \frac{du}{dx} = n a_n x^{n-1} + (n-1) a_{n-1} x^{n-2} + \cdots + a_1$$

where n is a positive integer (called the degree of the polynomial) and $a_n, a_{n-1}, \ldots, a_0$ can be any constants. Recall that the numeric value obtained when the derivative of a function is evaluated for a particular x is equal to the *slope* of the graph of the function at that value of x. In other words, the derivative represents the *rate of change* of the function with respect to x. For example, the rate of change of the polynomial

$$u(x) = 3x^3 - 4x^2 + 2x$$

at $x=2$ is found from

$$u'(2) = \left.\frac{du}{dx}\right|_{x=2} = \left. 9x^2 - 8x + 2 \right|_{x=2} = (9)(4) - 8(2) + 2 = 22$$

Example 9.1 Write a BASIC program that prompts the user to enter the degree n of a polynomial and the values of the coefficients $a_n, a_{n-1}, \ldots, a_1$ and then computes and prints the rate of change of the polynomial at a user-supplied value of the independent variable (x).

The program in Figure 9.1 computes the sum S of the terms $n a_n x^{n-1}$, $(n-1) a_{n-1} x^{n-2}$, etc., beginning with $n=N$ and ending with $n=1$. Statement 110 computes the value of each such term and adds it to the sum S. Note that the constant a_0 in the general form of the polynomial has no effect on the value of the derivative and is therefore not entered or used in the program. Figure 9.2 shows the results of two runs of the program. In the first run, we find that the rate of change of $3x^2 + 4x + 2$ at $x = 1.5$ is 13, and in the second run the rate of change of $5.2x^4 - 9.1x^3 + 7x$ at $x = -0.84$ is -24.5911 (indicating that the function is *decreasing* at $x = -0.84$). Note that $a_2 = 0$ in the second run, since x^2 does not appear in the polynomial.

If we know the rate of change of a function at a certain point, then we can find the equation of a line tangent to the function at that point. Since the rate of change of a function at a given point is by definition the slope of the line that is tangent at that point, we can use the *point-slope form* to find its equation:

$$y - y_1 = m(x - x_1)$$

where m is the slope and (x_1, y_1) is the point at which the line is tangent (see Figure 9.3, page 232). To illustrate, suppose that $y = f(x) = 3x^2 - 4x$ and we want to find the equation of a line tangent to $f(x)$ at $x=2$. Then

$$\frac{dy}{dx} = 6x - 4, \quad \text{and} \quad m = \left.\frac{dy}{dx}\right|_{x=2} = 12 - 4 = 8.$$

Now, $y_1 = f(x_1) = f(2) = 3(2)^2 - 4(2) = 4$. Therefore, using the point-slope form:

$$y - 4 = 8(x - 2)$$

or
$$y = 8x - 12$$

```
10 REM  A PROGRAM THAT FINDS THE RATE OF CHANGE OF
20 REM  AN NTH DEGREE POLYNOMIAL AT A SPECIFIED POINT
30 PRINT "ENTER DEGREE N OF POLYNOMIAL"
40 INPUT N
50 PRINT "ENTER POINT AT WHICH RATE OF CHANGE IS
   DESIRED"
60 INPUT X
70 LET S=0
80 FOR I=N TO 1 STEP-1
90 PRINT "ENTER COEFFICIENT OF X**"I
100 INPUT A
110 LET S=S+A*I*X**(I-1)
120 NEXT I
130 PRINT "RATE OF CHANGE="S
140 END
>
```

FIGURE 9.1 Program for Example 9.1

Example 9.2 Write a BASIC program to find the equation of a line that is tangent to the polynomial $y = a_3 x^3 + a_2 x^2 + a_1 x + a_0$ at a point x supplied by the user. The user also enters all coefficient values. The result should be displayed in the form $y = mx + b$, where m and b are (computed) constants.

We will use an approach similar to the program given in Example 9.1 to compute the derivative dy/dx and thus obtain the required slope of the tangent line. To find y_1, we will evaluate $y(x)$ at $x = x_1$, where x_1 is the x coordinate of the tangent point (supplied by the user). Finally, we will expand the point-slope form

$$y - y_1 = m(x - x_1)$$

```
ENTER DEGREE N OF POLYNOMIAL
?2
ENTER POINT AT WHICH RATE OF CHANGE IS DESIRED
?1.5
ENTER COEFFICIENT OF X** 2
?3
ENTER COEFFICIENT OF X** 1
?4
RATE OF CHANGE=13

ENTER DEGREE N OF POLYNOMIAL
?4
ENTER POINT AT WHICH RATE OF CHANGE IS DESIRED
?-.84
ENTER COEFFICIENT OF X** 4
?5.2
ENTER COEFFICIENT OF X** 3
?-9.1
ENTER COEFFICIENT OF X** 2
?0
ENTER COEFFICIENT OF X** 1
?7
RATE OF CHANGE=-24.5911
```

FIGURE 9.2 The results of two runs of the program in Figure 9.1

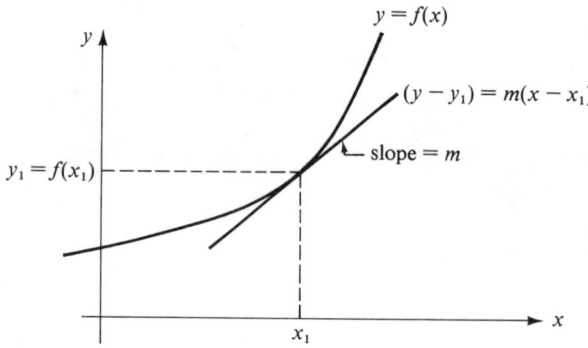

FIGURE 9.3 The point-slope form of a line tangent to a curve is $(y - y_1) = m(x - x_1)$

into the equivalent form

$$y = mx - mx_1 + y_1$$

Note that the constant b in the required output format is equal to $-mx_1 + y_1$.

The program to perform these computations is shown in Figure 9.4.

We did not assign the constant a_0 to A(0), since some versions of BASIC do not accept 0 as a value for the index of an array. Instead, the value of a_0 is assigned to the variable A0. Figure 9.5 shows the results of one run of the program, in which it is determined that the equation of the tangent to $2x^3 + 4x^2 + 3x + 15$ at $x = 1$ is $y = 17x + 7$.

The *second derivative* of a function $y = f(x)$ is the derivative of its derivative:

$$y''(x) = \frac{d^2y}{dx^2} = \frac{d}{dx}\left(\frac{dy}{dx}\right)$$

```
10 REM   PROGRAM THAT FINDS THE EQUATION Y=MX + B OF A
         LINE THAT IS
20 REM   TANGENT TO A 3RD DEGREE POLYNOMIAL AT A
         SPECIFIED POINT
30 PRINT "Y=A3*X**3 + A2*X**2+A1*X+A0"
40 DIM A(3)
50 PRINT "ENTER A3,A2,A1,A0"
60 INPUT A(3),A(2),A(1),A0
70 PRINT "ENTER X-COORDINATE OF TANGENT POINT"
80 INPUT X1
90 REM   COMPUTE SLOPE M:
100 LET M=0
110 FOR I=3 TO 1 STEP -1
120 LET M=M+A(I)*I*X1**(3-I)
130 NEXT I
140 REM   COMPUTE Y1 AND B:
150 LET Y1=A(3)*X1**3+A(2)*X1**2+A(1)*X1+A0
160 LET B=-M*X1+Y1
170 PRINT "Y="M"X+"B
180 END
>
```

FIGURE 9.4 Program for Example 9.2

```
Y=A3*X**3 + A2*X**2 + A1*X + A0
ENTER A3,A2,A1,A0
?2,4,3,15
ENTER X-COORDINATE OF TANGENT POINT
?1
Y=17X+ 7
```

FIGURE 9.5 Results of a run of the program in Figure 9.4

For example, if $y = 3x^3 - 5x + 7$, then

$$y'(x) = \frac{dy}{dx} = 9x^2 - 5$$

and

$$y''(x) = \frac{d^2y}{dx^2} = 18x$$

Similarly, the third derivative, d^3y/dx^3, is the derivative of the second derivative, and the nth derivative is the derivative of the $(n-1)$st derivative. In the above example,

$$\frac{d^3y}{dx^3} = 18 \quad \text{and} \quad \frac{d^4y}{dx^4} = 0$$

The expression $d^n y/dx^n$ or $y^{(n)}(x)$, is called an nth *order* derivative.

Example 9.3 Write a BASIC program that prompts the user to enter the degree D of a polynomial, the values of the coefficients, the order M of the derivative desired, and the value x at which the derivative is to be evaluated and then computes the value of the mth order derivative at the specified value of x.

In planning our approach to this problem, we first note that the mth order derivative of x^n is $(n)(n-1) \ldots (n-m+1)x^{n-m}$. For example, the second-order derivative of x^4 is $(4)(3)x^2$. Thus, each term of the mth order derivative of the polynomial will be multiplied by a product of m integers, each integer one less in value than the preceding one. To illustrate this, suppose that the polynomial is $x^5 + x^4 + x^3$. The third derivative of this polynomial is $(5)(4)(3)x^2 + (4)(3)(2)x + (3)(2)(1)x^0$, or $60x^2 + 24x + 6$. This observation suggests that we use a separate computation in our program to find the factor that multiplies each term in the mth order derivative of the polynomial. This computation can be performed effectively by using a subroutine that creates the product of an integer multiplied by itself minus 1. If we call that subroutine once for a second-order derivative, twice for a third-order derivative, and so forth, we can then generate the proper multiplying factor for each term in the mth order derivative of the polynomial.

Figure 9.6 is a flowchart for this program. We create arrays A, B, and N of fixed dimension 10 to store values of the coefficients A_i in the polynomial and to store the multiplying factors we have just discussed. Therefore, use of the program is limited to a 10th-degree polynomial and a tenth-order derivative (maximum). Arrays of some arbitrarily large dimension must be reserved, because we cannot set an array dimension equal to a variable. (To save memory, we would like to write DIM A(D),B(D),N(D), where D is the degree of the polynomial, but BASIC does not permit this.)

The subroutine that computes the multiplying factors requires some elaboration. Note that B(I) and N(I) are initialized in the main program so that B(1)=N(1)=1, B(2)=N(2)=2, . . . , etc. These are the correct multiplying factors if the order M

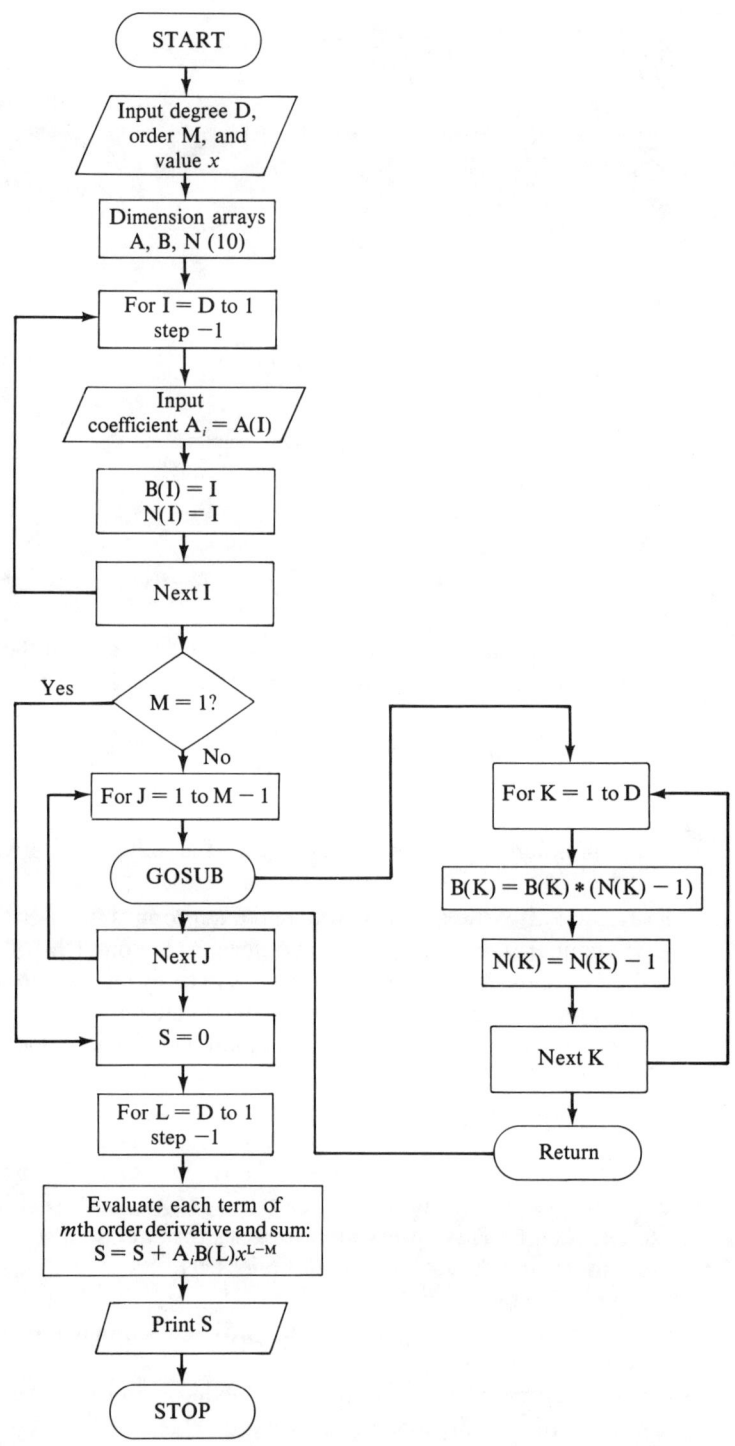

FIGURE 9.6 Flowchart for Example 9.3

of the derivative is 1. If this is so, the program then branches immediately to the evaluation of the derivative, using these factors. On the other hand, if the order of the derivative is 2 or greater, the subroutine is called. The first time that the subroutine is called, we set $B(1) = B(1)\{N(1) - 1\} = 1(0) = 0$, $B(2) = B(2)\{N(2) - 1\} = 2(1) = 2$, $B(3) = B(3)\{N(3) - 1\} = 3(2) = 6$, $B(4) = B(4)\{N(4) - 1\} = 4(3) = 12$, and so forth. The computed values of $B(K)$, for $K = 1, \ldots, D$, are therefore the multiplying factors that result when the second derivative is performed. For example, if the polynomial is $u(x) = a_4x^4 + a_3x^3 + a_2x^2 + a_1x$, then

$$u'(x) = 4a_4x^3 + 3a_3x^2 + 2a_2x + a_1$$

and

$$u''(x) = 12a_4x^2 + 6a_3x + 2a_2 + 0$$

The subroutine then diminishes each value N(K) by one. If the subroutine is called a second time (to compute the multiplying factors for a third derivative) it computes

$$\begin{aligned} B(1) &= B(1)\{N(1)-1\} = 0(0-1) = 0 \\ B(2) &= B(2)\{N(2)-1\} = 2(1-1) = 0 \\ B(3) &= B(3)\{N(3)-1\} = 6(2-1) = 6 \\ B(4) &= B(4)\{N(4)-1\} = 12(3-1) = 24 \end{aligned}$$

Thus, $u'''(x) = 24a_4x + 6a_3 + 0 + 0$. We see that the subroutine is called $M-1$ times to compute the multiplying factors for a derivative of order M.

Figure 9.7 shows a BASIC program to implement the logic shown in the flowchart of Figure 9.6.

```
10 REM  A PROGRAM THAT EVALUATES THE MTH DERIVATIVE
   OF AN NTH
20 REM  DEGREE POLYNOMIAL AT A SPECIFIED VALUE. M
   AND N  <=10.
30 PRINT "ENTER DEGREE OF POLYNOMIAL"
40 INPUT D
50 PRINT "ENTER ORDER OF DERIVATIVE DESIRED"
60 INPUT M
70 PRINT "ENTER VALUE AT WHICH DERIVATIVE IS TO BE
   EVALUATED"
80 INPUT X
90 DIM A(10),B(10),N(10)
100 FOR I=D TO 1 STEP -1
110 PRINT "ENTER COEFFICIENT OF A" I
120 INPUT A(I)
130 LET N(I)=I
140 LET B(I)=I
150 NEXT I
160 IF M=1 THEN 200
170 FOR J=1 TO M-1
180 GOSUB 260
190 NEXT J
200 LET S=0
210 FOR L=D TO 1 STEP -1
220 LET S=S+A(L)*B(L)*X**(L-M)
230 NEXT L
240 PRINT "THE DERIVATIVE OF ORDER" M "AT X=" X " IS"
    S
250 STOP
260 FOR K=1 TO D
270 LET B(K)=B(K)*(N(K)-1)
280 LET N(K)=N(K)-1
290 NEXT K
300 RETURN
310 END
```

FIGURE 9.7 Program for Example 9.3

```
ENTER DEGREE OF POLYNOMIAL
?4
ENTER ORDER OF DERIVATIVE DESIRED
?2
ENTER VALUE AT WHICH DERIVATIVE IS TO BE EVALUATED
?2
ENTER COEFFICIENT OF A 4
?2
ENTER COEFFICIENT OF A 3
?3
ENTER COEFFICIENT OF A 2
?2
ENTER COEFFICIENT OF A 1
?1
THE DERIVATIVE OF ORDER 2 AT X=2 IS 136

ENTER DEGREE OF POLYNOMIAL
?3
ENTER ORDER OF DERIVATIVE DESIRED
?4
ENTER VALUE AT WHICH DERIVATIVE IS TO BE EVALUATED
?2
ENTER COEFFICIENT OF A 3
?1.5
ENTER COEFFICIENT OF A 2
?5
ENTER COEFFICIENT OF A 1
?-7
THE DERIVATIVE OF ORDER 4 AT X= 2 IS 0
```

FIGURE 9.8 The results of two runs of the program in Figure 9.7

Figure 9.8 shows the results of two runs of the program in Figure 9.7. In the first run, we see that the second derivative of $2x^4 + 3x^3 + 2x^2 + x$ evaluated at $x=2$ is 136. In the second run, we see that the fourth derivative of $1.5x^3 + 5x^2 - 7x$ at $x=2$ is zero. Note that whenever the order of the derivative is greater than the degree of the polynomial, the result will be zero.

EXERCISES

9.1 Write a BASIC program to evaluate the derivative dy/dx of $y = Ax^{M/N}$ at a specified value of x. The user enters A, M, and N. Run your program to find the rate of change of the following:

(a) $y = -5x^{1/3}$ at $x=8$
(b) $y = 3.75x^{-3/4}$ at $x=17.63$

9.2 The slope of a line that is *normal* (perpendicular) to a curve $y=f(x)$ at a specified value of x is the *negative reciprocal* of the slope of the line that is tangent to the curve at that point. For example, if the slope of a tangent line is 2 at a certain point, the slope of the normal line is $-\frac{1}{2}$. Write a BASIC program to find the equation of a line normal to the curve $y(x) = a_2x^2 + a_1x + a_0 + a_{-1}x^{-1}$ at a user-specified value of x. The user enters the values of all coefficients. The solution should be presented in the form $y = mx + b$, where m and b are computed constants. Run the program to find the equation of a line normal to $y(x) = 6x^2 - 4x + 2 + 3x^{-1}$ at $x = 2$.

9.3 If the displacement s of a body is given as a function of time, $s(t)$, then its velocity and acceleration at any time t_0 can be found from $s'(t_0)$ and $s''(t_0)$, respectively. Thus:

$$v(t_0) = \left.\frac{ds}{dt}\right|_{t=t_0} \quad \text{and} \quad a(t_0) = \left.\frac{d^2s}{dt^2}\right|_{t=t_0}$$

The units of a derivative are the same as the units of the numerator (units of y in dy/dx) divided by the units of the denominator (units of x in dy/dx). Therefore, if s is given in meters and t is given in seconds, $v = ds/dt$ has the units of meters per second. Similarly, $a = d^2s/dt^2$ has units of m/sec^2.

Write a BASIC program to compute and print the velocity and acceleration of a body whose displacement is given by an nth degree polynomial: $s(t) = a_nt^n + a_{n-1}t^{n-1} + \ldots + a_1t + a_0$. The user enters the point in time at which these quantities are to be computed and the units of s and t. The user also enters all coefficients for the polynomial. The results should be displayed with the proper units. (Hint: Input the units to a string variable.)

Run your program to find the velocity and acceleration

(a) at $t = 1.4$ minutes, when $s(t) = 6.2t^3 - 4t^2 + t + 10$ feet;
(b) at $t = 0.25$ hours, when $s(t) = 3t - 7$ miles;
(c) at $t = 162$ seconds, when $s(t) = 0.004t^4 - t^2 + 3t$ km.

9.4 Write a BASIC program to evaluate the mth order derivative of the function

$$u(x) = a_px^p + a_{p-1}x^{p-1} + \ldots + a_1x + a_0 + a_{-1}x^{-1} + a_{-2}x^{-2} + \ldots + a_{-r}x^{-r}$$

at a value of x specified by the user. The user also enters values for m, p, r, and all coefficients.

Run your program to evaluate the third derivative of

$$2x^5 - 3x^4 + 2x^3 + 12x - 4x^{-1} + 50x^{-2} \text{ at } x = 2.5.$$

2 MAXIMUM AND MINIMUM PROBLEMS

A function $y = f(x)$ is said to have a *local maximum* at a point x_0 if the value of that function at x_0 is greater than all values it has when x is close to x_0. In other words, a local maximum occurs where the graph of the function has a "peak" in it. Similarly, a *local minimum* occurs at the point x_0 when the value of the function at x_0 is less than its surrounding values (when the graph shows a "dip"). As an example, consider the function $f(x) = 2x^3 - 24x$. The graph of this function is shown in Figure 9.9, where we see that $f(x)$ has a local maximum at $x = -2$. The value of $f(-2)$ is 32, and 32 is greater than other values of $f(x)$ in the vicinity of $x = -2$. However, note that 32 is not *the* maximum value of the function, since there are other values of x that make $f(x)$ quite large. For example, $f(6) = 72$. In fact, there is no *absolute* maximum for this function, since it becomes arbitrarily large as we allow x to increase indefinitely. Therefore, we use the term "local maximum" at $x = -2$. Similarly, we notice that $f(x)$ has a local minimum at $x = 2$, where $f(x) = -32$. Again, this is a

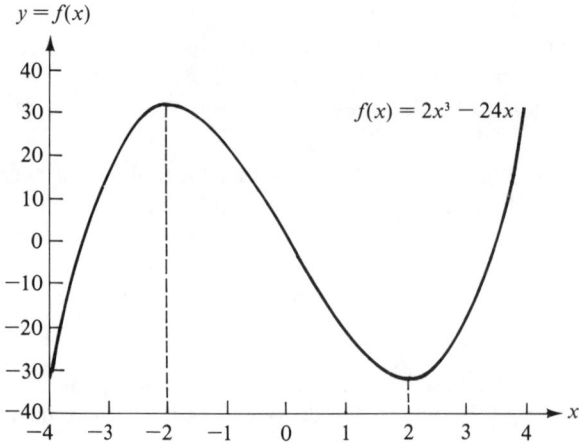

FIGURE 9.9 Graph of the function $f(x) = 2x^3 - 24x$, showing local minimum and maximum points

local phenomenon, since there are other values of x that make $f(x)$ much smaller than -32.

One characteristic of a local maximum or local minimum is that a line drawn tangent to either point is a horizontal line. In other words, tangent lines have zero slope at these points. But if the slope of a tangent line is zero, the derivative of the function must then be zero at the point of tangency. This fact provides us with the mathematical means for finding local minimum and local maximum points: Differentiate the function, set the derivative equal to zero, and solve for x. This way we can find values of x at which the derivative of x equals zero—that is, values of x at which the tangent line is horizontal and at which there must therefore be a local minimum or maximum.

To illustrate this, consider the function whose graph is shown in Figure 9.9: $f(x) = 2x^3 - 24x$. Following the procedure we have just described:

$$f'(x) = 6x^2 - 24$$
$$6x^2 - 24 = 0$$
$$x^2 = 4$$
$$x = \pm 2$$

These results confirm that local minimum and maximum points occur at $x = -2$ and $x = +2$.

The same procedure can of course be used to find the point at which an absolute maximum or minimum occurs. This point is called a *global* maximum or global minimum—a point where $f(x)$ is greater than (or less than) *all* other values of $f(x)$. For example, the parabola $f(x) = x^2 - 18x$ has a global minimum at $x = 9$. (As an exercise, sketch the graph and verify this.) Hereafter, we will refer to maximum and minimum points without distinguishing between local and global.

Once we have found a value of x at which the derivative of a function is zero, the next task is to determine whether we have found a minimum or maximum point. One method used to make that determination is called the *second derivative test*. This test requires that we find the second derivative of the function and evaluate that derivative at the x value being tested. If the result is *positive,* then we have a *minimum* point; if the result is negative, we have a *maximum* point. As an example, we will apply the test to the function whose graph is shown in Figure 9.9. Recall that we found $f'(x) = 6x^2 - 24$ and that $x = \pm 2$ are the points for which $f'(x) = 0$—that is, the candidates for minimum and maximum points. Now:

$$f''(x) = 12x$$

$$f''(+2) = 12(2) = +24$$

and

$$f''(-2) = 12(-2) = -24$$

Using the second derivative test, we conclude that $x = +2$ is a minimum point and $x = -2$ is a maximum point. The graph confirms this conclusion. We should note that the second derivative test can produce a value of zero rather than a positive or negative number, when applied to certain functions. In that case, we cannot form a conclusion about whether the point being tested is a maximum or minimum. (However, another test, the first derivative test, which we will not discuss in this book, can be used in such cases.)

Example 9.4 Write a BASIC program to find any minimum or maximum points in the function $y = a_3 x^3 + a_2 x^2 + a_1 x + a_0$. The user enters the coefficient values. The program should print the x-y coordinates of any such points and identify them as either minimum or maximum. If there are no such points or if the second derivative test is inconclusive, the program should print an appropriate message.

The general form of the first derivative of y is

$$\frac{dy}{dx} = 3a_3 x^2 + 2a_2 x + a_1$$

Equating dy/dx to zero and using the quadratic formula to solve for x, we find

$$3a_3 x^2 + 2a_2 x + a_1 = 0$$

$$x = \frac{-2a_2 \pm \sqrt{4a_2^2 - 12a_3 a_1}}{6a_3}$$

$$= \frac{-a_2 \pm \sqrt{a_2^2 - 3a_3 a_1}}{3a_3} \tag{1}$$

We see that a real solution exists only if $a_2^2 \geq 3a_3 a_1$, since only in that case is the quantity under the radical nonnegative. The general form of the second derivative of y is

$$\frac{d^2 y}{dx^2} = 6a_3 x + 2a_2$$

Thus, to apply the second derivative test, we evaluate $6a_3 x + 2a_2$ at the x values found in Equation 1, if they exist.

Figure 9.10 on page 240 shows a BASIC program to perform the required computations.

Figure 9.11 (page 241) shows the results of three runs of the progam in Figure 9.10. The first run shows that $y = 2x^3 - 24x$ has a minimum point at $x = 2$, $y = -32$ and a maximum point at $x = -2$, $y = 32$. This is the same example that we used in the preceding discussion. The second run shows that $y = 10x^3 + x^2 + 4x + 6$ has no minimum or maximum points. Note that $a_2^2 < 3a_3 a_1$ in this case. The third run shows that $y = 7.5x^3 + 12.1x^2 + x - 9.9$ has a minimum at $x = -0.043045$, $y = -9.92122$ and a maximum at $x = -1.03251$, $y = -6.28849$.

```
10 REM  A PROGRAM THAT FINDS ANY MIN OR MAX
20 REM  VALUES OF A THIRD DEGREE POLYNOMIAL
30 DIM A(3)
40 FOR I=3 TO 1 STEP -1
50 PRINT "ENTER A"I
60 INPUT A(I)
70 NEXT I
80 PRINT "ENTER A 0"
90 INPUT A0
100 IF A(2)*A(2)<3*A(3)*A(1) THEN 320
110 LET D=SQR(A(2)**2-3*A(3)*A(1))
120 LET X1=(-A(2)+D)/(3*A(3))
130 LET X2=(-A(2)-D)/(3*A(3))
140 LET T1=6*A(3)*X1+2*A(2)
150 LET T2=6*A(3)*X2+2*A(2)
160 DEF FNY(X)=A(3)*X**3+A(2)*X**2+A(1)*X+A0
170 IF T1<0 THEN 210
180 IF T1>0 THEN 230
190 PRINT "NO MIN OR MAX CAN BE FOUND AT X="X1
200 GOTO 240
210 PRINT "X="X1", Y="FNY(X1)" IS A MAX PT."
220 GOTO 240
230 PRINT "X="X1 ", Y="FNY(X1) "IS A MIN PT."
240 IF T2<0 THEN 280
250 IF T2>0 THEN 300
260 PRINT "NO MIN OR MAX CAN BE FOUND AT X="X2
270 GOTO 330
280 PRINT "X=" X2 ", Y="FNY(X2) "IS A MAX PT."
290 GOTO 330
300 PRINT "X=" X2 ", Y="FNY(X2) "IS A MIN PT."
310 GOTO 330
320 PRINT "NO MIN OR MAX POINTS CAN BE FOUND"
330 END
>
```

FIGURE 9.10 Program for Example 9.4

The power rule for functions states that

$$\frac{\mathrm{d}u^n}{\mathrm{d}x} = nu^{n-1}\frac{\mathrm{d}u}{\mathrm{d}x}$$

where u is an arbitrary function of x and n can be any real number. To illustrate how this rule can be used to find the derivative of a polynomial raised to a power, suppose we wish to find the derivative of $f(x) = (x^3 + 4x^2 + 2x)^4$. Letting $u = x^3 + 4x^2 + 2x$, we have $\frac{\mathrm{d}u}{\mathrm{d}x} = 3x^2 + 8x + 2$. Then, applying the power rule with $n=4$, $\frac{\mathrm{d}f(x)}{\mathrm{d}x} = 4(x^3 + 4x^2 + 2x)^3(3x^2 + 8x + 2)$. As another illustration, suppose that $f(x) = \{(x^2 + 1)^3 + 2\}^{1/2}$. Here we must apply the power rule twice. For the first application, let $u = (x^2 + 1)^3 + 2$ and $n = 1/2$. Thus,

$$\frac{\mathrm{d}f}{\mathrm{d}x} = \frac{1}{2}\{(x^2+1)^3+2\}^{-1/2}\frac{\mathrm{d}\{(x^2+1)^3+2\}}{\mathrm{d}x}$$

```
ENTER A 3
?2
ENTER A 2
?0
ENTER A 1
?-24
ENTER A 0
?0
X=2, Y=-32 IS A MIN PT.
X=-2, Y=32 IS A MAX PT.
>

ENTER A 3
?10
ENTER A 2
?1
ENTER A 1
?4
ENTER A 0
?6
NO MIN OR MAX POINTS CAN BE FOUND
>

ENTER A 3
?7.5
ENTER A 2
?12.1
ENTER A 1
?1
ENTER A 0
?-9.9
X=-4.30450E-02, Y=-9.92122 IS A MIN PT.
X=-1.03251, Y=-6.28849 IS A MAX PT.
```

FIGURE 9.11 The results of three runs of the program in Figure 9.10

or, since the derivative of $+2$ is zero,

$$\frac{df}{dx} = \frac{1}{2}\{(x^2+1)^3+2\}^{-1/2}\frac{d(x^2+1)^3}{dx}$$

The evaluation of the remaining derivative requires another application of the power rule. This time, we let $u = x^2 + 1$ and $n = 3$:

$$\frac{d(x^2+1)^3}{dx} = 3(x^2+1)^2\frac{d(x^2+1)}{dx} = 3(x^2+1)^2(2x)$$

Thus, finally we have

$$\frac{df}{dx} = \frac{1}{2}\{(x^2+1)^3+2\}^{-1/2}3(x^2+1)^2(2x) = \frac{3x(x^2+1)^2}{\{(x^2+1)^3+2\}^{1/2}}$$

If we want to set this derivative equal to zero and solve for x,

$$\frac{3x(x^2+1)^2}{\{(x^2+1)^3+2\}^{1/2}} = 0$$

then multiplication of both sides by the quantity $(1/3)\{(x^2+1)^3+2\}^{1/2}$ leads to the following equivalent equation:

$$x(x^2+1)^2 = 0$$

Since $(x^2+1)^2$ is never zero, the equation has only the solution $x=0$.

The *product rule* provides a means for finding the derivative of the product of two functions, u and v:

$$\frac{d(uv)}{dx} = \frac{udv}{dx} + \frac{vdu}{dx}$$

For example, to find the derivative of $f(x)=(x+1)^2 x^3$, we let $u=(x+1)^2$ and $v=x^3$. Then,

$$\frac{d(x+1)^2 x^3}{dx} = (x+1)^2 \frac{dx^3}{dx} + x^3 \frac{d(x+1)^2}{dx}$$

$$= (x+1)^2(3x^2) + x^3(2)(x+1)$$

The *quotient rule* is used to find the derivative of the quotient of two functions, u and v:

$$\frac{d(u/v)}{dx} = \frac{v(du/dx) - u(dv/dx)}{v^2}$$

For example, to find the derivative of $f(x) = \dfrac{x^2}{x^2+1}$, let $u=x^2$, $v=x^2+1$, and find

$$\frac{d\left(\dfrac{x^2}{x^2+1}\right)}{dx} = \frac{(x^2+1)\dfrac{dx^2}{dx} - x^2 d\dfrac{(x^2+1)}{dx}}{(x^2+1)^2}$$

$$= \frac{(x^2+1)(2x) - x^2(2x)}{(x^2+1)^2}$$

$$= \frac{2x}{(x^2+1)^2}$$

Example 9.5 The displacement s of a body as a function of time t is given by $s(t) = (At^2 + Bt + C)^{1/2}$. Write a BASIC program that prompts the user to enter values for the constants A, B, and C and then computes and prints a table of the values of t, displacement $s(t)$, velocity $v(t)$, and acceleration $a(t)$. Assume that A, B and C are positive constants.

The velocity $v(t)$ is given by

$$v(t) = \frac{ds}{dt} = \frac{d(At^2+Bt+C)^{1/2}}{dt} = \frac{1}{2}(At^2+Bt+C)^{-1/2}(2At+B)$$

$$= \frac{At + B/2}{(At^2+Bt+C)^{1/2}}$$

The acceleration $a(t)$ is found by applying the quotient rule to find the derivative of $v(t)$:

$$a(t) = \frac{dv}{dt} = \frac{d(At + B/2)/(At^2 + Bt + C)^{1/2}}{dt}$$

$$= \frac{(At^2 + Bt + C)^{1/2}\dfrac{d(At + B/2)}{dt} - (At + B/2)\dfrac{d(At^2 + Bt + C)^{1/2}}{dt}}{\{(At^2 + Bt + C)^{1/2}\}^2}$$

$$= \frac{(At^2 + Bt + C)^{1/2}(A) - (At + B/2)\dfrac{1}{2}(At^2 + Bt + C)^{-1/2}(2At + B)}{At^2 + Bt + C}$$

Since A, B, and C are all greater than zero, we know that the quantity $At^2 + Bt + C$ will always be greater than zero and, therefore, we do not need to be concerned that $(At^2 + Bt + C)^{1/2}$ might be undefined for some values of t. Furthermore, we know that division by $At^2 + Bt + C$ will not be responsible for any attempts to divide by zero. Without these restrictions on A, B, and C, we would have to make provisions in the program to avoid those error-generating situations.

To facilitate computation, we will define the functions

$$FNX(t) = At^2 + Bt + C$$
$$FNY(t) = (At^2 + Bt + C)^{1/2} = \{FNX(t)\}^{1/2}$$
$$FNZ(t) = (At^2 + Bt + C)^{-1/2} = 1/FNY(t)$$

Then,

$$s(t) = FNY(t)$$
$$v(t) = (At + B/2)FNZ(t)$$

and

$$a(t) = \frac{A\{FNY(t)\} - 0.5(At + B/2)(2At + B)FNZ(t)}{FNX(t)}$$

Figure 9.12 on page 244 shows a BASIC program that prints the required table of values for 21 values of t, ranging from $t=0$ to a maximum t specified by the user. Also shown in this figure are the results of one run for the case $A=10$, $B=2$, $C=5$ and time of computation $= 10$.

In the table of results shown in Figure 9.12, note that the acceleration is positive since the velocity is increasing. Also note that, as the velocity approaches a constant value, the acceleration becomes smaller in magnitude.

EXERCISES

9.5 Write a BASIC program to print a table of values of x, $f(x)$, and $f'(x)$, for 41 values of x from $x=0$ to $x=10$, inclusive, where

$$f(x) = (8x^3 + 3x^2 + 4x + 7)^{0.8}$$

```
10 PRINT "ENTER COEFFICIENTS A,B,C. ALL MUST BE >0."
20 INPUT A,B,C
30 PRINT "ENTER MAXIMUM TIME POINT FOR COMPUTATIONS"
40 INPUT T1
50 DEF FNX(T)=A*T**2+B*T+C
60 DEF FNY(T)=SQR(FNX(T))
70 DEF FNZ(T)=1/FNY(T)
80 PRINT "T","S(T)","V(T)","A(T)"
90 PRINT
100 FOR T=0 TO T1 STEP T1/20
110 PRINT T,FNY(T),(A*T+B/2)*FNZ(T),(A*FNY(T)-.5*(A*T+B/
    2)*(2*A*T+B)*FNZ(T))/FNX(T)
120 NEXT T
130 END
>

ENTER COEFFICIENTS A,B,C. ALL MUST BE >0.
?10,2,5
ENTER MAXIMUM TIME POINT FOR COMPUTATIONS
?10
```

T	S(T)	V(T)	A(T)
0	2.23607	.447214	4.38269
.500000	2.91548	2.05798	1.97728
1	4.12311	2.66789	.699073
1.50000	5.52268	2.89714	.290902
2	7	3.00000	.142857
2.50000	8.51469	3.05355	7.93760E-02
3	10.0499	3.08462	4.82741E-02
3.50000	11.5974	3.10414	3.14132E-02
4	13.1529	3.11717	2.15341E-02
4.50000	14.7139	3.12629	1.53819E-02
5	16.2788	3.13291	1.13587E-02
5.50000	17.8466	3.13786	8.62049E-03
6	19.4165	3.14166	6.69397E-03
6.50000	20.9881	3.14464	5.30002E-03
7	22.5610	3.14702	4.26697E-03
7.50000	24.1350	3.14895	3.48540E-03
8	25.7099	3.15053	2.88333E-03
8.50000	27.2855	3.15185	2.41212E-03
9	28.8617	3.15296	2.03811E-03
9.50000	30.4385	3.15390	1.73751E-03
10	32.0156	3.15471	1.49317E-03

FIGURE 9.12 Program for Example 9.5 and the results of one program run

9.6 Write a BASIC program to find any values of p at which the function

$$f(p) = \frac{A}{Fp^2 + Gp + H}$$

has a minimum or maximum value. The user enters values for A, F, G, and H, all of which must be greater than zero. (Also assume $G^2 \neq 4FH$.) Determine and print the values of $f(p)$ at any such points and identify them as either minimum or maximum points. Run your program for $A = 150$, $F = 2$, $G = 10$, and $H = 5$.

9.7 Recall that the derivative of a function evaluated at a given point represents the rate of change of the function at that point—that is, the slope of a line drawn tangent to the graph of the function at that point. The mathematical procedure for obtaining the derivative of a function can be envisioned as a process beginning with the calculation of the slope of a line drawn through the point in question and another point on the graph of the function. This slope is a rough approximation of the true value of the derivative.

In Figure 9.13, we can approximate the derivative of $y = f(x)$ at point P_1 by computing the slope $\Delta y/\Delta x$ of the line drawn through P_1 and P_2. If we then allow Δx to become smaller and smaller, so that P_2 approaches P_1, the approximation $\Delta y/\Delta x$ gets better and better until, in the limit, it is equal to the slope of the actual tangent line at P_1.

Mathematically, we write

$$\frac{dy}{dx} = \lim_{\Delta x \to 0} \frac{\Delta y}{\Delta x}$$

In Figure 9.13, note that $\Delta x = x_2 - x_1$ and $\Delta y = f(x_2) - f(x_1)$.

Suppose we want to find the derivative of the function

$$y = f(x) = (5x^2 - 3x)^3(x^2 - 2)$$

at $x_1 = 1$. To observe how the slope of a line through $x_1 = 1$ and $x_2 = 1.1$ approaches the true derivative of $f(x)$ at $x_1 = 1$, write a BASIC program to compute

$$\frac{\Delta y}{\Delta x} = \frac{f(x_2) - f(x_1)}{x_2 - x_1}$$

for 21 values of x_2 between $x_2 = 1.1$ and $x_2 = 1.01$. Also compute and display the true value of $f'(1)$.

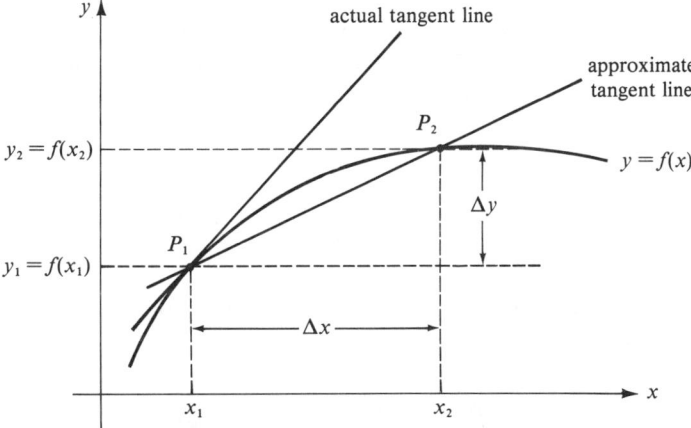

FIGURE 9.13 Exercise 9.7

3 DIFFERENTIATION OF TRANSCENDENTAL FUNCTIONS

Transcendental functions are "nonalgebraic" functions. The most important are the trigonometric, exponential, and logarithmic functions. For reference and review, we present the derivatives of a number of these functions in the following list. (In each case, $u(x)$ is an arbitrary function of x.)

$$\frac{d\sin(u)}{dx} = \cos(u)\frac{du}{dx}$$

$$\frac{d\cos(u)}{dx} = -\sin(u)\frac{du}{dx}$$

$$\frac{d\tan(u)}{dx} = \sec^2(u)\frac{du}{dx}$$

$$\frac{d\cot(u)}{dx} = -\csc^2(u)\frac{du}{dx}$$

$$\frac{d\sec(u)}{dx} = \sec(u)\tan(u)\frac{du}{dx}$$

$$\frac{d\csc(u)}{dx} = -\csc(u)\cot(u)\frac{du}{dx}$$

$$\frac{de^u}{dx} = e^u\frac{du}{dx}$$

$$\frac{d\ln(u)}{dx} = \frac{1}{u}\frac{du}{dx}$$

As an example, suppose we wish to find the derivative with respect to time of the function

$$f(t) = A\sin(\omega t + \theta)$$

where A is a constant (the *amplitude* of the sine wave),
 ω is the angular frequency (a constant) in radians per second, and
 θ is a constant phase angle in radians.

If we let $u(t) = \omega t + \theta$, then $du/dt = \omega$ and we find that

$$f'(t) = \frac{dA\sin(\omega t + \theta)}{dt} = A\omega\cos(\omega t + \theta)$$

Now consider the function $f(t) = Ae^{-at}\sin(\omega t + \theta)$, where a is a positive constant and all other quantities are as previously defined. This function is the damped sine wave that we studied in detail in Chapter 7. To differentiate the function, we must use the product rule, as follows:

$$\frac{d(uv)}{dx} = u\frac{dv}{dx} + v\frac{du}{dx}$$

In this case, we let $u(t) = Ae^{-at}$ and $v(t) = \sin(\omega t + \theta)$. Then, applying the product rule:

$$f'(t) = \frac{d\{Ae^{-at}\sin(\omega t + \theta)\}}{dt} = A\omega e^{-at}\cos(\omega t + \theta) - Aae^{-at}\sin(\omega t + \theta)$$

Figure 9.14 shows graphs of $f(t) = 10e^{-2t}\sin 3t$ and $f'(t) = 30e^{-2t}\cos 3t - 20e^{-2t}\sin 3t$ plotted on the same set of axes. Note how differentiation tends to *amplify* (increase the range of) a sine function. This amplification occurs in direct proportion to frequency ω, because the differentiated function is multiplied by ω. Also note in the figure that $f'(t)$ is largest where $f(t)$ is changing fastest and that $f'(t)$ is positive when $f(t)$ is increasing and negative when $f(t)$ is decreasing. These observations agree with our interpretation of a derivative as the rate of change of a function.

Example 9.6 Write a BASIC program to compute and print a table of values of $f(t) = e^{-\sin\omega t}$ and its derivative, for values of t between $t=0$ and $t=4\pi/\omega$, inclusive. The user enters the value of ω in radians per second. The program should also determine: the first value of t at which a minimum or maximum value of $f(t)$ occurs, the value of $f(t)$ at that point, and whether this is a minimum or maximum point.

Letting $u(t) = -\sin\omega t$, we find that $du/dt = -\omega\cos\omega t$, and therefore,

$$f'(t) = \frac{de^{-\sin\omega t}}{dt} = (-\omega\cos\omega t)e^{-\sin\omega t}$$

To determine $f''(t)$ (which we will need to perform the second derivative test), we use the product rule and find that

$$f''(t) = \frac{df'(t)}{dt} = -\omega\cos\omega t \frac{d(e^{-\sin\omega t})}{dt} + e^{-\sin\omega t}\frac{d(-\omega\cos\omega t)}{dt}$$
$$= (\omega^2\cos^2\omega t)e^{-\sin\omega t} + (\omega^2\sin\omega t)e^{-\sin\omega t}$$

To find a minimum or maximum value of $f(t)$, we set $f'(t)$ equal to zero:

$$-\omega\cos\omega t e^{-\sin\omega t} = 0$$

Now, $e^{-\sin\omega t}$ is never zero, so the only solution to this equation occurs for $\cos\omega t = 0$. Note that there are infinitely many solutions to this equation ($\omega t = \pi/2, 3\pi/2, 5\pi/2, \ldots$, etc.) but the *first* one is

$$\omega t = \pi/2$$

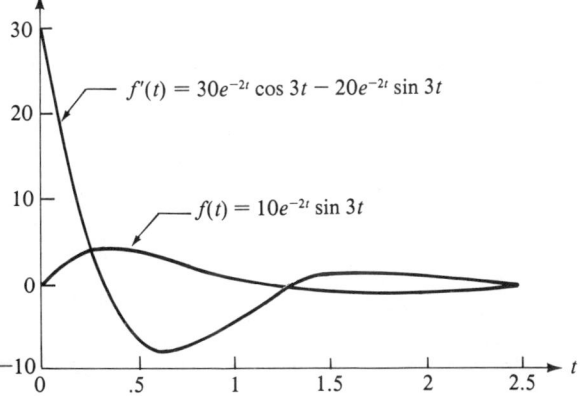

FIGURE 9.14 The graph of $f(t)$ and its derivative $f'(t)$—Note that $f'(t)$ is positive when $f(t)$ is increasing and negative when $f(t)$ is decreasing.

from which we conclude that the first minimum or maximum occurs at

$$t_0 = \pi/2\omega$$

Figure 9.15 shows a BASIC program that prints the required table at 21 different values of t between 0 and $4\pi/\omega$. Note that the program defines the following functions:

```
10 PRINT "ENTER FREQUENCY IN RAD/SEC"
20 INPUT W
30 DEF FNF(T)=EXP(-SIN(W*T))
40 DEF FND(T)=-W*COS(W*T)*FNF(T)
50 DEF FNS(T)=-W*COS(W*T)*FND(T)+W**2*SIN(W*T)*FNF(T)
60 PRINT "T","F(T)","F'(T)"
70 PRINT
80 LET N=12.56637/W
90 FOR T=0 TO N STEP N/20
100 PRINT T,FNF(T),FND(T)
110 NEXT T
120 LET T0=3.1415926/(2*W)
130 IF FNS(T0)>0 THEN 160
140 PRINT "T="T0", F(T)="FNF(T0)" IS A MAX PT."
150 GOTO 170
160 PRINT "T=" T0", F(T)="FNF(T0) " IS A MIN PT"
170 END
>

ENTER FREQUENCY IN RAD/SEC
?10
T                F(T)            F'(T)

 0                1               -10
 6.28318E-02     .555556          -4.49455
 .125664         .386333          -1.19383
 .188496         .386333           1.19383
 .251327         .555556           4.49454
 .314159        1.00000           10.0000
 .376991        1.80000           14.5623
 .439823        2.58844            7.99873
 .502655        2.58844           -7.99872
 .565487        1.80000          -14.5623
 .628318        1.00000          -10.0000
 .691150         .555556          -4.49455
 .753982         .386333          -1.19384
 .816814         .386333           1.19383
 .879646         .555556           4.49454
 .942478        1.00000           10.0000
1.00531         1.80000           14.5623
1.06814         2.58844            7.99874
1.13097         2.58844           -7.99872
1.19381         1.80000          -14.5623
1.25664         1.00000          -10.0000
T=.157080, F(T)=.367879 IS A MIN PT.
```

FIGURE 9.15 Program for Example 9.6 and the results of a program run

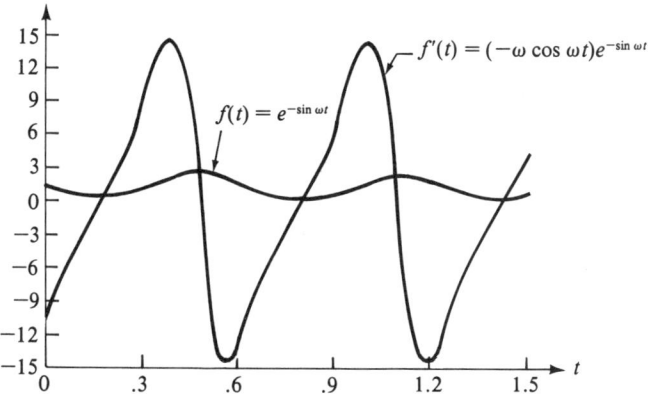

FIGURE 9.16 Graphs of $f(t)$ and its derivative $f'(t)$—Note that both functions are periodic (Example 9.6).

$$f(t) = \text{FNF}(t) = e^{-\sin\omega t}$$
$$f'(t) = \text{FND}(t) = (-\omega\cos\omega t)e^{-\sin\omega t} = (-\omega\cos\omega t)\text{FNF}(t)$$
$$f''(t) = \text{FNS}(t) = (\omega^2\cos^2\omega t)e^{-\sin\omega t} + \omega^2\sin\omega t\, e^{-\sin\omega t}$$
$$= (-\omega\cos\omega t)\text{FND}(t) + (\omega^2\sin\omega t)\text{FNF}(t)$$

The second derivative test is performed in line 130 of the program. Also shown in Figure 9.15 is the result of a run of the program for $\omega = 10$.

Note the oscillatory nature of $f(t)$ and $f'(t)$. No damping occurs because the values of $\sin\omega t$ range from -1 to $+1$, so the values of $e^{-\sin\omega t}$ range from e^{-1} to e^1. The period of both $f(t)$ and $f'(t)$ are seen to equal

$$1.0053 - 0.376992 = 0.628308, \text{ or } 2\pi/\omega$$

Figure 9.16 shows graphs of $f(t)$ and $f'(t)$ plotted on the same set of axes. Note that these functions are definitely *not* sinusoidal. Also, note again that $f'(t)$ is positive when $f(t)$ is increasing and negative when $f(t)$ is decreasing. The derivative is seen to pass through zero at the points where $f(t)$ reaches its minimum and maximum values.

EXERCISES

9.8 The horizontal displacement s of a body as a function of time t is given by $s(t) = \ln(at^2 + bt + c)$. Write a BASIC program that prompts the user to enter values for a, b, and c and then prints a table of values for time t, displacement s, and velocity v. The table should list 21 values for t, ranging from $t=0$ to $t=a$. Assume that $c > 0$. Run your program for $a=10$, $b=2$, and $c=1$.

9.9 Write a BASIC program to find the first value of t at which a maximum or minimum value of the function $f(t) = A_1\sin\omega t + A_2\cos\omega t$ occurs, after the user enters values for A_1, A_2 and ω. Assume that A_1 must be a positive number. Determine whether a maximum or minimum occurs at the computed value of t, and find the value of $f(t)$ at that point. (Hint: After equating the derivative to zero, divide through by $\sin\omega t$.)

Run your program for the following combinations of input:

(a) $A_1 = 1$, $A_2 = 2$, $\omega = 4$ rad/sec
(b) $A_1 = 1.57$, $A_2 = -2.03$, $\omega = 165$ rad/sec

4 DEFINITE INTEGRALS

Integration is generally a more challenging computational problem than differentiation. There are many continuous functions that can be differentiated but cannot be integrated in "closed form." That is, it is not possible to write expressions containing a finite number of terms to represent the integrals of such functions. A good example is the widely used, *normal* ("bell-shaped") probability density function

$$p(x) = \frac{1}{\sqrt{2\pi}} e^{-x^2/2}$$

whose integral represents a probability. Since this function cannot be integrated in closed form, tables containing values of its integral must be used. Every elementary statistics book contains such a table.

Computers are especially valuable for solving integration problems and are used extensively in scientific programming for that purpose. There are many techniques that can be used to obtain accurate values of integrals of functions that cannot be integrated in closed form. In the next section of this chapter, we will investigate two such methods. In this section, we will review standard, closed-form integration of functions and write some BASIC programs that are useful in the evaluation of integrals.

Recall that an integral is also called an *antiderivative*, because it is obtained by a process that is the inverse of differentiation. That is, if

$$\frac{dy}{dx} = f(x)$$

then

$$y = \int f(x) dx$$

In other words, given a function $f(x)$, integration of $f(x)$ is the process of finding a function $y(x)$ that, when differentiated, produces $f(x)$. For example, given $f(x) = x$, we know that

$$\int f(x) dx = \int x \, dx = x^2/2$$

is one solution, because

$$\frac{d(x^2/2)}{dx} = x = f(x)$$

The power rule for integration states that

$$\int u^n du = \frac{u^{n+1}}{n+1} + C$$

where u is a function, n can be any number except -1, and C is an arbitrary constant. In particular, if $u(x) = x$, then $du = dx$, and

$$\int x^n dx = \frac{x^{n+1}}{n+1} + C$$

where, again, n can be any number except -1. If $n = -1$, then

or, if $u(x) = x$,

$$\int u^{-1} du = \ln(u) + C$$

$$\int \frac{dx}{x} = \ln(x) + C$$

The presence of the arbitrary constant C in all these cases identifies each as an *indefinite* integral.

Integration is *linear* in the same sense that differentiation is: The integral of a sum of functions is the sum of the integrals of the functions, and the integral of a constant multiplied by a function equals the constant multiplied by the integral of the function. We can therefore find the integral of a polynomial in x by writing

$$\int (a_n x^n + a_{n-1} x^{n-1} + \ldots + a_1 x + a_0) dx$$
$$= \frac{a_n x^{n+1}}{n+1} + \frac{a_{n-1} x^n}{n} + \ldots + \frac{a_1 x^2}{2} + a_0 x + C$$

where $a_n, \ldots, a_0$ are constants. Notice especially that, for any constant a_0,

$$\int a_0 dx = a_0 x + C$$

For example,

$$\int 3 dt = 3t + C$$

Recall that the value of a *definite* integral is computed by finding the difference between the values of the corresponding indefinite integral at an *upper limit* and a *lower limit*. Thus, if $F(x)$ is the indefinite integral of $f(x)$—that is, if

$$\int f(x) dx = F(x)$$

then the definite integral between a lower limit a and an upper limit b is

$$\int_a^b f(x) dx = F(b) - F(a)$$

For example,

$$\int_1^2 x^2 dx = \frac{x^3}{3} \bigg|_1^2 = \frac{8}{3} - \frac{1}{3} = \frac{7}{3}$$

All the following examples and exercises concern the computation of definite integrals.

Example 9.7 If the displacement $s(t)$ of a body is given as a function of time, we know that its velocity and acceleration can be found from

$$v(t) = \frac{ds(t)}{dt}$$

$$a(t) = \frac{dv(t)}{dt} = \frac{d^2 s(t)}{dt^2}$$

It follows then that, given acceleration $a(t)$ as a function of time, we can find the velocity at any time t_0 by integrating $a(t)$ between $t=0$ and $t=t_0$:

$$v(t) = \int_0^{t_0} v(t)dt + v(0)$$

where $v(0)$ is the velocity at $t=0$, called the *initial velocity*. Similarly,

$$s(t) = \int_0^{t_0} v(t)dt + s(0)$$

where $s(0)$ is the initial displacement.

Write a BASIC program that prompts the user to enter the coefficients for the acceleration $a(t) = k_3 t^3 + k_2 t^2 + k_1 t + k_0$, as well as the initial velocity and displacement, and the time t_0 at which the displacement and velocity are desired. The program should compute and print $s(t_0)$ and $v(t_0)$.

Integrating the general expression for $a(t)$ to find $v(t)$, we obtain

$$v(t) = \int_0^{t_0} (k_3 t^3 + k_2 t^2 + k_1 t + k_0)dt + v(0)$$

$$= \left. \frac{k_3 t^4}{4} + \frac{k_2 t^3}{3} + \frac{k_1 t^2}{2} + k_0 t \right|_0^{t_0} + v(0)$$

$$= \frac{k_3 t_0^4}{4} + \frac{k_2 t_0^3}{3} + \frac{k_1 t_0^2}{2} + k_0 t_0 + v(0)$$

Similarly,

$$s(t) = \int_0^{t_0} \left[\frac{k_3 t^4}{4} + \frac{k_2 t^3}{3} + \frac{k_1 t^2}{2} + k_0 t + v(0) \right] dt + s(0)$$

$$= \left. \frac{k_3 t^5}{20} + \frac{k_2 t^4}{12} + \frac{k_1 t^3}{6} + \frac{k_0 t^2}{2} + v(0)t \right|_0^{t_0} + s(0)$$

$$= \frac{k_3 t_0^5}{20} + \frac{k_2 t_0^4}{12} + \frac{k_1 t_0^3}{6} + \frac{k_0 t_0^2}{2} + v(0)t_0 + s(0)$$

Figure 9.17 shows a BASIC program that performs the required computations. Also shown are the results of two runs of the program. The first listing shows that $s=95.833$ and $v=55$ at $t=5$, when $a(t)=4t+1$ and $v(0)=s(0)=0$, and the second shows that $s=351.524$ and $v=282.071$ at $t=4.72$, when $a(t)=1.7t^3 - 4.9t^2 + 16t + 22.3$, $v(0)=-40.6$, and $s(0)=17.9$.

Recall also that the definite integral of a function can be interpreted as the area enclosed by the graph of the function, the x axis, and vertical lines drawn through the two limit points. For example, the area A bounded by the parabola $y=3x^2$, the x axis, and the lines $x=0$, $x=4$ is found from

$$A = \int_0^4 3x^2 dx = \left. \frac{3x^3}{3} \right|_0^4 = 4^3 - 0^3 = 64$$

(See Figure 9.18.)

```
10 PRINT "ENTER COEFFICIENTS K3,K2,K1,K0"
20 INPUT K3,K2,K1,K0
30 PRINT "ENTER INITIAL VELOCITY, INITIAL DISPLACEMENT"
40 INPUT V0,S0
50 PRINT "ENTER TIME OF COMPUTATION"
60 INPUT T0
70 LET V=K3*T0**4/4+K2*T0**3/3+K1*T0**2/2+K0*T0+V0
80 LET S=K3*T0**5/20+K2*T0**4/12+K1*T0**3/6+K0*T0**2/2+V0*T0+S0
90 PRINT "THE VELOCITY AT T="T0" IS"V
100 PRINT "THE DISPLACEMENT AT T= "T0" IS"S
110 END
>

ENTER COEFFICIENTS K3,K2,K1,K0
?0,0,4,1
ENTER INITIAL VELOCITY, INITIAL DISPLACEMENT
?0,0
ENTER TIME OF COMPUTATION
?5
THE VELOCITY AT T=5 IS 55
THE DISPLACEMENT AT T=5 IS 95.8333

ENTER COEFFICIENTS K3,K2,K1,K0
?1.7,-4.9,16,22.3
ENTER INITIAL VELOCITY, INITIAL DISPLACEMENT
?-40.6,17.9
ENTER TIME OF COMPUTATION
?4.72
THE VELOCITY AT T=4.72000 IS 282.071
THE DISPLACEMENT AT T=4.72000 IS 351.542
```

FIGURE 9.17 Program for Example 9.7 and the results of two program runs

The area *below* the x axis is negative and thus subtracts from any area enclosed by the graph of a function above the x axis. Consider, for example, the area enclosed by the parabola $y = x^2 - 2x$ between $x = 0$ and $x = 4$ (see Figure 9.19).

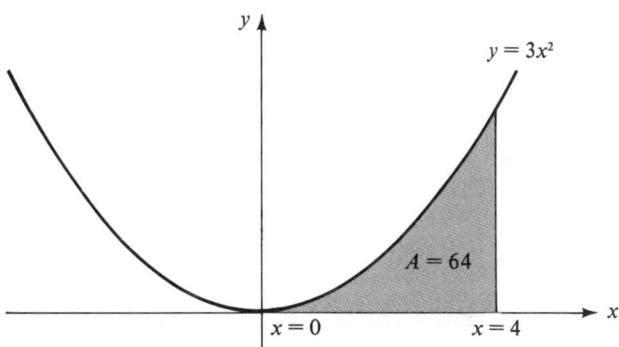

FIGURE 9.18 The definite integral of a function between two limits is the area under the graph of the function, bounded by the limits.

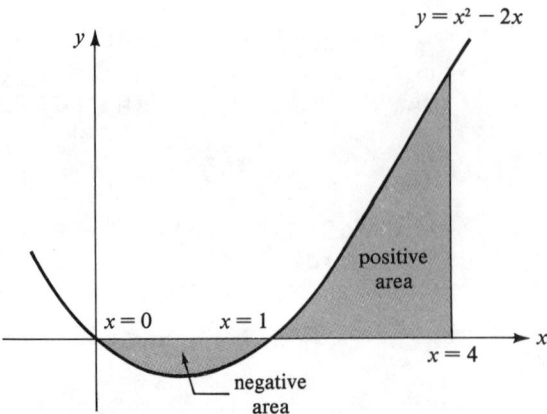

FIGURE 9.19 The area under the graph of a function can be either positive or negative.

We see that the area between $x=0$ and $x=1$ is negative, and the area between $x=1$ and $x=4$ is positive. The net area is

$$\int_0^1 (x^2 - 2x)dx + \int_1^4 (x^2 - 2x)dx$$

$$= \left(\frac{x^3}{3} - x^2\right)\Big|_0^1 + \left(\frac{x^3}{3} - x^2\right)\Big|_1^4$$

$$= \left(\frac{1}{3} - 1\right) + \left\{\left(\frac{64}{3} - 16\right) - \left(\frac{1}{3} - 1\right)\right\}$$

$$= -\frac{2}{3} + 6 = 5\frac{1}{3}$$

Thus, the area below the x axis is $-2/3$ and the area above the x axis is 6, yielding a net area of 5 1/3. We divided this computation into two separate integrals to demonstrate that part of the area is negative, but the same net area can be found by integrating $x^2 - 2x$ between the entire range from $x=0$ to $x=4$:

$$\int_0^4 (x^2 - 2x)dx = \frac{x^3}{3} - x^2\Big|_0^4 = \frac{64}{3} - 16 = 5\frac{1}{3}$$

To find the area bounded by *two* functions, it is often necessary to construct a differential rectangle whose area dA is equal to a differential width dx multiplied by an expression f(x) representing the height of the rectangle. We then integrate dA between limits determined by the points of intersection of the two functions. As an example, suppose we want to find the area enclosed between the points of intersection of the line $y=6x$ and the parabola $y=3x^2$. In these kinds of problems, it is wise to make a sketch of the region whose area is desired, as an aid in setting up the differential rectangle. Figure 9.20 shows a sketch for the present example.

As can be seen in the figure, the differential rectangle has a height equal to the length of a vertical line between the two functions. But this height equals the difference between the y coordinates of the two functions—namely, $6x - 3x^2$. Thus, the differential area of the rectangle is $dA = (6x - 3x^2)dx$. The points where the two curves intersect are found by equating the expressions for the curves and solving for x

$$3x^2 = 6x$$

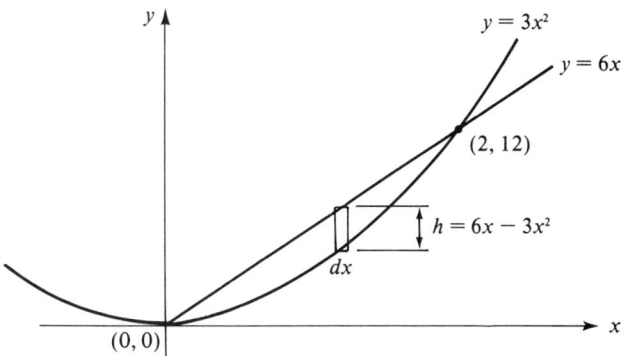

FIGURE 9.20 The area between the graphs of two functions

From this we find $x=0$ and $x=2$. The required area is therefore

$$A = \int_0^2 dA = \int_0^2 (6x - 3x^2)dx = 3x^2 - x^3 \Big|_0^2 = 4$$

Example 9.8 Write a BASIC program to find the area enclosed between the two parabolas $y = k_1 x^2$ and $y = -k_2 x^2 + k_3$. The user enters values for k_1, k_2, and k_3, all of which must be greater than zero.

Figure 9.21 shows the two parabolas and the differential rectangle within the region enclosed by the parabolas.

We see that $dA = \{k_1 x^2 - (-k_2 x^2 + k_3)\}dx = \{(k_1 + k_2)x^2 - k_3\}dx$. The limits x_1 and x_2 of integration are found by solving

$$k_1 x^2 = -k_2 x^2 + k_3$$

$$(k_1 + k_2)x^2 = k_3$$

$$x = \pm \sqrt{\frac{k_3}{k_1 + k_2}}$$

Thus

$$x_1 = -\sqrt{\frac{k_3}{k_1 + k_2}} \quad \text{and} \quad x_2 = +\sqrt{\frac{k_3}{k_1 + k_2}}$$

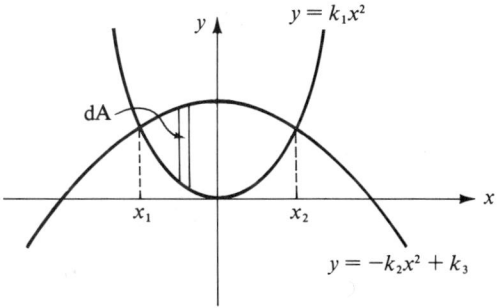

FIGURE 9.21 The area bounded by two parabolas (Example 9.8)

Therefore,

$$A = \int_{x_1}^{x_2} \{(k_1 + k_2)x^2 - k_3\}dx$$

$$= \frac{(k_1 + k_2)x_2^3}{3} - k_3 x_2 - \frac{(k_1 + k_2)x_1^3}{3} + k_3 x_1$$

$$= \frac{(k_1 + k_2)}{3}(x_2^3 - x_1^3) + k_3(x_1 - x_2)$$

or, since $x_1 = -x_2$,

$$A = \frac{2}{3}(k_1 + k_2)x_2^3 - 2k_3 x_2$$

The corresponding BASIC program is shown in Figure 9.22. Also shown is the result of a run that finds the area enclosed between $y = 1.75x^2$ and $y = -3.41x^2 + 7.92$. (Note that +3.41 is entered for K2.)

Listed below for reference and review are the indefinite integrals of some frequently encountered transcendental functions:

$$\int e^u du = e^u + C$$

$$\int \sin u\, du = -\cos u + C$$

$$\int \cos u\, du = \sin u + C$$

$$\int \sec^2 u\, du = \tan u + C$$

$$\int \csc^2 u\, du = -\cot u + C$$

$$\int \sec u \tan u\, du = \sec u + C$$

$$\int \csc u \cot u\, du = -\csc u + C$$

It is often necessary to perform algebraic manipulations on an integral, so it can be expressed in an equivalent form that can be integrated. Consider, for example, the integral

```
10 PRINT "Y=(K1)X**2, ENTER K1"
20 INPUT K1
30 PRINT "Y=(-K2)X**2+K3, ENTER K2 AND K3"
40 INPUT K2,K3
50 LET X=-SQR(K3/(K1+K2))
60 LET A=2*(K1+K2)*X**3/3-2*K3*X
70 PRINT "THE AREA IS" A
80 END
>

Y=K(1)X**2, ENTER K1
?1.75
Y=(-K2)X**2 + K3, ENTER K2 AND K3
?3.41,7.92
THE AREA IS 13.0828
```

FIGURE 9.22 Program for Example 9.8 and the results of one program run

$$\int \sin \omega t \, dt$$

where ω is a constant. Notice that this integral is not in any of the forms previously listed. We must perform algebraic manipulations to obtain the form

$$\int \sin u \, du$$

Toward that end, let $\omega t = u$. Then $du = \omega dt$. Now multiply dt in the original integral by ω to obtain ωdt and also divide the integral by ω. Multiplying and dividing the integral by the same constant ω does not affect its value.

$$\int \sin \omega t \, dt = \frac{1}{\omega} \int (\sin \omega t)(\omega dt) = \frac{1}{\omega} \int \sin u \, du$$

$$= -\frac{1}{\omega} \cos u + C = -\frac{1}{\omega} \cos \omega t + C$$

We see that this procedure creates the equivalent form $\sin u \, du$, which we can then integrate to obtain $-\frac{1}{\omega} \cos \omega t + C$.

It is often necessary to use trigonometric identities to change an integral into a form that can be integrated. Consider, for example,

$$\int \sin^2(\omega t + \theta) dt$$

Using the identity $\sin^2 x = \frac{1}{2}(1 - \cos 2x)$, we can rewrite the integral in the equivalent form

$$\int \frac{1}{2}\{1 - \cos 2(\omega t + \theta)\} dt = \int \frac{1}{2} dt - \frac{1}{2} \int \cos 2(\omega t + \theta) dt$$

Letting $u = 2(\omega t + \theta)$, so that $du = 2\omega dt$, we find

$$\frac{1}{2}\int dt - \frac{1}{2\omega}\int \{\cos 2(\omega t + \theta)\}(2\omega dt) = \frac{t}{2} - \frac{1}{2\omega} \sin 2(\omega t + \theta) + C$$

Example 9.9 Write a BASIC program to find the area of the region shown in Figure 9.23, bounded by the t axis, and the curves $y_1 = A_1 \sin \omega t$ and $y_2 = A_2 \sin(\omega t + \theta)$. The user enters values for A_1, A_2, ω, and θ. Assume that $0 < \theta < \pi/2$ radians.

As can be seen in this figure, the height of the differential rectangle depends on its location. Its height is y_1, if t is to the left of the point of intersection ($t < t_0$); and it has a height of y_2, if $t > t_0$. The required area is thus found by evaluating two integrals: one between 0 and t_0 and the other between t_0 and t_1.

$$A = \int_0^{t_0} A_1 \sin \omega t \, dt + \int_{t_0}^{t_1} A_2 \sin(\omega t + \theta) dt$$

$$= -\frac{A_1}{\omega} \cos \omega t \Big|_0^{t_0} - \frac{A_2}{\omega} \cos(\omega t + \theta) \Big|_{t_0}^{t_1}$$

$$= -\frac{A_1}{\omega}\{\cos(\omega t_0) - 1\} - \frac{A_2}{\omega}\{\cos(\omega t_1 + \theta) - \cos(\omega t_0 + \theta)\}$$

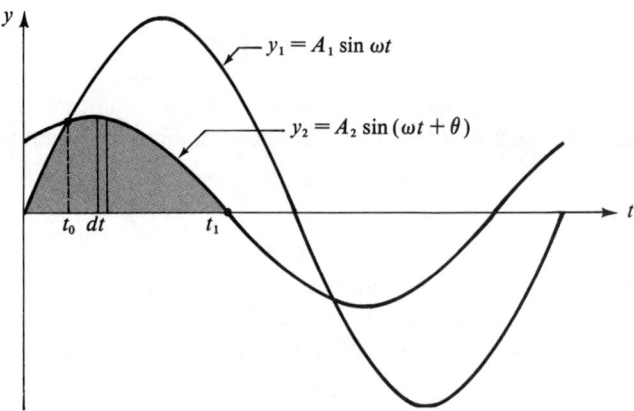

FIGURE 9.23 The area bounded by two sine waves (Example 9.9)

The point t_1 is the point where $A_2\sin(\omega t + \theta) = 0$ and is therefore found from

$$\omega t_1 + \theta = \pi$$

$$t_1 = \frac{\pi - \theta}{\omega}$$

The value t_0 where the two sine functions intersect is found by equating them and solving for t, as follows:

$$A_2\sin(\omega t_0 + \theta) = A_1\sin\omega t_0$$

Applying the trigonometric identity $\sin(x + y) = \sin x \cos y + \cos x \sin y$ to the left side of the equation, we obtain

$$A_2\sin(\omega t_0)\cos\theta + A_2\cos(\omega t_0)\sin\theta = A_1\sin(\omega t_0)$$

Therefore

$$\sin(\omega t_0)\{A_2\cos\theta - A_1\} = -A_2\sin\theta\cos(\omega t_0)$$

$$\tan(\omega t_0) = \frac{-A_2\sin\theta}{A_2\cos\theta - A_1} = \frac{A_2\sin\theta}{A_1 - A_2\cos\theta}$$

$$\omega t_0 = \arctan\left(\frac{A_2\sin\theta}{A_1 - A_2\cos\theta}\right)$$

$$t_0 = \frac{1}{\omega}\arctan\left(\frac{A_2\sin\theta}{A_1 - A_2\cos\theta}\right)$$

The corresponding BASIC program is shown in Figure 9.24, along with the results of two runs. In the first run, we find that the area between $10\sin 8t$ and $5\sin(8t + 45°)$ is 0.953984. In the second run, the area between $50\sin.01t$ and $50\sin(.01t + 89°)$ is seen to be 2990.91.

The technique called *integration by parts* is often helpful for expressing an integral in a form that makes it more readily integrable. Recall that

$$\int u\,dv = uv - \int v\,du$$

USING BASIC FOR PROBLEMS IN APPLIED CALCULUS

```
10 PRINT "Y1=(A1)SIN(WT), ENTER A1 AND W, (W IN RAD/
   SEC)."
20 INPUT A1,W
30 PRINT "Y2=(A2)SIN(WT+D), ENTER A2 AND D, (D IN
   DEGREES)."
40 INPUT A2,D
50 REM  CONVERT DEGREES TO RADIANS:
60 LET R=D/57.295781
70 LET T1=(3.1415927-R)/W
80 LET T0=(1/W)*ATN(A2*SIN(R)/(A1-A2*COS(R)))
90 LET I1=-(A1/W)*(COS(W*T0)-1)
100 LET I2=-(A2/W)*(COS(W*T1+R)-COS(W*T0+R))
110 PRINT "AREA=" I1+I2
120 END
>

Y1=(A1)SIN(WT), ENTER A1 AND W, (W IN RAD/SEC).
?10,8
Y2=(A2)SIN(WT+D), ENTER A2 AND D, (D IN DEGREES).
?5,45
AREA= .953984
>

Y1=(A1)SIN(WT), ENTER A1 AND W, (W IN RAD/SEC).
?50,.01
Y2=(A2)SIN(WT+D), ENTER A2 AND D, (D IN DEGREES).
?50,89
AREA=2990.91
>
```

FIGURE 9.24 Program for Example 9.9 and the results of two program runs

The key to successful application of this technique is a judicious choice of u and $\mathrm{d}v$. As an example, suppose we wish to evaluate

$$\int_0^1 x \mathrm{e}^x \mathrm{d}x$$

We choose to let $u = x$ and $\mathrm{d}v = \mathrm{e}^x \mathrm{d}x$. Based on these choices, it remains to determine $\mathrm{d}u$ and v. Clearly,

$$\mathrm{d}u = \mathrm{d}x \quad \text{and} \quad v = \int \mathrm{d}v = \int \mathrm{e}^x \mathrm{d}x = \mathrm{e}^x$$

Consequently, we obtain

$$\int_0^1 x\mathrm{e}^x \mathrm{d}x = uv \Big|_0^1 - \int_0^1 v\,\mathrm{d}u$$

$$= x\mathrm{e}^x \Big|_0^1 - \int_0^1 \mathrm{e}^x \mathrm{d}x$$

$$= x\mathrm{e}^x \Big|_0^1 - \mathrm{e}^x \Big|_0^1$$

$$= (\mathrm{e} - 0) - (\mathrm{e} - \mathrm{e}^0) = \mathrm{e}^0 = 1$$

EXERCISES

9.10 The *infinite series* expansion for e^{ax} is

$$e^{ax} = 1 + ax + \frac{(ax)^2}{2!} + \frac{(ax)^3}{3!} + \frac{(ax)^4}{4!} + \ldots$$

where a is a constant. The integral of a function is often approximated by integrating a certain number of terms of its infinite series expansion. To determine how well this technique works for the integral of e^{ax}, write a BASIC program to obtain an approximate value for the integral

$$\int_0^1 e^{ax} dx$$

by integrating a certain number of terms of the infinite series expansion. The user enters the value of a. Compute the actual (exact) value of this integral, and then integrate the series expansion for e^{ax}, using as many terms as necessary to obtain a value within 0.01 of the actual value. For example, if you use only three terms of the series expansion, you would then compute

$$\int_0^1 e^{ax} dx \simeq \int_0^1 dx + \int_0^1 ax\, dx + \int_0^1 \frac{(ax)^2}{2!} dx$$

Three terms may or may not be enough to produce a value within 0.01 of the actual value. Note that the actual value is

$$\int_0^1 e^{ax} dx = \frac{1}{a} e^{ax} \Big|_0^1 = \frac{1}{a}(e^a - 1)$$

Your program should print the actual value, the number of terms of the series that must be used for the approximation to be within 0.01 of the actual value, and the approximate value obtained when that number of terms is used. Run your program for each of the following: (a) a=10, (b) a=0.1, and (c) a=−2.

9.11 Given that

$$\int e^{au} \sin bu = \frac{e^{au}(a \sin bu - b \cos bu)}{a^2 + b^2} + C$$

Write a BASIC program to compute the area A_1 and *net* area $A_1 + A_2$ (see Figure 9.25) enclosed by the first half-cycle and first full cycle of the function

$$f(t) = Me^{-at}\sin(\omega t)$$

The user enters values for *M*, *a*, and *ω*. Run your program for M=10, a=2, and ω=12.

9.12 Write a BASIC program to compute and print a table of values of

$$\int_0^b Ax^2 e^{ax} dx$$

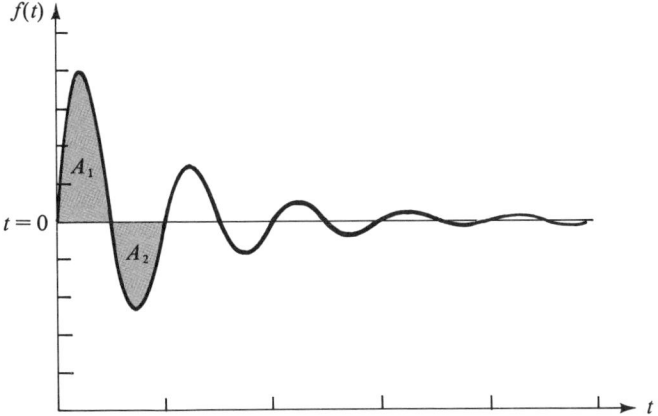

FIGURE 9.25 Exercise 9.11

The user enters values for a, b, and A. The table should contain 10 values of this integral, corresponding to the 10 upper limits of integration: $0.1b$, $0.2b$, $0.3b$, ..., b, with the values of a and A fixed. (Hint: Integrate twice by parts.) Run your program for the following combinations of input:

(a) $A = 10$, $a = 2$, $b = 2$
(b) $A = 10$, $a = -2$, $b = 2$

5 NUMERICAL INTEGRATION

As we indicated at the beginning of Section 9.4, computers are especially valuable for evaluating integrals that cannot be integrated in closed form. The techniques used to evaluate integrals without first obtaining explicit expressions for the indefinite integrals are called "numerical methods" of integration. In this section, we will discuss two such methods and show how we can write BASIC programs to implement them.

The first method of numerical integration that we will consider is called the *trapezoidal rule*. This method is based on the interpretation of a definite integral as the area under the graph of a function, between its limits of integration. Our approach is to divide this area into a number of trapezoids and then compute the sum of all trapezoids. Recall that a trapezoid is a figure that has two parallel sides, and the area of a trapezoid is

$$A = \frac{1}{2}h(s_1 + s_2)$$

where h is the height of the trapezoid (the perpendicular distance between its parallel sides) and s_1 and s_2 are the lengths of its two parallel sides. Here we are interested in the type of trapezoid shown in Figure 9.26, where we see that the lengths of the two parallel sides are y_1 and y_2 and the "height" is $x_2 - x_1 = \Delta x$. Thus:

$$A = \frac{1}{2}\Delta x(y_1 + y_2)$$

Hereafter, we will refer to Δx as the *width* of a trapezoid.

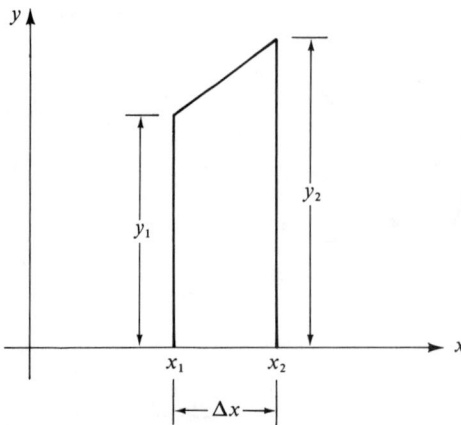

FIGURE 9.26 The area of the trapezoid is $(1/2)\Delta x(y_1 + y_2)$.

Now consider an arbitrary function $y = f(x)$. Under the graph of this function, we construct a number of trapezoids of equal width. These trapezoids are constructed between a lower limit $x = a$ and an upper limit $x = b$. See Figure 9.27, where four trapezoids are shown; each of their heights is a point on the graph of $y = f(x)$. We see that the area under the graph between $x_1 = a$ and $x_5 = b$ is approximated by the total area of the four trapezoids. Straight lines are drawn between the y coordinates where the trapezoids intersect the graph, so some trapezoids have more area (and some have less) than that appearing under the graph between those points. The total area of the trapezoids is

$$A = \frac{1}{2}\Delta x(y_1 + y_2) + \frac{1}{2}\Delta x(y_2 + y_3) + \frac{1}{2}\Delta x(y_3 + y_4) + \frac{1}{2}\Delta x(y_4 + y_5)$$

where

$$y_1 = f(x_1), y_2 = f(x_2), \ldots, \text{etc.}$$

and

$$\Delta x = (x_2 - x_1) = (x_3 - x_2), \ldots, \text{etc.}$$

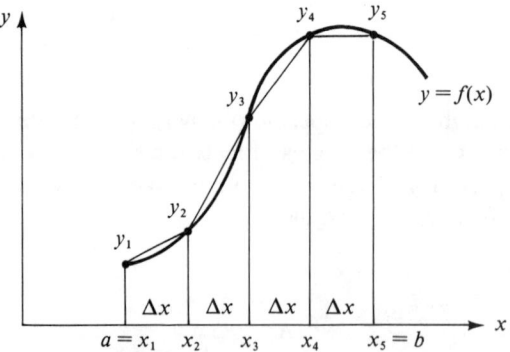

FIGURE 9.27 Four trapezoids whose total area approximates the area under the graph of $y = f(x)$ between $x = a$ and $x = b$.

Factoring out Δx, we obtain

$$A = \Delta x\left(\frac{1}{2}y_1 + \frac{1}{2}y_2 + \frac{1}{2}y_2 + \frac{1}{2}y_3 + \frac{1}{2}y_3 + \frac{1}{2}y_4 + \frac{1}{2}y_4 + \frac{1}{2}y_5\right)$$

Notice that every y value except the first and last (y_1 and y_5) has half its value added to half its value. Therefore,

$$A = \Delta x\left(\frac{1}{2}y_1 + y_2 + y_3 + y_4 + \frac{1}{2}y_5\right)$$

We repeat this derivation now for an arbitrary number (n) of trapezoids:

$$A = \frac{1}{2}\Delta x(y_1 + y_2) + \frac{1}{2}\Delta x(y_2 + y_3) + \ldots + \frac{1}{2}\Delta x(y_{n-1} + y_n) + \frac{1}{2}\Delta x(y_{n-1} + y_n)$$

$$= \Delta x\left(\frac{1}{2}y_1 + \frac{1}{2}y_2 + \frac{1}{2}y_3 + \ldots + \frac{1}{2}y_{n-1} + \frac{1}{2}y_{n-1} + \frac{1}{2}y_n\right)$$

$$= \Delta x\left(\frac{1}{2}y_1 + y_2 + y_3 + \ldots + y_{n-1} + \frac{1}{2}y_n\right)$$

where

$$\Delta x = \frac{x_n - x_1}{n} = \frac{b - a}{n}$$

This is the general version of the trapezoidal rule. The larger the value of n (the smaller the widths of the trapezoids), the more closely this area approximates the true area under $y = f(x)$ between $x_1 = a$ and $x_n = b$. In a limiting sense, if we let $\Delta x \to 0$, then the total area of the resulting (infinite) number of trapezoids would exactly equal the definite integral

$$\int_a^b f(x)dx$$

Using a high-speed computer, we can choose a very large number of trapezoids—that is, a small value of Δx, and thus obtain an excellent approximation for the integral.

To illustrate the trapezoidal rule, we will use it to approximate the value of a definite integral whose exact value is known. We can then compare the approximate value obtained with the true value. Suppose we want to evaluate

$$\int_0^2 x^{1/2}dx$$

Using n = 4 trapezoids, we have

$$\Delta x = \frac{2 - 0}{4} = 0.5$$

Therefore, the x values of the four trapezoids are $x_1 = 0$, $x_2 = 0.5$, $x_3 = 1.0$, $x_4 = 1.5$ and $x_5 = 2.0$. Note that there will always be one more x value (and one more y value) than the total number of trapezoids. Next, we tabulate the y coordinates corresponding to each x coordinate, using $y = x^{1/2}$, as follows:

x	y
0	0
0.5	0.7071
1.0	1.000
1.5	1.2247
2.0	1.4142

Applying the trapezoidal rule, we find

$$A = \Delta x \left(\frac{1}{2}y_1 + y_2 + y_3 + y_4 + \frac{1}{2}y_5\right)$$

$$= 0.5(0 + 0.7071 + 1.000 + 1.2247 + 0.7071)$$

$$= 1.8195$$

The exact value of the integral is

$$\int_0^2 x^{1/2}dx = \frac{2}{3}x^{3/2}\Big|_0^2 = (2/3)(2.82843) = 1.88562$$

Thus, we obtain a reasonably good approximation (3.5% error) using only four trapezoids.

Example 9.10 Write a BASIC program to compute and print the approximate value of

$$\int_a^b (Ax^2 + Bx + C)dx$$

using the trapezoidal rule. Use $n = 10$ trapezoids. The user enters values for a, b, A, B, and C. Assume that $a < b$. The program should also compute and print the exact value of the integral and the percent error between the approximate and exact values.

The BASIC program that performs these computations is shown in Figure 9.28.

In the program of Figure 9.28, the FOR-TO, NEXT loop in statements 100 through 140 computes the x and y coordinates of each intersection of a trapezoid with the function $Ax^2 + Bx + C$. For I = 1, we see that X = A1, the lower limit of integration; for I = 2, X = A1 + D, where D is the width of the trapezoid; for I = 3, X = A1 + 2D; and so forth, until I = 11 and X = A1 + 10D = B1, the upper limit of integration. For each of these x coordinates, the corresponding y coordinate is computed and stored in array Y, by statement 130. Statement 160 initializes the sum S of the y values to Y(1)/2 + Y(11)/2. Y(1) and Y(11) are the only two y values that are halved in the trapezoidal rule. The FOR-TO, NEXT loop in statements 170 through 190 then adds all the remaining values, and the final trapezoidal area, S * D (i.e., S * Δx), is printed by statement 200.

Figure 9.29 shows the results of three runs of the program in Figure 9.28. In the first run, we see that the trapezoidal rule finds an approximation for the value of

$$\int_0^1 (2x^2 + 3x + 4)dx$$

with a very small error of less than 0.06%. In the second run, using the same function, we see that the error is somewhat larger (but still less than 0.5 percent). However, in

```
10 PRINT "Y=AX**2+BX+C. ENTER A,B,C."
20 INPUT A,B,C
30 PRINT "ENTER LOWER, UPPER LIMITS OF INTEGRATION."
40 INPUT A1,B1
50 DIM Y(11)
60 REM   COMPUTE D=DELTA-X (TRAPEZOID WIDTH):
70 LET D=(B1-A1)/10
80 DEF FNY(X)=A*X**2+B*X+C
90 DEF FNZ(X)=A*X**3/3+B*X**2/2+C*X
100 REM   COMPUTE Y-VALUES
110 FOR I=1 TO 11
120 LET X=A1+(I-1)*D
130 LET Y(I)=FNY(X)
140 NEXT I
150 REM   COMPUTE SUM OF Y-VALUES
160 LET S=Y(1)/2+Y(11)/2
170 FOR J=2 TO 10
180 LET S=S+Y(J)
190 NEXT J
200 PRINT "THE APPROXIMATE VALUE IS" S*D
210 REM   COMPUTE EXACT VALUE E:
220 LET E=FNZ(B1)-FNZ(A1)
230 PRINT "THE EXACT VALUE IS" E
240 PRINT "THE ERROR IS" (S*D-E)*100/E "%"
250 END
>
```

FIGURE 9.28 Program for Example 9.10

```
Y=AX**2+BX+C. ENTER A,B,C.
?2,3,4
ENTER LOWER, UPPER LIMITS OF INTEGRATION.
?0,1
THE APPROXIMATE VALUE IS 6.17000
THE EXACT VALUE IS 6.16667
THE ERROR IS 5.40541E-02 %
>

Y=AX**2+BX+C. ENTER A,B,C.
?2,3,4
ENTER LOWER, UPPER LIMITS OF INTEGRATION.
?0,20
THE APPROXIMATE VALUE IS 6040
THE EXACT VALUE IS 6013.33
THE ERROR IS .443459 %
>

Y=AX**2+BX+C. ENTER A,B,C.
?0,5,2
ENTER LOWER, UPPER LIMITS OF INTEGRATION.
?0,10
THE APPROXIMATE VALUE IS 270
THE EXACT VALUE IS 270
THE ERROR IS 0 %
>
```

FIGURE 9.29 The results of three runs of the program in Figure 9.28

the second run, the limits of integration are 0 and 20, so the ten trapezoids represent a much coarser division of the region. In the third run, the trapezoidal rule finds the value of

$$\int_0^{10} (5x + 2)dx$$

with zero error. This result is expected, since $y = 5x + 2$ is a straight line. The straight line segments joining the tops of the trapezoids therefore coincide exactly with the line $5x + 2$.

The second method for numerical integration we will discuss is called *Simpson's rule*. This method is also based on finding an approximate area under the graph of a function, between the limits of integration. Again, we partition the region by constructing a number of equally spaced vertical lines intersecting the curve $y = f(x)$. However, instead of joining the points of intersection by straight-line segments, as when using the trapezoidal rule, we will join the points with *parabolic arcs*. (See Figure 9.30.)

Given three points on a parabola, it is possible to write the equation for the parabola. The essence of Simpson's rule is to fit a parabola between each successive set of points: (y_1, y_2, y_3), (y_2, y_3, y_4), and so forth, as shown in Figure 9.30. Based on this curve-fitting technique, it can be shown that the area A_1 under the parabolic arc between y_1 and y_3 in Figure 9.30 is

$$A_1 = \frac{\Delta x}{3}(y_1 + 4y_2 + y_3)$$

Similarly, the area A_2 is

$$A_2 = \frac{\Delta x}{3}(y_3 + 4y_4 + y_5)$$

Thus, the total area is

$$A_1 + A_2 = \frac{\Delta x}{3}(y_1 + 4y_2 + 2y_3 + 4y_4 + y_5)$$

If we divide the region between $x = a$ and $x = b$ into an arbitrary number n of intervals, then the general form of Simpson's rule is

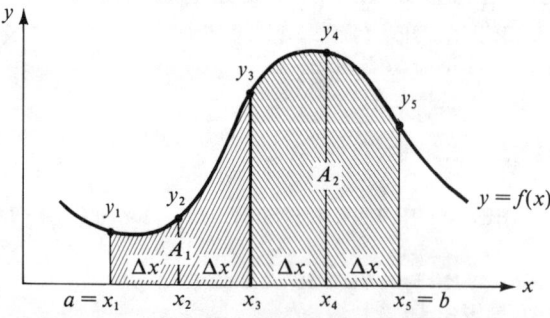

FIGURE 9.30 The sum of the areas A_1 and A_2 approximates the area under the graph of $y = f(x)$ between $x = a$ and $x = b$.

$$A = \frac{\Delta x}{3}(y_1 + 4y_2 + 2y_3 + 4y_4 + 2y_5 + \ldots + 4y_n + y_{n+1})$$

where

$$y_1 = f(x_1), \ y_2 = f(x_2), \ldots, y_{n+1} = f(x_{n+1})$$

and

$$\Delta x = \frac{x_n - x_1}{n} = \frac{b - a}{n}$$

Notice that the application of Simpson's rule requires that the number of intervals n be an *even* number. In many cases, Simpson's rule produces a better approximation for a definite integral than does the trapezoidal rule.

Example 9.11 Repeat Example 9.10 using Simpson's rule instead of the trapezoidal rule.

The way that the program in Figure 9.28 computes the sum S of y values must be modified. Note that Simpson's rule requires every other y value, except the first and last, to be multiplied by either 4 or 2. Also, the final sum is multiplied by $\Delta x/3$ instead of Δx. Figure 9.31 shows the modified program.

Figure 9.32 on page 268 shows the results of two runs of the program in Figure 9.31. In the first run, Simpson's rule computes the value of

$$\int_{-5}^{25} (50x - 75) dx$$

```
10  PRINT "Y=AX**2+BX+C, ENTER A,B,C."
20  INPUT A,B,C
30  PRINT "ENTER LOWER, UPPER LIMITS OF INTEGRATION."
40  INPUT A1,B1
50  DIM Y(11)
60  REM   COMPUTE DELTA-X
70  LET D=(B1-A1)/10
80  DEF FNY(X)=A*X**2+B*X+C
90  DEF FNZ(X)=A*X**3/3+B*X**2/2+C*X
100 REM   COMPUTE Y-VALUES
110 FOR I=1 TO 11
120 LET X=A1+(I-1)*D
130 LET Y(I)=FNY(X)
140 NEXT I
150 REM   COMPUTE SUM OF Y-VALUES
160 LET S=Y(1)+4*Y(10)+Y(11)
170 FOR J=2 TO 5
180 LET S=S+4*Y(2*J-2)
190 LET S=S+2*Y(2*J-1)
200 NEXT J
210 LET V=S*D/3
220 PRINT "THE APPROXIMATE VALUE IS" V
230 REM   COMPUTE EXACT VALUE E:
240 LET E=FNZ(B1)-FNZ(A1)
250 PRINT "THE EXACT VALUE IS" E
260 PRINT "THE ERROR IS" (V-E)*100/E " %"
270 END
>
```

FIGURE 9.31 Program for Example 9.11

```
Y=AX**2+BX+C. ENTER A,B,C.
?0,50,-75
ENTER LOWER, UPPER LIMITS OF INTEGRATION.
?-5,25
THE APPROXIMATE VALUE IS 12750
THE EXACT VALUE IS 12750
THE ERROR IS 0 %
>

Y=AX**2+BX+C. ENTER A,B,C.
?2,3,4
ENTER LOWER, UPPER LIMITS OF INTEGRATION.
?0,20
THE APPROXIMATE VALUE IS 6013.33
THE EXACT VALUE IS 6013.33
THE ERROR IS 0 %
>
```

FIGURE 9.32 The results of two runs of the program in Figure 9.31

with zero error. The results of the second run show that it also computes

$$\int_0^{20} (2x^2 + 3x + 4)\,dx$$

with zero error. These results are expected, since the function $Ax^2 + Bx + C$ is a parabola!

One precaution should be observed when applying either of the two numerical methods of integration we have discussed: Avoid integrating over a large range of x in which there is very little contribution to the area under the curve. If the majority of the area under the graph of a function occurs over a limited range of x, then the value used for Δx should be small in that range. Suppose, for example, that we use $n = 10$ intervals to compute the value of

$$\frac{1}{\sqrt{2\pi}} \int_{-10}^{0} e^{-x^2/2}\,dx$$

In this case, $\Delta x = 1.0$, and the majority of the intervals occur in a range where the function is negligibly small (from $x = -10$ to about $x = -3$). Consequently, only three intervals are used to approximate the area between $x = -3$ and $x = 0$, where the majority of the entire area occurs. The approximation obtained could therefore be expected to have substantial error. A much closer approximation could be obtained by using ten intervals in the range from -3 to 0 and omitting the range from -10 to -3 entirely. Alternatively, we could increase the total number of intervals from ten to, say, 50.

EXERCISES

9.13 Using the trapezoidal rule, write a BASIC program to compute and print a table of approximate values of

$$\int_a^b xe^{kx}\,dx$$

The table should contain the approximate values computed when the number of intervals is 2, 4, 6, 8, and 10. Each entry in the table should be accompanied by the percent error between the approximate and exact values. The user enters values for a, b, and k.

Run your program for

(a) $a=1$, $b=4$, $k=1$
(b) $a=1$, $b=4$, $k=-1$

9.14 The probability that a normally distributed random variable x having mean μ and standard deviation σ has a value in the interval between a and b is given by

$$P(a \leq x \leq b) = \frac{1}{\sqrt{2\pi}\,\sigma} \int_a^b e^{-1/2[(x-\mu)/\sigma]^2} dx$$

This integral cannot be evaluated in closed form.

Write a BASIC program that can be used to print a table of approximate values of $P(0 \leq x \leq b)$ for $b = 0.2, 0.4, 0.6, \ldots, 3$, when $\mu = 0$. The user enters the value of σ. Use Simpson's rule and $n = 20$ intervals.

Run your program for each of the following: (a) $\sigma = 0.1$, (b) $\sigma = 1$, and (c) $\sigma = 10$.

9.15 Write a BASIC program to compare the trapezoidal and Simpson's rules by determining the number of intervals required to compute a value of

$$\int_1^b \frac{dx}{x}$$

within 0.1 percent of the exact value. The user enters the value of b. The program should print the number of intervals required to meet the accuracy criterion for each rule.

Run your program for

(a) $b = 1.5$
(b) $b = 10$

EXERCISES FOR ARCHITECTURAL/CIVIL/CONSTRUCTION TECHNOLOGY

9.16 Figure 9.33 shows a beam supported at each end and a load that increases linearly from one end to the other. The load increases at the rate of k lbs/ft, and the total length of the beam is l feet.

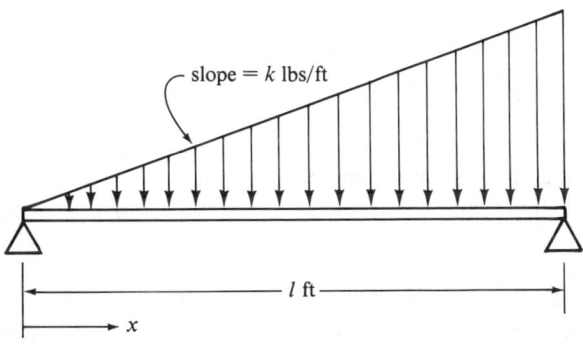

FIGURE 9.33 Exercise 9.16

The bending moment M is given by

$$M = \frac{kl^2 x}{6}\left(1 - \frac{x^2}{l^2}\right)$$

The shear force V is found by differentiating M with respect to x:

$$V = \frac{dM}{dx}$$

Write a BASIC program to perform the following:

(1) Prompt the user to enter values for k in pounds per foot and l in feet.
(2) Compute and print a table of values of M and V for $x = 0, 0.1l, 0.2l, \ldots, l$.
(3) Compute and print the value of x at which the maximum moment occurs and the value of the maximum moment.
(4) Compute and print the value of the maximum shear force and the value of x at which it occurs. (Find the second derivative of M and equate it to zero.)

Run your program for

(a) $k = 200$ lbs/ft, $l = 10$ ft
(b) $k = 87.42$ lbs/ft, $l = 21.3$ ft

9.17 When the origin of an x-y coordinate system is set at the beginning of a certain highway curve, the equation for the curve is approximated by the elliptical function

$$y = b\sqrt{1 - ax^2}$$

The *radius of curvature* ρ of the curve is

$$\rho = \frac{\{1 + (dy/dx)^2\}^{1/2}}{d^2y/dx^2}$$

Write a BASIC program that prompts the user to enter values for a and b and then prints a table of values for ρ at values of x equal to $0, 0.1/\sqrt{a}, 0.2/\sqrt{a}, \ldots, 1/\sqrt{a}$.

Run your program for $b = 200$ and $a = 10^{-4}$.

9.18 The moment of inertia I of a right triangle with respect to its vertical leg is

$$I = \int_0^a y^2\left(b - \frac{b}{a}y\right)dy$$

where a = the length of the horizontal leg, and
b = the length of the vertical leg.

(See Figure 9.34.)

Write a BASIC program that prompts the user to enter values for a and b and then computes and prints the moment of inertia. Run your program for $a = 3.725$ and $b = 5.091$.

9.19 By using a computer program designed to fit coordinate data to a curve, a surveyor found that one boundary of a certain plot of land is approximated by the equation

$$y = 82{,}680 e^{-.5(x - 120)^2}$$

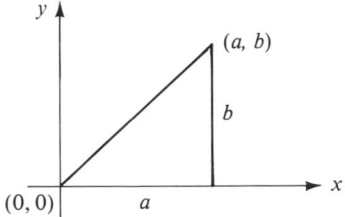

FIGURE 9.34 Exercise 9.18

where x and y are given in feet. The other boundaries of the property include a river and two parallel roads. (See Figure 9.35.) The origin of the coordinate system is the intersection of Peanut Road and the river. The distance between Peanut Road and Route 10 is 267 feet.

Write a BASIC program to compute the area of the property, using the trapezoidal rule and again using Simpson's rule, each with 20 intervals. Print the results and the percent difference between them, using the Simpson's-rule result as your reference

$$\text{percent difference} = \frac{|A_S - A_T|}{A_S} \times 100\%$$

where $A_S =$ the Simpson's-rule result and $A_T =$ the trapezoidal rule result.

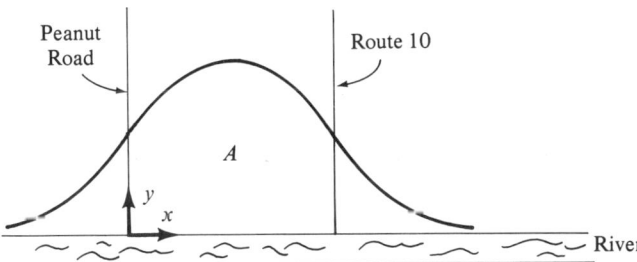

FIGURE 9.35 Exercise 9.19

EXERCISES FOR ELECTRICAL/ELECTRONICS/COMPUTER TECHNOLOGY

9.20 The voltage across an inductor having inductance L is given by

$$v = L\frac{di}{dt}$$

where L is the inductance in henries. (See Figure 9.36.)

Write a BASIC program to determine the maximum (positive) voltage across an inductor when the current is

$$i = Ae^{-at}\cos(\omega t + \theta) \text{ amps}$$

The user enters values for A, a, ω, θ, and the value of L in henries. Run your program for $A = 10$, $a = 2$, $\omega = 15$ rad/sec, $\theta = 45°$, and $L = 20$ mH.

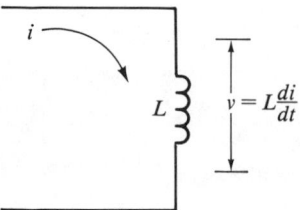

FIGURE 9.36 Exercise 9.20

9.21 Electrical current is, by definition, the rate of flow of electrical charge Q. Mathematically,

$$i = \frac{dQ}{dt}$$

Therefore,

$$\frac{di}{dt} = \frac{d^2Q}{dt^2}$$

Given an expression for the amount of charge at a particular point in a circuit as a function of time, $Q(t)$, you can therefore find the current and the rate of change of the current in the circuit. You can also find the time at which the *maximum rate of change of current* occurs by solving the equation

$$\frac{d^3Q}{dt^3} = 0$$

Given that $Q(t) = te^{-at}$, write a BASIC program that prompts the user to enter a value for a and then computes and prints the time at which di/dt is maximum and the values of Q, i, and di/dt at that time. Run your program for $a = 0.2$.

9.22 The voltage across a capacitor C at time t_0 is

$$v = \frac{1}{C}\int_0^{t_0} i(t)dt + v(0)$$

where C is the capacitance in farads, and $v(0)$ is the initial voltage on the capacitor (the voltage at $t=0$). (See Figure 9.37.)

$$v = \frac{1}{C}\int_0^{t_0} i(t)dt + v(0)$$

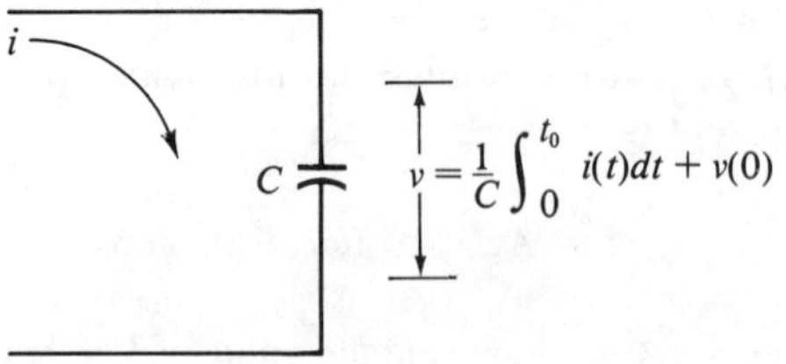

FIGURE 9.37 Exercise 9.22

Write a BASIC program to compute and print a table of values of v when $i(t) = Ae^{-at}\sin\omega t$ at 41 values of t from $t=0$ to $t=5/a$. The user enters values of A, a, C and ω.

Run your program for $A=10$, $a=2$, $\omega=150$ rad/sec, and $C=250\mu F$.

9.23 The *RMS* (or *effective*) *value* of a periodic voltage is

$$V_{\text{RMS}} = \sqrt{\frac{1}{T}\int_0^T v(t)dt}$$

where T is the period of the periodic voltage $v(t)$.

Write a BASIC program to compute and print the RMS value of the voltage

$$v(t) = \{A_1 + A_2\sin(\omega t + \theta)\}\{B_1 + B_2\sin(\omega t + \phi)\}$$

using both the trapezoidal and Simpson's rule and $n=12$ intervals. The period of $v(t)$ is π/ω. The user enters values for A_1, A_2, B_1, B_2, ω, θ, and ϕ. The program should also print the percent difference between the values obtained from Simpson's rule and the trapezoidal rule, using the Simpson's rule result as the reference

$$\text{percent difference} = \frac{|V_S - V_T|}{V_S} \times 100\%$$

where $V_S =$ Simpson's rule value and $V_T =$ trapezoidal rule value.

Run your program for the following combinations of input:

(a) $A_1=0$, $A_2=10$, $B_1=0$, $B_2=5$, $\omega=10$ rad/sec, $\theta=0$, $\phi=0$
(b) $A_1=10$, $A_2=5$, $B_1=2$, $B_2=6$, $\omega=377$ rad/sec, $\theta=30°$, $\phi=-45°$

EXERCISES FOR INDUSTRIAL/MANUFACTURING/PRODUCTION TECHNOLOGY

9.24 The angular rotation $\theta(t)$ of the arm of an industrial robot as it sweeps periodically through the same trajectory is given by

$$\theta(t) = (A_1 + A_2\sin Ct)(B_1 + B_2\sin Ct) \text{ radians}$$

Calibration adjustments made by an operator determine the values of the constants A_1, A_2, B_1, and B_2, and the speed of operation determines the constant C. The angular velocity ω of the arm is given by

$$\omega = \frac{d\theta}{dt} \text{ rad/sec}$$

and the angular acceleration α is

$$\alpha = \frac{d\omega}{dt} = \frac{d^2\theta}{dt^2} \text{ rad/sec}^2$$

The period T of the rotation is $T = \pi/C$ seconds.

Write a BASIC program that prompts the user to enter values for A_1, A_2, B_1, B_2, and C and then prints a table of values of θ, ω, and α at 21 times between $t=0$ and $t=T$, inclusive. Run your program for the following values: $A_1=1$, $A_2=2$, $B_1=0.5$, $B_2=3$, and $C=10$.

9.25 The profit P per unit produced in a certain manufacturing process is low when the rate R at which units are produced is too small and also low when the rate is too large. An approximate equation relating profit per unit to rate of production was determined to be

$$P = aRe^{-(bR^2 + 1)}$$

where a and b are constants determined by the particular product being manufactured.

Write a BASIC program that prompts the user to enter values for a and b and then computes and prints the value of R required for maximum profit per unit. (Note that only positive values of R are meaningful.) The program should also print the maximum unit profit. Run your program for $a = 20$ and $b = 0.02$.

9.26 The probability that the lifetime L of a certain machine will be less than H hours is given by the exponential probability distribution function

$$P_\beta(L < H) = \int_0^H \frac{1}{\beta} e^{-x/\beta} dx$$

where β is the mean (average) lifetime of similar machines. If a production process depends on the continued operation of n machines having average lifetimes of $\beta_1, \beta_2, \ldots, \beta_n$, the probability that the process will operate for less than H hours is

$$\left[P_{\beta_1}(L < H)\right]\left[P_{\beta_2}(L < H)\right] \cdots \left[P_{\beta_n}(L < H)\right]$$

Write a BASIC program that prompts the user to enter the average lifetimes $\beta_1, \beta_2, \beta_3$ of three machines used in a production process and then prints the probability that the process will operate for less than H hours, where H is supplied by the user.

Run your program for the following combinations of input:

(a) $\beta_1 = 110$ hrs, $\beta_2 = 175$ hrs, $\beta_3 = 205$ hrs, $H = 75$ hrs
(b) $\beta_1 = 110$ hrs, $\beta_2 = 175$ hrs, $\beta_3 = 205$ hrs, $H = 750$ hrs

9.27 The probability that the diameter D of a manufactured machine shaft will be between D_1 and D_2 inches is

$$P(D_1 < D < D_2) = \frac{1}{\sqrt{2\pi}\sigma} \int_{D_1}^{D_2} e^{-1/2[(D - \mu)/\sigma]^2} dD$$

where μ is the nominal (average) diameter and σ is the standard deviation of the diameters.

Write a BASIC program to compute $P(D_1 < D < D_2)$ using the trapezoidal rule and again using Simpson's rule, each for $n = 8$ intervals. The user enters values for μ, σ, D_1, and D_2. The program should print the probabilities computed by each method and the percent difference between them, using the Simpson's rule probability as the reference:

$$\text{percent difference} = \frac{|P_S - P_T|}{P_S} \times 100\%$$

where P_S = Simpson's-method probability and P_T = trapezoidal-method probability.

Run your program for the following combinations of input:

(a) $\mu = 1.00$ inch, $\sigma = 1$, $D_1 = 0.9$ inches, $D_2 = 1.1$ inches
(b) same as (a), except that $\sigma = 0.1$
(c) same as (a), except that $\sigma = 10$

EXERCISES FOR MECHANICAL TECHNOLOGY

9.28 The force F developed by a dashpot having damping factor B is given by

$$F = B\frac{dx}{dt}$$

where x is the displacement of its movable element. (See Figure 9.38.)

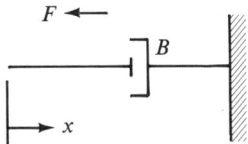

FIGURE 9.38 Exercise 9.28

Given that $x(t) = Ae^{-at}\cos(\omega t + \theta)$, write a BASIC program to find the maximum force developed by the dashpot. The user enters values for A, a, ω, θ, and the damping constant B. If x is expressed in feet, B in pounds per foot per second, and ω in radians per second, then F is in pounds.

Run your program for $A = 10$, $a = 2$, $\omega = 15$ rad/sec, $B = 0.04$ lb/ft/sec, and $\theta = -45°$.

9.29 If the horizontal and vertical displacements of a body are given in *parametric* form—that is, as individual functions of time $x(t)$ and $y(t)$—the horizontal and vertical velocities v_x and v_y are given by

$$v_x(t) = \frac{dx(t)}{dt}$$

$$v_y(t) = \frac{dy(t)}{dt}$$

The *resultant* velocity v has magnitude

$$v = \sqrt{v_x^2 + v_y^2}$$

and angle θ, measured counterclockwise from the positive x axis, given by

$$\theta = \arctan(v_y/v_x)$$

(See Figure 9.39.)

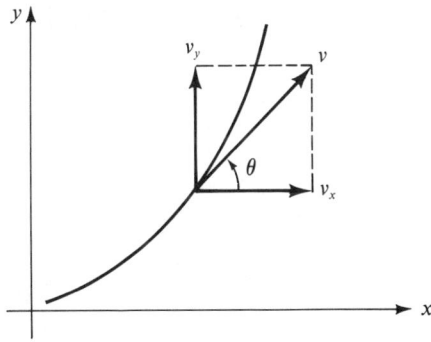

FIGURE 9.39 Exercise 9.29

Given that $x(t) = Ate^{-at}$ and $y(t) = Bte^{-bt}$, write a BASIC program to compute and print the maximum value of $v_x(t)$, the maximum value of $v_y(t)$, and the resultant velocity (magnitude and angle) at *each* of the values of time where v_x is maximum and v_y is maximum. The user enters values for A, B, a, and b.

Run your program for $A = 12$, $B = 9$, $a = 0.2$, $b = 0.1$.

9.30 If a force $F(t)$ varies periodically with time, the *average* force F_{AVG} is defined by

$$F_{AVG} = \frac{1}{T}\int_0^T F(t)dt$$

where T is the *period* of the force (the time required to complete one cycle of the periodic variation).

Given that $F(t) = \{A_1\sin(\omega t + \theta)\}A_2\sin\omega t$, write a BASIC program to compute and print the average force, after the user enters values for A_1, A_2, ω, and θ. The period of $F(t)$ is π/ω.

Run your program for the following combinations of input:

(a) $A_1 = 5$, $A_2 = 10$, $\omega = 40$ rad/sec, $\theta = 0$
(b) $A_1 = 5$, $A_2 = 10$, $\omega = 40$ rad/sec, $\theta = \pi/2$ radians

9.31 The acceleration of a certain body is given by

$$a(t) = Ate^{-t^2/b} \text{ ft/sec}^2$$

Therefore, its velocity at time t_0 is

$$v(t_0) = \int_0^{t_0} Ate^{-t^2/b}dt + v(0)$$

where $v(0)$ is the initial velocity.

Write a BASIC program to compute and print $v(t_0)$, using the trapezoidal rule and again using Simpson's rule, each with $n = 14$ intervals. The user enters values for A, b, $v(0)$, and t_0. Your program should print $v_T(t_0)$: the velocity computed using the trapezoidal rule; $v_S(t_0)$: the velocity computed using Simpson's rule; and the percent difference between these values.

$$\text{percent difference} = \frac{|v_S(t_0) - v_T(t_0)|}{v_S(t_0)} \times 100\%$$

Run your program for $A = 120$, $b = 16$, $t_0 = 1$, and $v(0) = 0$.

GLOSSARY

This glossary is a dictionary of terms frequently encountered in the fields of computer science and computer technology. Many of these terms are not mentioned or discussed elsewhere in the book. Readers who want to further their knowledge of computers and computing, read and understand popular periodicals devoted to these subjects, or who are interested in purchasing their own computer system will find this summary of definitions a useful reference.

Absolute value. 1. The value of a number, without regard to its algebraic sign. The mathematical symbol for the absolute value of the number N is $|N|$. For example, $|3.7| = 3.7$ and $|-12| = 12$. 2. The absolute value of a *complex* number $|a + jb|$ is the positive square root of the sum of the squares of its real and imaginary parts: $|a + jb| = +\sqrt{a^2 + b^2}$. (In both cases, the term *magnitude* is also used.)

Account number. A number assigned to a programmer—usually in a large, time-shared computer system—to authorize access to a computer. The programmer enters the account number from a keyboard as part of the *log on* procedure, and the computer keeps track of the total computer time used on each account. This time record is then used for reporting or billing purposes.

Acoustic coupler. An electronic device used for transmitting data between a computer and a peripheral device (usually a terminal) by telephone. In this process, tones are transmitted through a telephone line (see **FSK**), and an acoustic coupler is used to convert the audible tones to electrical signals, and vice versa. The coupler is designed so that the ear- and mouthpiece of an ordinary telephone can be placed into specially designed circular cups. The acoustic coupler is an example of an *interface* device.

Address. A number assigned to each location where a word is stored in computer memory. A memory address is like a postal address, or box number, and should not be confused with the contents (data) stored therein. The computer generates an address and transmits it to memory every time it reads a word from memory or writes a word into memory. In some systems, peripheral devices are also assigned addresses, so the computer can distinguish one device from another.

Algorithm. A procedure to be followed to achieve a certain programming goal. A series of steps that lead to a certain result. An algorithm is most often expressed in a set of English-language statements and then converted to computer-language statements, for incorporation into a program. As a simple example, the algorithm for squaring a number would be: (1) Obtain the number to be squared. (2) Multiply the number by itself.

Alphanumeric. Pertaining to letters, numbers, and usually also special symbols. A keyboard has keys for entering alphanumeric data (letters, decimal numbers, and special symbols) into the computer.

AND. 1. A logical operator used in programming to signify that both of two conditions must be true before a specified action can occur. For example, in the BASIC statement: IF X < 4 AND Y > 10 THEN 200, both of the conditions "X is less than 4" *and* "Y is greater than 10" must be true, for a branch to statement 200 to occur. 2. In binary number processing and Boolean algebra, an operator that produces the result binary 1, if *all* bits it operates on are binary 1; and produces binary 0 otherwise. For example, 1 AND 1 = 1, whereas 1 AND 0 = 0. 3. In logic circuitry, a type of electronic gate whose output is binary 1, if and only if all inputs are binary 1; and whose output is binary 0 otherwise.

ANSI (American National Standards Institute). An organization formed to develop manufacturing and industrial standards, including standard forms or versions of programming languages.

APL (from "A Programming Language"). A computer language that uses a large number of mathematical operators, each represented by a special symbol. Special APL keyboards containing these symbols are available for writing APL programs. The language was developed primarily for scientific programming.

Arc. A prefix used with a trigonometric function. An arc trigonometric function produces the angle corresponding to a particular value of the function itself. For example, sin(30°) = 0.5, so arcsin(0.5) = 30°. Also called an *inverse* and written $\sin^{-1}(0.5) = 30°$.

Architecture. In connection with computer hardware, the structure and design of the fundamental electronic components from which a computer is constructed. Also refers to the way that these components are interconnected and the sequences in which they are actuated to process machine-language instructions.

Argument. The quantity operated on by a function to produce a numeric result. For example, when the argument of the sine function is 30°, the result is sin(30°) = 0.5. The argument is the quantity enclosed within parentheses following the symbol representing a function. In BASIC, arguments can be numbers, variable names, or expressions, as in LET X = ABS(X − 3). The more general term *operand* is often used to mean argument, when used in connection with functions.

Array. A set of data values arranged like a list or table. Each value has a certain position in the array and can be referenced by specifying the position it occupies. (See also **Matrix** and **Vector**.)

ASCII (American Standard Code for Information Interchange). A code in which alphanumeric data, including special symbols, is represented by a set of standard eight-bit words. This code is commonly used for transmitting data between a computer and a peripheral device, and it ensures compatibility between equipment produced by different manufacturers. The acronym is pronounced "Askee."

Assembler. A computer program that converts an assembly-language program into machine language, for execution by the computer.

Assembly language. A computer language using alphanumeric abbreviations (see **Mnemonic**) to represent the most fundamental instructions a certain computer can execute. Assembly language uses English-language characters to represent machine-language instructions and is therefore said to be one level higher than machine language. Assembly language is a machine-dependent language.

Asynchronous. The opposite of synchronous. A sequence of events having no regular pattern of occurrence. Not regulated or controlled in a predictable or systematic way. In asynchronous data transmission, bits are generated without timing constraints (independent of the clock); bits are not required to occur at regular time intervals.

Auxiliary memory. Memory other than the main, internal memory of a computer. Examples include magnetic disk and magnetic tape memory. Auxiliary memory is provided by *peripheral devices* and is generally slower than main memory—that is, it takes longer to store or retrieve data.

Base. In connection with a number system, a decimal number whose various powers can be summed to represent any desired magnitude (same as *radix*). The base of the decimal number system is 10, since any decimal number can be represented as a sum of powers of 10. For example: $527.3 = 5 \times 10^2 + 2 \times 10^1 + 7 \times 10^0 + 3 \times 10^{-1}$. The base of the binary number system is 2, the base of the octal number system is 8, and the base of the hexadecimal number system is 16.

BASIC. Acronym for Beginner's All-purpose Symbolic Instruction Code. A programming language, developed at Dartmouth College, that is easy to learn and easy to use.

BASIC-PLUS. A powerful version of BASIC that includes statements and statement variations not generally available in microcomputer versions of the language. It is used on a large, time-shared computer system and has, for example, statements for performing matrix (MAT) operations.

Batch processing. The operations performed by a computer when a program is submitted (entered) in its entirety (for example, on a deck of punched cards). Results of program runs are retrieved later in the form of printouts (hard copy). By contrast, in time-shared operation the programmer can enter programs, view results, and *interact* directly with the computer during a program run, by use of a remote terminal. Contrast with **Interactive**.

Baud. A unit used to specify the rate or speed of data transmission between a computer and a peripheral device. The term *baud rate* is also commonly used. For practical purposes, one baud is a rate of one bit per second.

BCD (binary-coded decimal). A means for representing decimal numbers by encoding each decimal digit with a single binary code group. For example, the most common (8-4-2-1) BCD code for 27 is 0010 0111. Contrast this with the pure binary number equivalent to 27: 1011. (There are also other kinds of BCD codes).

Binary number system. The number system (base 2) in which all digits are either 0 or 1. This number system is widely used in design and analysis of computer hardware, since computer components (such as logic gates) operate by changing back and forth between the two states ON and OFF. Machine-language programs are written in binary code.

Bit. Abbreviation for *binary digit*. A bit can be either 0 or 1 (see **Binary number system**).

Boolean algebra. A symbolic mathematical system used to represent logical relationships. Named after mathematician George Boole (1815–1864), who first proposed using symbols to express logical thought processes. The system is widely applied to the design and analysis of computer hardware, especially combinations of logic gates. Boolean algebra involves the manipulation of binary variables (those that can only have values of 0 or 1) and logical operators, such as AND, OR, and INVERT.

Bootstrap. A procedure used by a computer to initiate loading a program. A bootstrap *loader* is a program or subroutine, often stored in read-only memory, that initiates the reading of a program that causes the rest of itself to be loaded into the computer.

Branch (also called a *jump*). The execution of a statement that is not the next one in the sequence in which statements appear in a program. Unless a branch occurs, the computer will execute statements in the same order as they are written in the program. A branch statement causes the computer to skip over a number of statements to find the next one to be executed.

Breakpoint. A point in the sequence of program execution where the computer is instructed to stop, so that a programmer can examine intermediate results and take appropriate action based on those results. This is used in program *debugging*, especially in assembly-language programs.

Bug. 1. A programming error that prevents a program from achieving a desired goal. This can be an error in *syntax* (in this case, an *error message* is generated by the computer) or an error in programming logic. (See also **Run-time error**.)   2. A defect in the computer hardware.

Byte. Eight bits. In many microcomputer systems, a data *word* is equivalent to an eight-bit byte. In so-called 16-bit microcomputers or microprocessors, a word is composed of two bytes.

Calculator mode (also called direct or immediate mode). A way of using BASIC to perform calculations, without writing a program or using statement numbers. Interactive computation using the keyword PRINT. For example, to compute $40(5-3)$ in the calculator mode, the user need only enter the statement

PRINT 40 * (5 − 3)

The result, 80, is displayed immediately.

Call. A branch to the first statement in a *subroutine*. In BASIC, a subroutine call is initiated by the statement GOSUB *sn*, where *sn* is the statement number of the first statement in a subroutine. Also used in connection with user-defined functions.

Card. 1. (Also called an "IBM card.") A rectangular piece of stiff paper, approximately $7\frac{3}{8}'' \times 3\frac{1}{4}''$. Holes are punched in various patterns on the card to represent alphanumeric data or program instructions. Used as a permanent storage medium for data and programs. The *Hollerith* code defines the hole patterns used to represent data and program instructions.   2. A circuit board on which are mounted electronic components such as integrated circuit chips. Cards are inserted in connectors mounted inside a card cage. Cards are commonly used for assembling the hardware components of a computer.

Card punch. A machine used to punch hole patterns in cards. Card-punch operators depress alphanumeric keys, causing holes to be punched in patterns specified by the *Hollerith* code. Another kind of card-punch machine is designed for use as an *output* device for a computer; the results of program runs are automatically punched on a deck of cards.

Card reader. An electronic device designed to convert the punched-hole patterns on a deck of cards to electrical signals that are entered into a computer. A card reader is an example of an *input* device.

Cassette. A magnetic tape cartridge used for storing data and programs. Small, portable tape cassette units are often used with microcomputer systems to provide auxiliary memory.

Cathode ray tube (CRT). The "picture tube" on which the video display appears at a terminal or television monitor. A beam of electrons (cathode rays) is accelerated within the tube and strikes a special phosphor coating that emits light. The beam is deflected and positioned electronically to create images on the screen.

Central Processing Unit (see **CPU**).

Character. Any alphanumeric symbol. Data is in "character form" when a binary code such as ASCII is used to represent each alphanumeric symbol.

Chip. (See also **Integrated circuit**.) A small piece of silicon (containing many transistors and diodes) in an integrated circuit. The term "chip" is used synonymously with integrated circuit. For example, the 7400 chip (integrated circuit) contains four logic gates; each of these gates has two inputs and one output. A microprocessor is an example of a much more complex device fabricated on a single chip.

Clock. An electronic signal consisting of pulses occurring at regular intervals to synchronize hardware operations in a computer. A typical microcomputer clock generates one pulse every microsecond.

Code. 1. A means for representing numbers and/or alphabetic characters. In computer operations, codes are formed by using groups of binary digits (zeros and ones) to represent alphanumeric characters. Examples are BCD and ASCII. 2. Used as a verb, to write a program or to prepare data in a form that can be accepted by a computer.

Color graphics (see also **Graphics**). Visual displays created by a computer program that controls the color(s) of the patterns it generates. Color television sets are typically used as the display medium.

Command. A means (other than a programming statement) for controlling the operation of a computer. Commands can be used to initiate program runs (the RUN command), to interrupt program runs, to store (save) programs in external memory, to load programs from memory, and to perform other tasks that do not directly affect program execution. Commands are typed in from a keyboard, and their formats are machine-dependent.

Comment (see also **Remark**). A statement inserted in a program for the sole purpose of explaining steps or procedures in the program. In BASIC, comments are inserted using REM (remark) statements. Comments are *nonexecutable* statements in the sense that they neither affect the way a program processes other statements nor alter the results of a program run. Comments are printed out as part of a program listing.

Compilation. The act of compiling. (See **Compile** and **Compiler**.)

Compile. To convert a program written in a high-level language (such as BASIC or FORTRAN) into a machine-language program, so it can be executed by a computer. The computer itself does the compiling by executing a program called the *compiler*.

Compiler. A computer program that converts a program written in a programming language to machine language, so it can be executed by the computer. (See also **Source program** and **Object program**.) Contrasted with an interpreter, a compiler converts the entire source to an object program, before execution begins.

Complement (see **Invert**).

Computer-aided design (CAD). The use of computers to solve engineering problems and prepare documents necessary for construction or manufacturing. Computer graphics are widely applied in this field.

Computer language. A means for enabling a human to communicate with a computer. A set of alphanumeric expressions and symbols, with rules for combining them to cause desired operations to be performed in a computer. (See also **Low-level** and **High-level language**.)

Concatenate. To juxtapose or place adjacent to. Often used in connection with strings. For example, if the strings "DOG" and "GONE" are concatenated, the result is "DOGGONE."

Conditional branch (also called conditional *jump*). A branch that occurs in a program only if a specified condition is satisfied. For example, the BASIC conditional branch statement IF X < 0 THEN 200 causes a branch to statement 200 only if the condition X < 0 is satisfied—that is, only if the value of the variable X is negative at the time the statement is executed.

Connector. 1. A symbol used on a flowchart. A number or letter is written inside the symbol, and another corresponding symbol is located somewhere else on the chart and given the same letter or number. The first connector shows an exit point from the chart, and the second one shows the reentry point. Connectors are used to avoid drawing a line between the two points, or when portions of a flowchart appear on different pages. 2. In hardware, a device (such as a socket or plug) used to join electrical circuits together. For example, a card connector makes electrical connections to the components on a printed-circuit card, when the card is inserted into the connector.

Console. The unit in a large computer system where the operating controls, switches, and monitoring devices are located. A computer operator controls the overall computer system from the console.

Control character. A special character used to interrupt, restart, or otherwise alter computer operations. Many keyboards have a *control key* that can be depressed simultaneously with another key to cause a prescribed action to occur.

Core. In large computer systems, the main, internal computer memory. Also called core memory. It consists of small magnetic cores, each of which can store one bit. Most modern computers, including all microcomputers, have a solid-state main memory rather than a core memory. (See also **Magnetic core**.)

Counter variable. A variable whose value is changed in accordance with the number of times a statement or group of statements is executed, particularly in a programming loop. For example, in the BASIC statement FOR I = 1 to 10, the variable I is the counter variable and is automatically increased by one each time the loop is executed.

cpi. Abbreviation for characters per inch.

CPM. Abbreviation for cards per minute.

CP/M. An *operating system* developed for use with microcomputers having an Intel 8080 or Zilog Z-80 microprocessor.

CPU (Central Processing Unit). That part of a computer where arithmetic, logic, and control functions are performed. In a microcomputer, the *microprocessor* serves as the CPU.

Cross assembler (see also **Assembler**). An assembler written so it can be run on one type of computer to create machine code for another type of computer.

CRT (see **Cathode ray tube**).

Current loop. A means of interfacing computer peripherals. The presence or absence of electrical current signifies binary ones or zeros.

Cursor. An illuminated marker (such as a bar, rectangle, or underline) showing where the next character will appear on a video display. Some terminals have cursor-control keys allowing the operator to move the cursor to a position where display of the next character should occur. In some terminals, the cursor is a dot or rectangle that flashes on and off.

Cycle. 1. The interval of time between successive pulses of a clock signal. 2. The time required for a computer to perform a certain kind of operation. For example, a memory cycle is the total time required to read a word from or write a word into memory.

Daisy-wheel printer. A printer that uses a circular plastic or metal wheel containing "fingers" (similar to petals on a daisy) that strike the paper and thus make impressions of characters. The wheel rotates to whatever position is required to print one of the characters it contains. This generally produces a high-quality ("letter quality") printout.

Data. 1. In general, any collection of information, instructions, or numbers. 2. Data is also used to mean *numbers* in contrast to program statements. The word is commonly used both as a singular and plural noun. (It is therefore correct to say either "the data is . . . " or "the data are . . . .")

Data base (also database). The entire collection of data available in computer memory for retrieval of information related to a certain task or for accomplishing a specific programming goal. For example, the data base used in a computerized airline-reservation system includes all data related to schedules, aircraft capacities, occupancies, and so forth.

Data processing. Use of a computer to handle large quantities of data while performing relatively simple computational tasks. Contrast with **Scientific programming**, wherein complex mathematical computations are performed on relatively small amounts of data.

Debounce. To eliminate the effect of "switch bounce." A mechanical switch (or key) makes and breaks contact several times when it is depressed. It therefore generates a short series of electrical pulses that can cause erroneous input to a computer. Debouncing can be accomplished by using either special hardware or a program that ignores the short-duration pulses.

Debug. 1. To locate and correct errors in a program (see **Bug**). 2. To locate and correct hardware faults.

Deck. A set of punched cards. Usually represents either a program, data, or both (see **Card**).

Decrement. To decrease by 1. For example, if the variable X has a value of 13 and it is decremented, its new value is 12.

Dedicated. Designed to perform a specific function and incapable of being altered to perform other functions. For example, a portion of computer memory can be dedicated to the storage of a compiler program and therefore will not be available for general-purpose storage.

Development system. A hardware system with appropriate software, used to aid in the design and development of microcomputer systems. The system is capable of emulating one or more microprocessors and thus permits a designer to check and debug microcomputer programs, interface logic, timing and synchronization, and I/O operations. (See also **Emulator**.)

Digit. One of the symbols 0 through 9 used to represent a number. Some number systems only permit certain digits to be used for number representation. For example, in the octal number system, only the digits 0 through 7 are permitted. In number systems having a base greater than 10, additional symbols are used. For example, the digits in the hexadecimal number system (base 16) are 0 through 9 and A through F.

Digital. Pertains to devices that perform computations and process data using *discrete* number values. By contrast, *analog* devices operate on continuous variables, such as voltage levels, that can have any of an infinite number of values.

Dimension. 1. For an array whose entries are referenced by a single index variable, the total number of entries or data values that can be stored in the array. 2. The number of index variables required to reference any entry in an array. These two usages of the word *dimension* are quite different and, since both definitions are associated with arrays, the two meanings are easily confused. For example, the BASIC array A containing entries A(1) through A(20) has a dimension of 20 in the sense of the first definition, but it is a one-dimensional array in the sense of the second definition. 3. For a matrix, the number of rows and columns—for example, 6×2.

Direct mode (see **Calculator mode**).

Disassembler. A program that converts machine code to assembly language. The machine code is often expressed in hexadecimal rather than binary form, and the disassembler produces assembly-language mnemonics corresponding to the machine code.

Discrete. 1. Having only specific, distinct values, as opposed to continuous (or analog) in nature. For example, the digits 0 through 9 comprise a set of discrete integers: There are no integers between any of these. On the other hand, an infinite number of values exist in the continuum between 0 and 9. 2. In relation to hardware, *discrete* means made of individual electronic components, separately mounted and individually accessible, as opposed to fabricated in an integrated circuit.

Disk. 1. In regard to external memory, a means of auxiliary storage wherein data bits are written and read magnetically on a flat, circular plate (disk). The disk rotates, and magnetic read/write heads move across it to access data. 2. The flat, circular plate itself: the physical storage medium in disk memory. (See also **Floppy disk**.)

Disk drive. The mechanism into which a disk is inserted or upon which a disk is mounted.

Diskette (see **Floppy disk**).

Disk operating system (DOS). A program or group of programs stored on disk and used to control program runs, input/output operations, debugging, and similar tasks. The system was originally developed by IBM for its System 360 equipment.

Documentation. 1. Used in connection with programs, all information pertaining to the purpose of a program, how it is used, the logical sequences or algorithms it incorporates, input/output requirements, and any other facts relevant to its use. Documentation includes comments or remarks in the program listing and flowcharts to show the logic being implemented. 2. Used in connection with hardware, the schematic diagrams, logic diagrams, troubleshooting guides, operating manuals, and similar information pertaining to the operation and maintenance of a system.

DOS (see **Disk operating system**).

Dot-matrix printer. A printer that creates images of characters by striking a pattern of dots on paper. The matrix of dots available for creating a pattern is typically 5×7 or 5×9. Dot-matrix printers produce a lower-quality image than daisy-wheel printers but are less expensive and can be used to synthesize special symbols.

Double-density disk. A disk that can store twice as much data as a conventional disk of the same dimensions. A typical double-density floppy disk can store 256 characters per sector.

Down. Malfunctioning or not in operation. For example, a computer system could be "down" for routine maintenance or because of a hardware failure.

Dummy variable. A variable used only to define what a function does. For example, a function f that squares numbers can be defined using the dummy variable x, by writing $f(x) = x^2$. When the function is used, the quantity to be squared need not be x but can be any other variable or number.

Dump. To obtain a readout of the contents of all memory locations. The term is also used when referring to the readout itself, as a *memory dump*.

Duplex (see **Full duplex** and **Half duplex**).

Editor, text (see **Text editor**).

E format. A notation similar to scientific notation, used to represent numbers. The digits following the letter E signify the power of 10 that multiplies the digits preceding E. The power can be negative. For example, $17E2 = 1700$ and $395E-2 = 3.95$.

Emulator. A program or hardware system that allows programs written for one type of computer to be run on another type. An emulator accepts programs just as they are written to be run on the original system and produces the same results. (See also **Development system**.)

Error message. A message generated by a computer and printed out or displayed on a video terminal, to notify the user of a programming error. Error messages are caused by either syntax errors or run-time errors.

Error, run-time (see **Run-time error**).

Exclusive OR. 1. In binary number processing and Boolean algebra, a logic operator that produces binary 1, if one and only one of the two bits it operates on is binary 1, and produces 0, otherwise. The symbol $\oplus$ is used to denote the exclusive OR operation. For example, $1 \oplus 0 = 1$ and $1 \oplus 1 = 0$. 2. In relation to hardware, a type of electronic gate (or combination of gates) whose output is binary 1, if one and only one of its two inputs is binary 1, and whose output is binary 0, otherwise.

Execute. A computer *executes* a programming statement by performing whatever action(s) that statement is meant to cause. A BASIC program is executed in response to a RUN command, which causes the computer to execute each statement in the program.

Exponent. The power to which a number is raised. For example, in the numbers 2^6 and $12^{-1/3}$, the exponents are 6 and $-1/3$, respectively. In scientific notation, the exponent is an integer power of 10, as for example in 1.76×10^4.

Exponential (E) format (see **E format**).

Exponentiation. Raising a number to a power. Exponentiation is a mathematical operator having a variety of symbolic representations in different programming languages and in different versions of the same language. Examples include $**$ and $\uparrow$. Thus 5^3 might be expressed in one version of BASIC as $5 ** 3$; and in another, as $5 \uparrow 3$.

External memory (also called external *storage*). Memory made available to a computer system through peripheral devices, such as magnetic tape and disk units. External memory is slower than internal, or main, memory, since it takes longer to store data in or retrieve data from it. (See also **Mass storage**.)

Field. A portion of a program line or line of data, usually identified by the column numbers it occupies. Fields are specified to reserve certain part(s) of a line for specific kinds of information.

File. A collection of either program statements, data, or both. Any collection of related *records* that together comprise an identifiable entity. The whole file is treated as a single unit for storage or retrieval purposes. A programmer assigns a name to the file, so it can be saved in memory and recalled by name when needed.

Firmware. A program stored permanently in a computer system. Software that is *hard-wired* in a computer or stored in *read-only memory* and thus cannot be altered or removed.

Fixed point. A means of number representation wherein the decimal point has a fixed position, and number magnitudes are expressed relative to that position. Contrast with **Floating point**.

Floating point. A means of number representation wherein magnitudes are expressed by specifying a value and an exponent. Within a computer, the exponent bits occupy a certain portion of a data *word* and represent the power of 2 by which the value bits are multiplied to obtain the magnitude of the number. Floating-point representation permits the expression of a much larger range of numbers than does fixed-point representation.

Floppy disk (also called diskette). A small, flexible, plastic disk used for external memory in microcomputer systems. (See also **Disk**.) A floppy disk is a smaller and less expensive version of the "hard" disk unit used in large computer systems.

Flowchart. A drawing that shows the logical processes, branches, and computations performed in a program. Special symbols are used to show each different kind of action that a program can take, and lines are drawn between the symbols to show the sequence in which they are performed. This is a valuable tool in program design and an important part of program *documentation*.

Formatted disk (see also **Hard-sectored** and **Soft-sectored disk**). A disk whose sectors are identified either by codes recorded on the disk or holes punched into it. Formatting is necessary to specify the locations where data is stored.

FORTRAN. (derived from "formula translation"). A high-level programming language. This was one of the earliest high-level languages developed and has been widely used for both scientific programming and data processing. The formats required for writing statements and performing I/O operations in FORTRAN are much more rigid than those of BASIC. Later versions include FORTRAN IV and FORTRAN '77.

Friendly. A computer or programming language that is easy to use (also said to be "forgiving"). Tolerates variations rather than requiring strict adherence to a set of rules. Generates helpful prompting messages and understandable error messages to assist the user.

FSK. Abbreviation for frequency-shift keying. A method of binary data communication wherein bits are transmitted by using one tone (frequency) to represent a binary 0 and another tone to reprsent a binary 1. This method is widely used for data communication using telephone lines between a computer and remote terminals.

Full duplex. Describes a type of data communication in which information can be transmitted and received *simultaneously*. Voice communication in ordinary telephone service is an example of full duplex communication, since a caller can speak and hear at the same time. Contrast with **Half duplex**.

Function. 1. Mathematically, a (numeric) *function* is a rule or procedure applied to a number to produce another number. A function can be defined by giving the equation that expresses the mathematical operations (the rule) that must be performed on the *argument* of the function (the number to which it is applied) to produce the result. For example, a function $f(x)$ can be defined by $f(x) = x^2 - 4$, so if $x = 3$, then $f(3) = 3^2 - 4 = 5$. Many functions cannot be defined in equation form and are thus given special names; one example is the sine function, $f(x) = \sin(x)$. 2. In a programming language, a function can be defined by writing the appropriate statement(s), in a specified format, expressing the mathematical operations to be performed on any number to which the function is applied, to produce a new number. The function is assigned a name that can be used wherever the programmer wishes the operations it represents to be performed in a program. In BASIC, a function is defined using the DEF statement. (See also **Library function**.)

Gate (also called logic gate). An electronic device in computer hardware, used to perform a logical operation such as *AND, OR,* or *EXCLUSIVE OR*. The inputs to a gate are voltages representing binary zeros and ones, and the output is a similar voltage that depends on the logic implemented by the gate. Gates are the fundamental building blocks of computers, and a single integrated circuit chip can contain thousands of them.

General-purpose computer. A computer that can be programmed to perform a variety of computational tasks. Contrast with **Special-purpose computer**.

Giga (G). A Greek prefix used with numbers to mean 10^9. For example, 4 gigabits $= 4 \times 10^9$ bits.

GIGO. An acronym for "Garbage In, Garbage Out." This motto means that erroneous programming techniques or data will produce erroneous results.

Graphics. A specialty field in computer science concerned with the use of a computer to generate diagrams, patterns, graphs, etc., as opposed to alphanumeric displays. (See also **Color graphics**.)

Half duplex. Describes a type of data communication wherein information can be either transmitted or received—but not both at the same time. Contrast with **Full duplex**.

Hard copy. Program listings, program results, or data produced on paper or in another permanent form. Printers produce hard copy. Contrast with the display produced on a video terminal, which is lost when the terminal is shut down.

Hard-sectored disk (see also **Sector** and **Formatted disk**). A disk whose sectors are identified by permanently punched holes, as contrasted with **Soft-sectored disk**.

Hardware. The electronic devices, wiring, and other physical components used to construct a computer system. Contrast with **Software**.

Hard-wired. Permanently connected in a system's hardware. Usually refers to a control function or programming procedure that is created by using hardware and that might otherwise be performed using software.

Hexadecimal (hex). Pertaining to the base-16 number system, which represents numbers using the 16 symbols 0 through 9 and A through F. The symbols A through F carry weights 10 through 15. For example, A3 in hex is equivalent to $10 \times 16^1 + 3 \times 16^0 = 163$ in decimal. Hex is widely used in assembly-language programming, because it can be easily converted to equivalent binary (machine) code.

Hierarchy (of operators). The order in which mathematical and/or logic operations are performed by a computer, when evaluating an expression involving two or more of them. For example, in BASIC, multiplication takes precedence over addition in the hierarchy, so the expression $2 * 3 + 4$ is evaluated as $6 + 4 = 10$ rather than $2(7) = 14$.

High-level language. A computing language whose statements are similar to English-language statements. BASIC and FORTRAN are examples. Programmers having no knowledge of computer hardware or internal computer operations can learn to program using a high-level language. Contrast with a **Low-level language**, such as machine or assembly language.

Hollerith code. The code used for punching holes in cards to represent alphanumeric characters.

IC (see **Integrated circuit**).

Identity matrix. A square matrix having the number 1 in all diagonal positions, and zeros elsewhere.

Immediate mode (see **Calculator mode**).

Increment. To increase by one. For example, if the variable X has a value of 12 and it is incremented, it then has a new value of 13.

Index. In an array, the number that specifies the position occupied by a particular entry in the array. (See also **Index variable**.)

Index variable. An integer-valued variable whose value specifies the position of an entry in an array. In BASIC, the name of an index variable is enclosed in parentheses following the name of the array whose entries it identifies. For example, if the index variable I has a value of 7, then A(I) refers to entry number 7 in array A.

Infinite loop (see also **Loop**). A set of program statements repeatedly executed an indefinite number of times. A loop that either contains no provisions for termination or has a terminating condition that can never be satisfed. Once program execution is in an infinite loop, it can only be terminated by external intervention—for example, by an "escape" or "break" command. A simple example of an infinite loop is the BASIC statement

10 GOTO 10

Programmers must be careful to avoid constructing infinite loops inadvertently, particularly when doing *iterative* programming.

Initialize.   To assign a value to a variable before using its name in an expression. In many versions of BASIC, all variables are assumed to have a value of zero unless they are initialized—that is, unless they are given another value by assignment statements occurring before their names are used. For example, the BASIC statement LET Y = X + 5 would assign Y the value of 5, if no previous statement had assigned a value to X. On the other hand, if X had been previously initialized by a statement such as LET X = 2, Y would then be assigned the value of 7.

Input.   To supply data (in its most general sense) *to* a computer—more precisely, to the CPU of a computer. The word *input* is also used as a noun to mean the data supplied to the computer. The computer is always the point of reference when speaking of input or output.

Input device.   Any device that supplies input *to* a computer. One example is a terminal, which is used to input programming statements, data, and commands to a computer. A terminal also serves as an *output* device, since it receives and displays data from the computer.

Instruction.   1. The fundamental component of a computer program. A coded sequence of bits or alphanumeric characters prescribing a certain operation for the computer to perform. In machine- and assembly-language programs, instructions prescribe very elementary operations, such as "add" or "read."   2. Often used synonymously with "statement," although the latter is a more general term and does not necessarily prescribe an executable operation.

Integer.   A "whole" number, such as 0, 1, 2, 3, or 4. An integer can be positive, negative or zero. Also used as an adjective to mean "having no fractional part."

Integer BASIC.   A version of BASIC without the capability of performing mathematical operations on numbers other than integers. In integer BASIC, all numeric results are expressed as integers. This version is obviously not very useful for scientific programming, but it requires little memory space for its compiler or interpreter.

Integrated circuit (IC) (see also **Chip**).   A small electronic device (the size of a fingernail or smaller) containing many interconnected logic *gates* or similar components. ICs are classified by "family," depending on their design. Two families widely used in computer hardware are "TTL" and "MOS"; the latter is used in the construction of *microprocessors*.

Interactive.   In connection with a programming language, permitting spontaneous communication with a computer before, during, and after a program run. BASIC is an example of an interactive language, since it permits the user to observe the results of a program run, interrupt and modify program execution, perform editing, and input data from a terminal. Contrast with **Batch processing**.

Interface.   Hardware required to make one electronic component or system compatible with another. Interface circuitry is often required to change the format or characteristics of electrical signals representing binary data, so that two devices of different design can communicate with each other. For example, data generated by a computer in *parallel* form must be transmitted to some peripheral devices in *serial* form. Interface is also used as a verb, as " . . . the hardware necessary *to interface* a computer with a terminal."

Interpreter.   A computer program designed to convert program statements written in a high-level language to machine-language statements that can be executed by the computer. In contrast to a *compiler,* an interpreter converts each high-level statement to machine language, as needed during program execution. This means that a complete *object* program is not created prior to program execution. BASIC, with its *interactive* capabilities, is said to be an interpretive language, and many computers have BASIC interpreters.

Intrinsic function (see **Library function**).

Inverse. 1. Pertaining to a matrix, another matrix that, when multiplied by the first, produces the *identity* matrix. 2. Pertaining to a function, another function that, when applied to the first, produces the argument of the first. An example is the inverse sine function, arcsin: $\arcsin(\sin(\theta)) = \theta$.

Invert (to complement). In logic circuitry, to change to the opposite binary state. The output of a logic *inverter* is binary 0 when its input is binary 1, and vice versa. (See also **NOT**.)

I/O (Input/Output). Pertains to the process of supplying data to or obtaining data from a computer. For example, an I/O device either inputs data to the computer, receives output data from the computer, or both. (See also **Input** and **Output**.)

Iteration. One repetition of the execution of a set of program statements. (See also **Iterative programming** and **Loop**.) The word *iteration* is generally used to imply one repetition of a computation, the result of which is a more accurate approximation to the exact solution of a problem than that produced by a previous repetition.

Iterative programming. A technique whereby a programming goal is achieved through repeated executions (iterations) of a set of statements. Often used in *numerical analysis* to implement an algorithm; the result of each iteration is a closer approximation to the exact solution of a problem. Iterations are allowed to continue until a certain accuracy criterion is satisfied.

JCL. Abbreviation for job control language. A language having a limited number of statements and used only to impart general information, such as the language in which a program is written, the name and account number of the programmer, how the program results are to be output, and so forth. JCL is used in *batch processing* on large computer systems. The first few statements in a batched program are usually JCL statements; these can also appear at the beginning of a set of data statements and at the end of a program.

Job control language (see **JCL**).

Joystick. A movable lever used to control the position of an image on a video display or a pen on a plotter. Used in *graphics* and video games.

Jump (see **Branch**).

k. Abbreviation for kilo, a Greek prefix used with numbers to mean 10^3.

K. Used to describe memory capacity in bytes. One K is usually regarded as 1000 bytes but is actually $2^{10} = 1024$ bytes. Similarly, $16K = 16 \times 1024 = 16,384$ bytes.

Key. A push button on a *keyboard*. Performs the function of a switch. Each key is labeled with an alphanumeric symbol and, when a key is depressed, a code corresponding to the symbol is generated.

Keyboard. A set of keys, or pushbutton switches, arranged like those on a typewriter. Terminals, card punches, and similar devices have keyboards permitting an operator to "key in" (enter or type in) alphanumeric data.

Keypunch (see **Card punch**).

Keyword. A special word, abbreviation, mnemonic, or other set of alphanumeric characters that serves to distinguish a particular kind of programming statement from other kinds in the language being used. For example, in BASIC, the keyword LET identifies an assignment statement. The characters in a keyword are usually the first characters to appear in a statement, apart from a label or statement number.

Kilo (see **k**).

Label. A set of alphanumeric characters chosen by a programmer as a unique identifier for a particular statement. The label is a means for one statement to reference another, as in branching. In BASIC, every statement has a label: its statement number. However, in some languages, it is not necessary for every statement to have a label, and labels are not always restricted to numbers.

Language, computer (see **Computer language**).

LCD. Abbreviation for liquid crystal display. A means for displaying alphanumeric symbols. Consumes very little power.

LED. Abbreviation for light-emitting diode, a device that produces light when electrical current flows through it. It is available in various colors and is used in combination with other such diodes to display alphanumeric symbols.

Library function (also called intrinsic function). A function with a specific name, defined as an inherent part of a computer language. A programmer can cause this function to be evaluated anywhere in a program, by simply writing the function name and the argument for which it is to be evaluated. An example is the BASIC *absolute value* function, which has the name ABS. Different versions of a language sometimes use different sets of library functions.

Light pen. A photosensitive electronic device used to sense images in a video display and to transmit electrical signals to a computer.

Line printer. A large, high-speed printer used most often in large-scale computer systems. Typically, a continuously rotating chain with metal impressions of alphanumeric characters is struck at appropriate points to produce images on paper. Unlike other printers, characters are not printed in succession from left to right across a line, but instead, whenever a chain character is in the correct location on the line.

Listing. A display of all statements or lines in a program. In most versions of BASIC, the command "LIST" will cause a listing to be generated.

Load. To read a file stored in external memory and copy it into the main memory of a computer. For example, a program can be loaded from a disk, in preparation for execution by the computer.

Logic circuitry (see also **Gate**). Electronic components designed to produce voltage levels in accordance with logic relations, such as AND, OR, and combinations of these. In a typical computer, a large number of such components are connected together to implement many complex logic expressions. All voltage levels are either "low" (corresponding to binary 0) or "high" (corresponding to binary 1). Logic circuitry is constructed using both *integrated circuits* and *discrete* components.

Logic gate (see **Gate**).

Logical operator (or logic operator). Any Boolean algebra function, such as AND, OR, Exclusive OR, and Invert. Used for expressing conditions in *conditional branch* statements. In some languages, Boolean algebra relations can be programmed using logical operators and logical variables.

Log on. To gain access to a time-shared computer system, by supplying an account number or other authorization. This is usually accomplished by the user at a remote terminal.

Loop (see also **Iteration**). A set of statements that is executed repeatedly. Loops are set up in a programming language using special statements, such as the FOR-TO and NEXT statements in BASIC. A loop is usually executed a predetermined number of times in a program, whereas an *iteration,* which is also a loop, is generally performed until a specific criterion is satisfied.

Low-level language (see also **Computer language**). Machine language or a language similar to machine language, such as *assembly* language. A computer language whose statements include, or are limited to, the most fundamental instructions to which a computer can respond. To program in a low-level language, a programmer must understand the internal operations (at the word or byte level) performed by the hardware when the computer executes instructions.

M (see **Mega**).

Machine-dependent. Having variations that depend upon the computer being used. For example, *machine language* is machine-dependent, but high-level languages such as BASIC are not.

Machine language (see also **Computer language**). Used to cause a computer to respond to the most fundamental instructions for which it is designed. Machine-language instructions are expressed in binary code.

Magnetic core (see also **Core**). A toroid (doughnut-shaped) component used in core memory to store a binary 1 or a binary 0. By controlling the direction of current flow in the wires passing through a core, the polarity of the core's magnetization can be changed to correspond to one of the two binary states. Other wires are used to sense the polarity and thus *read* the bit stored by a core.

Magnetic disk (see **Disk**).

Magnetic tape. A means for storing data by magnetizing sequential sections of a specially coated plastic tape. Used for *external* computer memory.

Magnitude (of a number).   See **Absolute value**.

Mainframe. The *console* of a large computer system. This term is also used to distinguish a large computer system (*mainframe* computer) from a mini or microcomputer.

Mass storage. External computer memory—such as magnetic tape, disk, or punched cards. Unlike internal memory, it is capable of storing unlimited quantities of data.

Mathematical operator. Any fundamental mathematical function, such as addition, multiplication, or exponentiation. An operator is like a function defined for two arguments. For example, applying the multiplication operator to the numbers 7 and 4 produces the result 28.

MAT operations. In BASIC, mathematical operations that can be performed on *matrices*. The MAT prefix is used to identify such operations. Matrices can be added, multiplied, inverted, etc. Available only in powerful versions of BASIC.

Matrix. An *array* having rows and columns; the values of two *index* variables must be specified, to identify the position (row and column) occupied by any data value. (Note also that a matrix can have only one column; in this case, it is the same as a one-dimensional array and is called a *vector*.)

Mega (M). A Greek prefix meaning 10^6 or one million. For example, 4 Megabytes = 4 M bytes = 4×10^6 bytes.

Memory. A means for storing data. (See also **RAM** and **ROM**.)

Memory map. A diagram showing how internal memory is organized. Shows which portions of memory are reserved for special functions and which are available for general-purpose storage. The various portions are identified by the range of *addresses* that each represents.

Micro (μ). 1. A Greek prefix meaning 10^{-6}, or one one-millionth. For example, 20 microvolts = $20\mu V = 20 \times 10^{-6}$ volts.   2. Commonly used as a prefix to mean extremely small, as in *microprocessor, microelectronics,* etc.

Microcomputer (see also **Microprocessor**). A computer whose CPU is a microprocessor. So-called home computers, hobby computers, and personal computers are microcomputers. The word is sometimes used loosely to mean a microprocessor. Some microprocessors are constructed on one IC chip with a limited amount of memory and are therefore rightfully called "microcomputers on a chip." The maximum internal memory of most current microcomputers is 64K (64,536 bytes).

Microprocessor (see also **Microcomputer**).   A single integrated circuit containing all the electronic logic circuitry necessary to perform the arithmetic and control functions of a small computer. Also, the CPU of a microcomputer. Microprocessors are said to be 8-bit or 16-bit, depending on their basic *word* size. Besides serving as the CPU for a microcomputer, microprocessors are used in many *dedicated* applications, such as traffic-light controllers, microwave-oven controllers, and video games.

Minicomputer. A computer whose speed, memory, capacity, and power generally fall between those of a microcomputer and those of a large *mainframe* computer. The word size in a minicomputer is typically 32 bits, as opposed to 8 or 16 bits in most microcomputers. As microprocessors and microcomputers become more sophisticated, the distinction between "minis" and "micros" has become less distinct.

Minimal BASIC. An early ANSI standard version of BASIC with all the most commonly used statements and a reasonably large set of library functions. More powerful than *tiny* BASIC but less powerful than BASIC PLUS and other extended versions.

MKS system. The meter-kilogram-second system of units. Very similar to the SI system of units (the proposed international standard metric system).

Mnemonic. A memory aid. In a computer language, a set of characters that abbreviate a computer instruction. For example, in many assembly languages SUB is the mnemonic for the subtract instruction.

Modem (Modulator/Demodulator). An electronic device used to convert bits to FSK code (modulate) and to convert FSK code to bits (demodulate). Used for data transmissions by telephone line between a computer and a terminal. A modem is often an integral part of an *acoustic coupler*.

Monitor. 1. An operating system. 2. The video display portion of a terminal. Also, a television receiver.

Multiline function. A user-defined function requiring more than one statement to define. A capability available in some powerful versions of BASIC.

Multiplex. To permit multiple use through sharing. For example, a time-shared computer system multiplexes the electrical signal lines to the CPU, so that each user is allocated periodic time intervals to transmit data to or receive data from the computer. (See also **Time-sharing**.)

n (see **Nano**).

NAND. 1. A logic operator equivalent to an AND operation followed by an INVERT operation. 2. A type of logic gate that implements the NAND logic operator. Its output is binary 0 if and only if all its inputs are binary 1.

Nano (n). A Greek prefix meaning 10^{-9}. For example, 16 nsec = 16×10^{-9} seconds.

Nesting. Describes a programming technique wherein a particular kind of operation or procedure is performed as part of another operation or procedure of the same kind. In other words, one operation is embedded in another of the same kind. For example, a subroutine can contain a statement calling another subroutine; in this case, the latter is nested in the former. Also, a loop can contain statements that themselves constitute an "inner" loop, which can in turn contain another loop, and so forth.

Nibble. Four bits. Half of a byte.

Nonvolatile memory. A type of memory that does not lose its contents when electrical power is removed. ROM (read-only memory) is an example.

NOR. 1. A logic operator equivalent to the OR operation followed by the INVERT operation. 2. A type of logic gate that implements the NOR operation. The output is binary 1 if and only if at least one input is binary 0.

NOT. In logic, the same as invert or complement. If the logical variable A equals binary 1, then "NOT A" equals binary 0.

Numeric. Data consisting of numbers, as opposed to characters or strings. A numeric variable has values equal to numbers.

Numerical analysis. A branch of mathematics wherein problems are solved by performing computations on numbers. For example, numerical analysis provides a means for evaluating a definite integral in calculus or finding the roots of an equation, by using strictly numerical computations. Because of the speed and power that computers have to perform numerical computations, numerical analysis has become an important specialty area in computer science. (See also **Iterative programming**.)

Object program (also called object code). The machine-language program produced by an assembler or compiler. (See also **Source program**.)

Octal. Pertains to the base-8 number system, which uses only digits 0 through 7. Sometimes used in assembly-language programming, because it can be easily converted to binary.

OEM. Abbreviation for original equipment manufacturer.

Off line. Describes operation or equipment not under the control or in communication with the CPU of a computer. For example, "smart" terminals are designed to operate on line but can be used off line to edit programs, provide program listings, and display previously generated program results—all without communicating with the central computer.

One-pass assembler. An assembler that can create a machine-language program, after having been supplied only once with the assembly-language program. Contrast with **Two-pass assembler**.

On line. The opposite of *off line*. In direct, interactive communication with (or under the control of) a central computer.

Operating system (see also **Monitor**). A program or group of programs that control the operation of a computer. Often includes provisions for editing programs, debugging, input and output file management, and similar tasks. (See also **Disk operating system**.)

Operator, logical (see **Logical operator**).

Operator, mathematical (see **Mathematical operator**).

Operator, relational (see **Relational operator**).

OR. 1. In Boolean algebra and binary number processing, a logical operator that produces binary 1, if at least one bit that it operates on is 1, and binary 0, otherwise. 2. In logic circuitry, a gate whose output is binary 1, if at least one input is binary 1, and binary 0, otherwise.

Output. 1. To transmit data (in the most general sense) *from* a computer to an external device. 2. Used as a noun to mean what is transmitted out of a computer. The computer is always the point of reference, when referring to either *input* or *output*.

Output device. Any external device that receives data transmitted from a computer. One example is a printer.

Overflow. A numeric result (of a computation) that is too large to be represented in a specific computer.

p (see **Pico**).

Paper tape. A strip of paper containing punched-hole patterns representing a code for alphanumeric data. An early means for storage, input, and output of data. Also used with teletypewriters.

Parallel. Describes a binary data form wherein every bit in a word is generated or transmitted simultaneously. In hardware that processes binary data in parallel form, it is possible to examine, display, and test every bit in a word simultaneously. There must therefore be one signal line for each bit, so data is not normally transmitted through phone lines in parallel form. However, most computers move, process, and operate on words internally in parallel form. Contrast with **Serial**.

Pascal. A high-level programming language with a complex *syntax*. Lacks the exponentiation operator.

Peripheral devices (also called peripherals). Devices that are used with a computer but that are not an integral part of the CPU or internal memory—that is, devices that are not necessary for it to function as a computer. Examples include I/O devices such as printers, terminals, magnetic disk units, and card readers.

Pico (p). A Greek prefix meaning 10^{-12}. For example, 500 picofarads = 500 × 10^{-6} farads.

Port. A termination point (or set of termination points) in the electronic circuitry of a device used especially for interconnecting with other devices. A port can have certain *interface* circuitry associated with it, for that purpose. A port provides external access to important signal lines in the hardware of a computer system. An input port provides a means for a device to receive signals generated by an external source, whereas an output port makes internally generated signals available to external devices. A parallel data port acccepts or presents data bytes in parallel form; a serial port does the same for serial data.

Power. 1. Describing a computer or computer language, a relative term characterizing computational capabilities. For example, one computer is said to be more powerful than another if it can perform computations faster and possibly with greater precision; if it can be programmed to perform more complex computations; and/or if it has (or can access) larger memories. A computer language, or a version of a language, is more powerful than another, if it contains statements that can be used to perform more complex computations, has a more sophisticated branching structure, and/or has a larger set of library functions. 2. The rate of energy dissipation, measured in watts. An important consideration in designing computer hardware because of the heat generated by components requiring a large amount of power. 3. Mathematically, an exponent.

Precision. A measure of the number of significant decimal places by which a number is or can be represented. For example, if the true value of a number is 17.6430, 17.64 is a more precise representation than 17.6. Do not confuse precision with *accuracy*: In this example, 17.84 is a more precise representation of the true number than 17.8, but neither one is very accurate. In connection with programming and computer systems, precision is limited by word size. In *double*-precision arithmetic, two words are used to represent each number.

Printer (see also **Daisy-wheel printer**, **Dot-matrix printer**, and **Line printer**). A device that produces a printed copy of data supplied to it. A printer is an example of an *output* device in a computer system.

Print out (see also **Hard copy**). 1. To cause to be printed on a printer. 2. As a noun (*printout*), anything that is printed by a printer.

Program. A set of statements written in a computer language, which results in the achievement of a computational goal, when executed by a computer.

PROM (see also **ROM**). Abbreviation for programmable read-only memory. A nonvolatile, read-only memory whose contents are established electronically ("burned in"), using a device called a PROM programmer. Blank, integrated-circuit PROM chips can be purchased and programmed by a user who has a PROM programmer (unlike factory-programmed or mask-programmable ROMs).

Proportional spacing. A technique used by printers to alter the spacing between adjacent characters, to improve the appearance of printed copy.

Pseudo-random number (see also **Random number**). A number obtained from an algorithm that simulates the generation of random numbers. Such algorithms are frequently called "random-number generators." However, the sequence of numbers they produce must eventually repeat, so the numbers are not truly random.

Pseudo-random numbers are used in statistical investigations and computer studies that simulate random phenomena in nature.

Punched paper tape (see **Paper tape**).

Queue. A waiting line. In programming, a queue is a kind of "data structure" created by a programmer to save and retrieve data. Data is stored and removed from the queue in a manner similar to the way that members of a waiting line are served. ("Queueing" is the only word in the English language that contains five vowels in succession.)

Radix (see **Base**).

RAM. Abbreviation for random-access memory. Memory for which it is possible to gain immediate access to any location by specifying the *address* of that location. The main, internal memory of a computer is one example. In contrast, data on magnetic tape memory can be accessed only by unwinding a reel of tape until the location where it is stored on the tape is reached. The word RAM is also frequently used to mean random-access memory that can be both read and written into, as opposed to ROM.

Random number (see also **Pseudo-random number**). A number whose value cannot be predicted before its occurrence. Depending on the probability function governing its occurrence, the range of possible values can be either limited (for example: between 0 and 1) or unlimited. Unlike pseudo-random numbers, truly random events can occur only in a few natural processes, such as radioactive decay.

Read. To acquire the contents of memory or a specific memory location. For example, a computer reads memory to obtain the statements of a stored program.

Read-only memory (see **ROM**).

Readout. Contents acquired from memory as a result of a read operation. Also, a *display* of the contents of a memory location or the state(s) of another electronic device. Used as a verb to mean to obtain or measure the contents or state of a device.

Real-time solution. A problem solution that is obtained instantaneously and whose value changes without delay, as the problem parameters change. Many computer problems are designed to simulate physical processes in which the variables are changing with time. A solution to this kind of problem is said to be in real time if it changes in unison with changes in the variables. A high-speed digital computer can produce some solutions with very little delay and thus closely approximate real time. However, some delay always occurs, because of the time required to perform computations on discrete numbers. Only an analog computer can produce solutions that are truly in real time. From a practical standpoint, any solution can be considered real time if it is produced quickly enough to permit the user to react to it before parameter variations change the solution appreciably.

Record. A fundamental unit of data, usually synonymous with a line (for example, one line in a program listing). A record is identified for purposes of storage and retrieval in external memory, and a *file* is a group of such records.

Recursion. A procedure that refers to itself in its own definition. For example, we must resort to recursion to define the word "ancestor": "a parent or the parent of a parent." Examples of computational recursion include user-defined functions or subroutines that call themselves. *Recursive* describes a procedure involving recursion. Some programming languages (but not BASIC) can perform recursive computations.

Relational operator. A function that, when applied to two numbers, produces a logical result 1 or 0 (true or false), depending on the relationship it is designed to test. Relationships associated with these operators include "less than," "equal to," and "greater than or equal to." For example, applying the equality operator to the numbers 3 and 7 produces a result of 0 (false).

Remark (see also **Comment**). A nonexecutable BASIC statement serving only to explain other statements, procedures, or logic in a program. A remark statement is identified by the keyword REM and is an important part of program documentation.

Remote terminal (see **Terminal**).

Resident (in). Stored in, usually implying permanence. For example, "a compiler resident in ROM"

Return. 1. To produce or generate, as in the action of a function when it is evaluated. For example: "The ABS function returns the absolute value of its argument." 2. After execution of a subroutine, to branch back to the point where the subroutine was called.

Reverse display. A video display wherein characters appear black on a light background.

Reverse Polish. A method of performing numeric computations. The mathematical operator is acquired after both numbers to be operated on are acquired. For example, the computation 4 + 7 would be entered, stored, and processed in the sequence 4, 7, +.

Rollover. On a keyboard, the retention of characters corresponding to keys depressed simultaneously or within a very short interval of time.

ROM. Abbreviation for read-only memory. Memory whose contents are permanent. This means that the contents can be read but cannot be altered by write operations. ROM is an example of nonvolatile memory. Contrast with **RAM**.

Round off. To delete less significant digits. The least significant digit retained is increased by 1, if the most significant digit deleted is greater than 5. Round-off algorithms differ when the most significant digit deleted is equal to 5.

Round-off error. Computational error caused by rounding numbers. For example, suppose that the exact values of the variables x and y are 3.764 and 3.892, respectively. The exact value of $(y - x)/(0.01)$ is then 13.8. On the other hand, if the numbers were rounded to 3.8 and 3.9, then $(y - x)/(.01) = 10$, causing a round-off error of 27.5 percent.

Routine. A computational procedure usually having a limited goal and used as part of a larger program. A subroutine is one example.

RS 232. A widely used standard for interfacing serial binary data-communications equipment. Specifies voltage levels used for binary zeros and ones and other relevant information. Published by the Electronic Industries Association.

Run. To execute the statements in a program. The *command* RUN is used in BASIC to run a program. Also used as a noun to mean execution, as in "a program run."

Run time. The time during which a program is being run, as opposed to being compiled.

Run time error. A programming error detected during run time. For example, the BASIC statement LET Y = 2/(X − X) has no syntax errors and thus would not be detected as an illegal attempt to divide by zero until run time, when the computer would attempt to execute the statement. (See also **Bug**.)

Scientific notation. A method of number representation wherein a magnitude is expressed as a number with one digit left of the decimal point multiplied by a power of 10. For example, 20763 is expressed in scientific notation as 2.0763×10^4 and 0.0095 is expressed as 9.5×10^{-3}.

Scientific programming. Computer programming with the objective of solving problems in the natural sciences, mathematics, technology, or engineering. Characterized by the application of mathematical algorithms in which complex computations are performed on a relatively small amount of data. Contrast with **Data processing**.

Scroll. To move the display on a video screen up or down. When a display is scrolled up, the lines at the top of the screen disappear; when a display is scrolled down, the lines at the bottom disappear.

Sector (see also **Track**, **Hard-sectored disk** and **Soft-sectored disk**). A portion of a track on a disk. A sector on a typical floppy disk can store 128 characters (256, on a double-density floppy disk).

Serial. Describes a method of binary data transmission in which bits are transmitted one at a time. A word is transmitted on a single line (wire or conductor) by making the voltage on that line high or low at succeeding instants of time, in accordance with the sequence of binary ones and zeros in the word. Serial transmission is used for data transmission on telephone lines. Contrast with **Parallel**.

Silicon Valley. A region in northern California with a high concentration of industry related to the design, manufacture, and application of integrated circuits.

Simplex. A type of communications wherein information can be transmitted in only one direction. Contrast with **Half duplex** and **Full duplex**.

Simulator. A program that permits the user to write and execute programs that were designed to be run on a different computer. For example, a large computer system can have a simulator that permits the user to write and run assembly-language programs designed for a particular microprocessor.

Smart terminal (see also **Terminal**). A terminal that can perform some functions normally performed by a computer. A built-in microprocessor and some internal memory permit the user to perform (off line) tasks such as storing, writing, and editing programs.

Soft-sectored disk (see also **Sector** and **Formatted disk**). A disk with sectors identified by codes that a program has caused to be recorded on the surface of the disk. A disk formatted by software.

Software. Programs. (See also **Firmware**.)

Solid state. Constructed from semiconductor material, as opposed to vacuum tubes. Transistors and integrated circuits are examples. The principal semiconductor material used in solid-state hardware is silicon.

Source program. A program written in a programming language. The original form of a program before it is converted to machine language by an assembler or compiler. (See also **Object program**.)

Special-purpose computer. A computer designed to perform a specific task and not generally programmable. A computer *dedicated* to a specific application. Examples include a navigation computer aboard a spacecraft and a computer designed to control traffic lights. Contrast with **General-purpose computer**.

Square matrix. A matrix having the same number of rows as columns. Only square matrices can have *inverses,* but not all square matrices do.

Storage. Memory space. Also, the act of storing, as "the storage of data in memory."

Store (see also **Write**). To cause to be inserted (written) into memory. To save in memory.

Stored-program computer. A type of computer that executes programs stored in memory. The CPU acquires (reads) program statements from memory and executes them in succession. All modern general-purpose computers fall in this category.

String. A set of alphanumeric symbols treated by the computer as characters, rather than numeric data or program statements. A string is identified in BASIC by enclosing it within quotation marks. For example, in BASIC the statement

PRINT 7 − 6/3

will result in the display 5, since 7 and 6/3 are treated as numeric data; but the statement

PRINT "7 − 6/3"

will result in the display 7 − 6/3, since "7 − 6/3" is treated as a string of characters.

String variable. A variable whose values are strings. In BASIC, a string variable is distinguished from a numeric variable by the suffix $ in its variable name.

Structured programming. Programming that allows computational tasks, routines, conditional branch logic, and similar components of a program to be isolated—that is, placed in certain well-defined program blocks. The component blocks are then interconnected to form the structure of the program. PASCAL is said to be a language that lends itself to structured programming.

Subroutine. A set of program statements set aside from the main part of a program and designed to achieve a limited goal or perform some intermediate computation leading to the overall goal of the program. A subroutine is often used when a program must perform the same kind of computations several times during the course of program execution.

Subscript. A number used to distinguish one data value or variable from a related one having the same identifying symbol or name; usually an integer. For example, the subscripts 1, 2, and 3 distinguish three entries in array A symbolized in BASIC by A(1), A(2), and A(3) and expressed mathematically as A_1, A_2, and A_3.

Supervisor. Same as a monitor or operating system.

Support. In relation to software, programs available from a manufacturer for a particular computer. These programs can be used with that computer to perform a variety of programming tasks. Software support is used as a criterion by prospective computer purchasers, when deciding which model to buy.

Synchronous data transmission. A type of data transmission in which bits are generated only at prescribed intervals of time (for example, in unison with the pulses of a clock signal). Contrast with **Asynchronous data transmission**.

Syntax. Rules prescribed in a particular computer language for expressing and forming statements. Includes rules governing: the use of special symbols to identify different kinds of statements, keywords, and labels; the formation of variable names; and the use of parentheses in expressions, symbols for operators, and so forth.

Tape drive. The mechanism that controls the winding and unwinding of magnetic tape. The electromechanical hardware in which spools or reels of tape are mounted. This term is often used to describe the entire unit in which the tape is installed.

Tape, magnetic (see **Magnetic tape**).

Teletypewriter (TTY) (also teleprinter). A machine having a keyboard and a printer and designed for transmitting and receiving data. A Teletype (a trademark of the Teletype Corporation) is an example. Used in communication systems and is often adapted for use as a computer terminal.

Terminal. A device used by a programmer to communicate with a computer. In a time-shared computer system, programmers send programs to the computer by entering them at a terminal and receive program results at the terminal. A terminal is equipped with a keyboard and usually a video display (CRT), so it is often called a video terminal. Terminals are usually located in areas away from the central computer installation and communicate with the computer through telephone lines, so they are also called remote terminals. (See also **Smart terminal**.)

Text editor. A program that permits the user to prepare and, in particular, to modify programs and data files. Allows the user to insert or delete characters, statements, or whole blocks of statements; rearrange sequences of lines; search programs or data files for specific characters; and perform similar tasks that cannot ordinarily be performed by a programming language.

Time-sharing. The process that allocates successive intervals of computer time to a number of users. A high-speed computer with substantial memory capacity can serve many programmers in such rapid succession that all appear to be using the computer simultaneously. (See also **Multiplex**.)

Tiny BASIC. A version of BASIC having limited capabilities. This version is not as powerful as other versions of BASIC but requires less memory to store the interpreter or compiler.

Track. A ring of storage area on a disk. A disk contains a number of concentric tracks, each of which is assigned a number, so it can be accessed for reading or writing data.

Transparent. Describes an operation that occurs automatically, as a result of a programmer's action, that the programmer need not be aware of or take into account. For example, the code numbers assigned to disk tracks and sectors when a programmer formats a soft-sectored disk are transparent; the programmer need not know what they are or be concerned with them.

Truncate. To delete the fractional part of a number. For example, if the numbers 2.76 and -3.85 are truncated, the results are 2 and -3 respectively. Note that truncation is not equivalent to rounding, nor the result produced by the INT function (in most versions of BASIC).

TTY (see **Teletypewriter**).

Two-pass assembler. An assembler that must be supplied twice with an assembly-language program, to complete the process of converting it to machine language. (Contrast with **One-pass assembler**.)

Unconditional branch (see also **Branch**). A statement that causes a branch every time it is executed, irrespective of any conditions. In BASIC, the GOTO statement is an unconditional branch.

Underflow. A nonzero result (of a computation) that is too small to be represented in a specific computer.

Unfriendly. Describes a computer or computer language that is difficult to use. An unfriendly language has a demanding syntax that does not permit variations (such as arbitrary insertion of blanks in a statement) and does not provide helpful error messages. (See also **Friendly**.)

Up. In operation. Neither disconnected nor under repair.

Variable. A quantity that can acquire or be assigned different values during the course of program execution. Every variable has a name and subsequently acquires values whenever its name appears in conjunction with an equal sign.

Vector. In matrix theory, a matrix or array having a single column. An $n \times 1$ matrix, where n is an integer.

Video terminal (see **Terminal**).

Volatile memory. Memory whose contents are lost when electrical power is removed. Most semiconductor RAM memory is volatile. Contrast with **Nonvolatile memory**.

Word. A group of bits that together constitute the fundamental unit for representing data in binary form. In a typical microcomputer, a word is the same as a byte (8 bits). In large computers, however, a word can consist of 32 or more bits, depending on the design. An instruction word has the same number of bits as a data word but is interpreted differently by the computer.

Word processor. A program used to create and edit text material, such as manuscripts, business forms, and letters. (*Word* is not used in the context of a group of bits, but in the sense of an English-language word.) Word processors permit the user to insert and delete words, sentences, paragraphs, and pages in written text; set margins; number pages; and perform similar tasks that are helpful to authors. Similar to a *text editor*, except the latter is oriented to the editing of material written in a programming language.

Workspace. A portion of internal memory that a programmer can use on line to build, edit, run, and debug programs. A program cannot be retained indefinitely in a workspace, so it must be stored in external memory if the programmer wishes to save it. Typically, a programmer uses a workspace to debug and develop the final form of a program and then saves it in external memory. A program can be recalled to the workspace anytime.

Write. To store in memory.

ANSWERS TO SELECTED EXERCISES

Complete solutions are given for selected programming exercises. Both the program and the results it produces are given for some odd-numbered exercises. Only the results of program runs are shown for other odd-numbered exercises.

CHAPTER 1

1.1 Microcomputers and time-shared (mainframe) computers (see Section 2).

1.3 Magnetic disk and magnetic tape. Older forms include punched paper tape and punch cards.

1.5 $2^8 - 1 = 255$.

1.7 (a) As part of the program, using special statements identifying data values; (b) By the programmer, using a keyboard; (c) From external, mass-storage devices.

1.9 Values for V_1, V_2, and V_3.

1.11 Data processing involves relatively simple computations on large quantities of data; scientific programming involves complex computations on small quantities of data.

CHAPTER 2

2.1 (a) 4.8399E2 (e) $-1.0E-6$
  (b) $-1.077625E4$ (f) 1.439E7
  (c) $4.2E-4$ (g) 1.0E5
  (d) $-2.00383E2$ (h) $7.518E-6$

2.3 (a) Correct.
  (b) Incorrect; has noninteger exponent.
  (c) Correct.
  (d) Correct.
  (e) Incorrect; has noninteger exponent.
  (f) Incorrect; BASIC does not use π.

2.5 (a) Valid.
  (b) Valid.
  (c) Invalid; has constant in place of variable to the left of the equal sign.
  (d) Valid.
  (e) Invalid; BASIC does not have π.
  (f) Invalid; has expression in place of variable on left of equal sign.
  (g) Invalid; has invalid variable name.

2.7 (a) 6
  (b) I =
    600
    DONE
  (c) X=25
    25

2.9
```
10 PRINT "ENTER X1"
20 INPUT X1
30 PRINT "ENTER X2"
40 INPUT X2
50 PRINT "ENTER X3"
60 INPUT X3
70 LET S=X1+X2+X3
80 PRINT "THE SUM IS:"
90 PRINT S
```

2.11 Statement 10: missing quotes; statement 20: invalid variable name; statement 50: invalid variable name.

2.13 The actual error messages depend on the brand of computer being used. The reasons for errors are:
  (a) 3X is not a valid variable name.
  (b) 4 is not a valid variable name.
  (c) $X+Y$ is not a valid variable name.

(d) Quotes should enclose "THE VALUE IS".
(e) 35 is not a valid variable name.
(f) 52 E 1.5 has a fractional exponent.
(g) A1 + B1 is not a valid variable name.
(h) WRITE is not a BASIC keyword.
(i) A quote mark should precede X =.
(j) Only one variable can be assigned in a LET statement.
(k) 40 * F5 is not a valid variable name.

2.15
```
10 PRINT "ENTER RADIUS IN INCHES"
20 INPUT R
30 LET R=2.54*R
40 LET A=3.14159*R*R
50 PRINT "AREA IN SQ. CM. IS:"
60 PRINT A
70 END
```

(a) ENTER RADIUS IN INCHES
? 10
AREA IN SQ. CM. IS:
2026.82821

(b) ENTER RADIUS IN INCHES
? .3937008
AREA IN SQ. CM. IS:
3.14159021

(c) ENTER RADIUS IN INCHES
? 1.968
AREA IN SQ. CM. IS:
78.4995429

2.17
```
10 PRINT "ENTER SPEED IN MPH"
20 INPUT V
30 PRINT "ENTER TIME IN MINUTES"
40 INPUT T
50 LET X=26.834*V*T
60 PRINT "DISTANCE IN METERS IS:"
70 PRINT X
80 END
```

ENTER SPEED IN MPH
? 45
ENTER TIME IN MINUTES
? 10
DISTANCE IN METERS IS:
 12075.3

2.19 (a) ENTER FORCE IN NEWTONS
? 8064.3
ENTER DISTANCE IN METERS
? 1.95

MOMENT IN FOOT-POUNDS
 11598.5534

(b) ENTER FORCE IN NEWTONS
    ? 15200
ENTER DISTANCE IN METERS
? 4.083
MOMENT IN FOOT-POUNDS
 45774.6999

2.21 10 PRINT "ENTER RESISTANCE IN OHMS"
    20 INPUT R
    30 PRINT "ENTER CURRENT IN AMPS"
    40 INPUT I
    50 LET V=I*R
    60 PRINT "VOLTAGE DROP IS:"
    70 PRINT V
    80 END

(a) ENTER RESISTANCE IN OHMS
    ? 330.94
    ENTER CURRENT IN AMPS
    ? .26045
    VOLTAGE DROP IS:
     86.193323

(b) ENTER RESISTANCE IN OHMS
    ? 68000
    ENTER CURRENT IN AMPS
    ? .0152
    VOLTAGE DROP IS:
     1033.6

2.23 (a) ENTER INPUT POWER IN WATTS
    ? 793.617
    ENTER EFFICIENCY
    ? .94
    OUTPUT HORSEPOWER:
     1.00001297

(b) ENTER INPUT POWER IN WATTS
    ? 54220
    ENTER EFFICIENCY
    ? .85
    OUTPUT HORSEPOWER:
     61.7796235

2.25 10 PRINT "ENTER AVG. WEEKLY WORKER WAGE"
    20 INPUT W
    30 PRINT "ENTER AVG. ANNUAL SUPERVISOR SALARY"
    40 INPUT S
    50 LET T=657.9*W+2.667*S

```
60 PRINT "TOTAL MONTHLY PAYROLL:"
70 PRINT T
80 END
```

(a) ENTER AVG. WEEKLY WORKER WAGE
? 325
ENTER AVG. ANNUAL SUPERVISOR SALARY
? 26500
TOTAL MONTHLY PAYROLL:
284493

(b) ENTER AVG. WEEKLY WORKER WAGE
? 368
ENTER AVG. ANNUAL SUPERVISOR SALARY
? 25800
TOTAL MONTHLY PAYROLL:
310915.8

2.27
```
10 PRINT "ENTER W1 IN RAD/SEC"
20 INPUT W1
30 PRINT "ENTER N1/N2"
40 INPUT N
50 LET W2=W1*N
60 PRINT "W2="
70 PRINT W2
80 END
```

(a) ENTER W1 IN RAD/SEC
? 2.54
ENTER N1/N2
? 1.52
W2=
3.8608

(b) ENTER W1 IN RAD/SEC
? 13704
ENTER N1/N2
? 2.5E-05
W2=
.3426

2.29 (a) ENTER SPRING CONSTANT IN N/M
? .063
ENTER DISPLACEMENT IN INCHES
? .25
FORCE IN POUNDS:
9.0011E-05

(b) ENTER SPRING CONSTANT IN N/M
? 3.779
ENTER DISPLACEMENT IN INCHES
? 1.2E-03
FORCE IN POUNDS:
2.5916E-05

CHAPTER 3

3.1 (a) 1 (e) −0.15
(b) 1 (f) 32
(c) 0.5 (g) −1.666
(d) 3.5 (h) 0

3.3 (a) 72 (c) 8E6
(b) 0.5 (d) 3

3.5 (a) −50 (f) 500
(b) 10 (g) .016667
(c) 17 (h) 1E5
(d) 64 (i) 0.1
(e) 30 (j) 35

3.7 (a) (4*4+3*3)**0.5/5
(b) 3*(2+4)**2/(2*2+5)
(c) (4**3+200**0.5)**0.25
(d) 3.29*7.41/1.09/4.66
(e) (39.36-12)**(1/5)/7**0.5/(8-4.05)
(f) (1E6-(5E5-20E4))/1E7/(1.55-3.92)
(g) (2.025)**3*(2**7-11)/(13.88-4.709**2)

3.9
```
10 PRINT "ENTER N1, N2"
20 PRINT "N1 AND N2 MUST BE POSITIVE"
30 PRINT "N2 MUST BE GREATER THAN N1"
40 INPUT N1,N2
50 LET S=N2*(N2+1)*(2*N2+1)-(N1-1)*(N1)*(2*N1-1)
60 LET S=(S/6)**.5
70 PRINT "SQUARE ROOT OF SUM OF SQUARES FROM N1 TO N2:"
80 PRINT S
90 END
```

(a) ENTER N1, N2
N1 AND N2 MUST BE POSITIVE
N2 MUST BE GREATER THAN N1
   1 , 10
SQUARE ROOT OF SUM OF SQUARES FROM N1 TO N2:
19.6214169

(b) ENTER N1, N2
N1 AND N2 MUST BE POSITIVE
N2 MUST BE GREATER THAN N1
   5 , 50
SQUARE ROOT OF SUM OF SQUARES FROM N1 TO N2:
207.111081

3.11 (a) ENTER A, B
? .5 , 2

```
            RESULT IS:
            .302083333

        (b) ENTER A, B
            ?-.1 , 1
            RESULT IS:
            1.10516667
```

3.13
```
    10 PRINT "ENTER P, K, B"
    20 INPUT P,K,B
    30 LET H=4*P*B**2/K/(B+1)**2
    40 PRINT "SETTLEMENT IS:"
    50 PRINT H
    60 END

    (a) ENTER P, K, B
        ? 8 , 400 , 6
        SETTLEMENT IS:
        .0587755102

    (b) ENTER P, K, B
        ? 6, 100 , 8
        SETTLEMENT IS:
        .18962963
```

3.15
```
    10 PRINT "ENTER Q1, Q2, R"
    20 INPUT Q1,Q2,R
    30 LET F=9E9*Q1*Q2/R**2
    40 PRINT "THE FORCE IS:"
    50 PRINT F
    60 END

    (a) ENTER Q1, Q2, R
        ? 2.5E-05 , 1E-04 , .03
        THE FORCE IS:
        25000

    (b) ENTER Q1, Q2, R
        ? 1 , 1 , .015
        THE FORCE IS:
        4E+13
```

3.17
```
    (a) ENTER R1, R2, R3
        ? 33000 , 33000 , 33000
        EQUIVALENT RESISTANCE:
        11000

    (b) ENTER R1, R2, R3
        ? 1 , 100 , 10000
        EQUIVALENT RESISTANCE:
        .99000099
```

3.19 10 PRINT "ENTER C, E, R"
20 INPUT C, E, R
30 LET Q=(20*C/E/(1-C/R))**.5
40 PRINT "ECONOMIC LOT SIZE:"
50 PRINT Q
60 END

(a) ENTER C, E, R
? 600 , 1E-03 , 1000
ECONOMIC LOT SIZE:
5477.22556

(b) ENTER C, E, R
? 250 , 2E-03 , 400
ECONOMIC LOT SIZE:
2581.98889

3.21 10 PRINT "ENTER D0, D1"
20 INPUT D0,D1
30 LET J=3.14159*(D0**4-D1**4)/32
40 PRINT "POLAR MOMENT OF INERTIA:"
50 PRINT J
60 END

(a) ENTER D0, D1
? .1 , .09
POLAR MOMENT OF INERTIA:
3.376E-06

(b) ENTER D0, D1
? .595 , .42
POLAR MOMENT OF INERTIA:
9.25E-03

3.23 (a) ENTER K, B, M
? 12 , 6 , 1
FREQUENCY:
1.73205081

(b) ENTER K, B, M
? 105 , 4 , .5
FREQUENCY:
13.9283883

CHAPTER 4

4.1 (a) A1= 16   A2= 35
(b) X**2= 81 X**.5= 3
(c) 248
    1632

(d) FORCE= 20 NEWTONS
    MASS= 40 KG

4.3 (a) 64   8   2
      64   2

(b) ROOT A=   100
    SIZE B=   2

(c) -6   30 VOLTS
        36

(d) ONE   TWO   THREE
    1   2   3

4.5
```
10 INPUT A,B,A$,B$
```

4.7 (a) ANSWERS 98 9.8
(b) 35
(c) 30 15 0

4.9
```
10 READ X1,X2,X3
20 LET X1=4-X1
30 LET X2=X2-5
40 LET X3=7*X3
50 PRINT X1,X2,X3
60 RESTORE
65 READ X1,X2,X3
70 LET X1=X1+1
80 LET X3=X2*X3
90 PRINT X1,X2,X3
100 DATA 7,-4,2
110 END
```

-3   -9   14
 8   -4   4

4.11
```
10 PRINT "HORIZONTAL" TAB(20) "VERTICAL" TAB(40)
"RESULTANT"
15 PRINT "----------" TAB(20) "--------" TAB(40) "---
------"
20 READ H,V
30 LET R=(H*H+V*V)**.5
40 PRINT H TAB(20) V TAB(40) R
50 READ H,V
60 LET R=(H*H+V*V)**.5
70 PRINT H TAB (20) V TAB(40) R
80 READ H,V
90 LET R=(H*H+V*V)**.5
100 PRINT H TAB(20) V TAB(40) R
```

```
110 READ H,V
120 LET R=(H*H+V*V)**.5
130 PRINT H TAB(20) V TAB(40) R
140 READ H,V
150 LET R=(H*H+V*V)**.5
160 PRINT H TAB(20) V TAB(40) R
170 DATA 153.4,171.9,208,188.6
180 DATA 537.4,600.2,1082.2,824.5
190 DATA 1925.5,1098
200 END
```

HORIZONTAL	VERTICAL	RESULTANT
153.4	171.9	230.393511
208	188.6	280.773859
537.4	600.2	805.629443
1082.2	824.5	1360.49884
1925.5	1098	2216.56361

4.13

TIME	DIST.	ACCEL.	% DIFF.
1.03	16.8	31.6712225	1.02742959
2.11	72.5	32.5689001	-1.7778127
3.84	245.8	33.3387587	-4.18362082
5.26	421.9	30.4977664	4.69448019
9.05	1402	34.2358292	-6.98696622

4.15
```
10 PRINT "VOLUME" TAB(20) "TEMP. DIFF." TAB(40) " HEAT"
15 PRINT "(CU. FT.)" TAB(20) "(FAHRENHEIT)" TAB(40)
"(BTU/HR)"
16 PRINT "--------" TAB(20) "------------" TAB(40)
"--------"
30 READ C,T
40 PRINT C TAB(20) T TAB(40) .027*C*T
50 READ C,T
60 PRINT C TAB(20) T TAB(40) .027*C*T
70 READ C,T
80 PRINT C TAB(20) T TAB (40) .027*C*T
90 READ C,T
100 PRINT C TAB(20) T TAB(40) .027*C*T
110 READ C,T
120 PRINT C TAB(20) T TAB(40) .027*C*T
130 DATA 240,42.5,112,39.1,466
140 DATA 57.7,357,49.8,265,29.9
150 END
```

VOLUME (CU. FT.)	TEMP. DIFF. (FAHRENHEIT)	HEAT (BTU/HR)
240	42.5	275.4
112	39.1	118.2384
466	57.7	725.9814
357	49.8	480.0222
265	29.9	213.9345

4.17
```
10 PRINT "ENTER E, V1, V2, R3"
20 INPUT E,V1,V2,R3
30 PRINT "THE CURRENT IS" (E-V1-V2)/R3 "AMPS."
40 END
```

(a) ENTER E, V1, V2, R3
? 25 , 6 , 9 , 4700
THE CURRENT IS 2.128E-03 AMPS.

(b) ENTER E, V1, V2, R3
? 6 , 1 , 2.8 , 1500000
THE CURRENT IS 1.467E-06 AMPS.

4.19

VOLTAGE	CURRENT	RESIS.	DIFF.
5	4.87	1.02669404	9.08157551
10	11.09	.901713255	9.0827117
15	14.42	1.04022191	9.08145253
20	22.11	.904568069	9.08268575

4.21
```
10 PRINT " N" TAB(20) " SUM" TAB(40) "STANDARD
   DEVIATION"
20 PRINT "----" TAB(20) "-----" TAB(40) "-------------
   -----"
30 READ N,S
40 PRINT N TAB(20) S TAB(40) (S/(N-1))**.5
50 READ N,S
60 PRINT N TAB(20) S TAB(40) (S/(N-1))**.5
70 READ N,S
80 PRINT N TAB(20) S TAB(40) (S/(N-1))**.5
80 READ N,S
100 PRINT N TAB(20) S TAB(40) (S/(N-1))**.5
110 READ N,S
120 PRINT N TAB(20) S TAB(40) (S/(N-1))**.5
130 DATA 30,185.5,47,201.9,25
140 DATA 62.6,139,277,52,118.4
150 END
```

N	SUM	STANDARD DEVIATION
30	185.5	2.52914051
47	201.9	2.09502516
25	62.6	1.61503354
139	277	1.41677323
52	118.4	1.52366946

4.23
```
10 PRINT "ENTER AREA, VELOCITY, VOLUME"
20 INPUT A,V,S
25 PRINT
30 PRINT "FLOW RATE=" A*V/S "LB/SEC."
40 END
```

```
          ENTER AREA, VELOCITY, VOLUME
          ? 1.7 , 30.2 , 13.09

          FLOW RATE= 3.92207792 LB/SEC.
```

4.25

P	V	P*V	% DIFF.
1050000	5.2E-03	5460	.727272719
2110000	2E-03	4220	23.2727273
4770000	1.7E-03	8109	-47.4363637
8630000	8E-04	6904	-25.5272728

CHAPTER 5

5.1 10   100
     101
     102

5.3 (a) 100   (d) 150
     (b) 90   (e) 40
     (c) 60

5.5 (a) 80   (c) 80
     (b) 90   (d) 150

5.7 (a)
```
10 PRINT "ENTER X, Y"
20 INPUT X, Y
30 PRINT
40 IF X<Y THEN 50  ELSE 80
50 PRINT "LARGER  : " Y
60 PRINT "SMALLER : " X
70 GOTO 100
80 PRINT "LARGER  : " X
90 PRINT "SMALLER : " Y
100 END
```

(b)
```
100 PRINT "ENTER NUMBER :"
110 INPUT N
120 IF N>=-9 AND N<-3 THEN 160
130 IF N>=3 AND N<9 THEN 160
140 PRINT " NO"
150 GOTO 170
160 PRINT " YES"
170 END
```

(c)
```
100 PRINT "EVEN OR ODD ?"
110 INPUT T$
```

```
120 IF T$="EVEN" THEN LET I=0
130 IF T$="ODD" THEN LET I=1
140 PRINT I
150 LET I=I+2
160 IF I<=10 THEN 140
170 END
```

5.9
```
10 PRINT "ENTER A POSITIVE INTEGER"
20 INPUT N
25 IF N<=0 THEN 100
30 LET S=0
40 LET J=1
50 LET S=S+J
60 LET J=J+1
70 IF J<=N THEN 50
80 PRINT "THE SUM OF THE FIRST" N "INTEGERS IS" S
90 GOTO 120
100 PRINT "ILLEGAL ENTRY"
110 GOTO 10
120 END
```

5.11
```
100 PRINT "ENTER TWO INTEGERS M AND N,"
110 PRINT "POSITIVE, NEGATIVE, OR ZERO,"
120 PRINT "WITH M<N:"
130 INPUT M,N
140 IF M<N THEN 170
150 PRINT "M MUST BE LESS THAN N . . ."
160 GOTO 100
170 LET I=M
180 LET S=0
190 LET S=S+I
200 LET I=I+1
210 IF I<=N THEN 190
220 PRINT "THE SUM OF THE INTEGERS BETWEEN"
230 PRINT "M AND N, INCLUSIVE, IS:"
240 PRINT S
250 END
```

5.13
```
100 REM : THIS PROGRAM PICKS AND
110 REM : PRINTS THE LARGEST OF
120 REM : X, Y, AND Z.
130 INPUT X,Y,Z
140 IF X>Y THEN 170
150 M=Y
160 GOTO 180
170 M=X
180 IF M<=Z THEN LET M=Z
190 PRINT M
200 END
```

5.15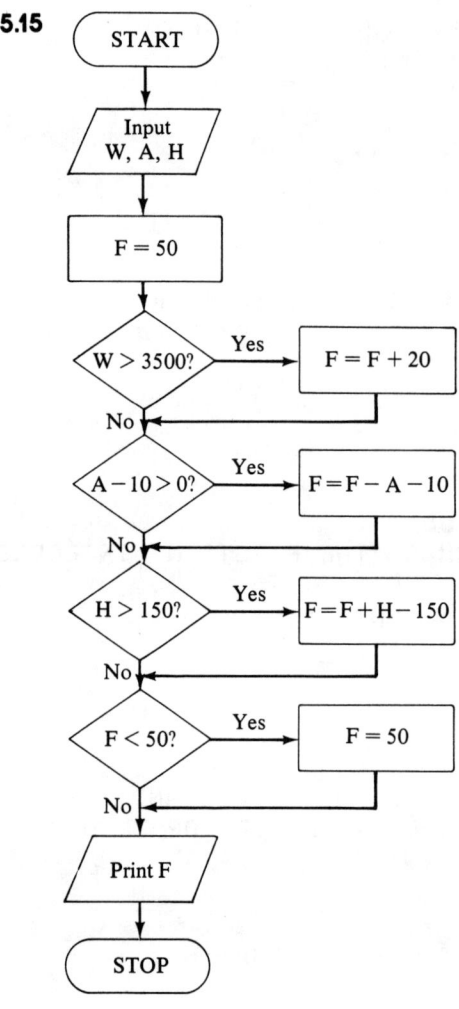

```
100 INPUT "WEIGHT : ";W
110 INPUT "   AGE : ";A
120 INPUT "  H.P. : ";H
130 LET F=50
140 IF W>3500 THEN F=70
150 IF A>10 THEN LET F=F-A+10
160 IF H>150 THEN LET F=F+H-150
170 IF F<50 THEN F=50
180 PRINT " FEE : $"F
190 END
```

5.17 (a) 3 (b) 8 (c) 720 (d) .2
3 12 .6
3 16 1
3 20 1.4
3 24 7

5.19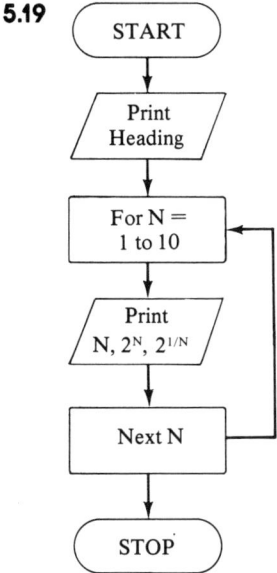

```
100 PRINT " N" TAB(10)"2**N" TAB(25)"2**(1/N)"
110 FOR N=1 TO 10
120 PRINT N TAB(10) 2**N TAB(25) N**(1/N)
130 NEXT N
140 END
```

```
N         2**N                 2**(1/N)
1         2                    1
2         4                    1.41421356
3         8                    1.44224957
4         16                   1.41421356
5         32.0000001           1.37972966
6         64.0000001           1.34800616
7         128                  1.32046925
8         256.000001           1.29683956
9         512.000002           1.27651801
10        1024                 1.25892541
```

5.21

STUDENT	AVERAGE
A	82.3333334
B	79
C	94.3333334
D	47.3333333
E	52.6666667
F	84
G	64.3333334
H	94

OVERALL AVG.: 74.75

5.23
```
10 PRINT "ENTER N"
20 INPUT N
30 ON N GOTO 40, 60, 60, 90, 90
40 PRINT "N=";
```

```
50 GOTO 130
60 PRINT "(N**2)!=";
70 LET N=N*N
80 GOTO 100
90 PRINT "N!=";
100 FOR I=N-1 TO 2 STEP -1
110 LET N=N*I
120 NEXT I
130 PRINT N
140 END
```

5.25

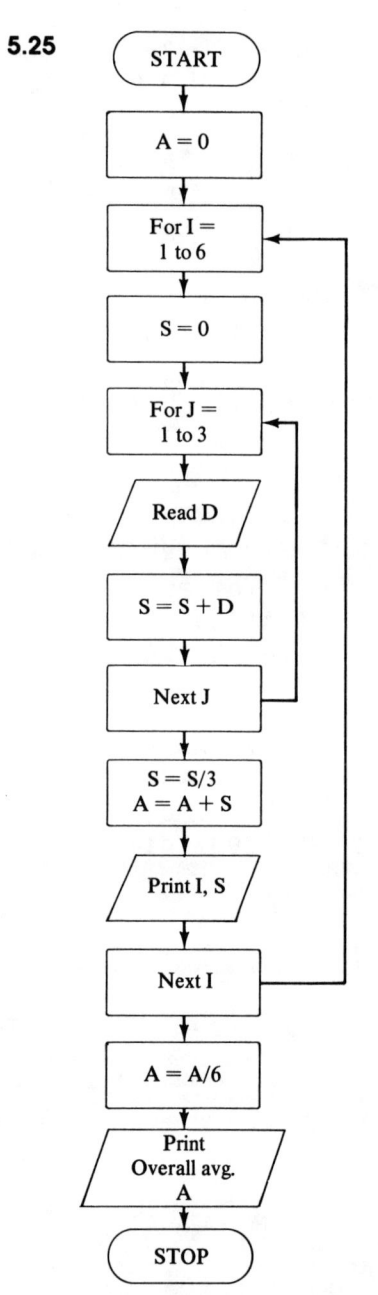

```
100 PRINT "ASTRONOMER","AVERAGE DIST."
110 PRINT
120 LET A=0
130 FOR I=1 TO 6
```

```
140 LET S=0
150 FOR J=1 TO 3
160 READ D
170 LET S=S+D
180 NEXT J
190 LET S=S/3
200 LET A=A+S
210 PRINT TAB(4);I,S
220 NEXT I
230 LET A=A/6
240 PRINT
250 PRINT "OVERALL AVERAGE="A
260 DATA 93.07,92.89,94.6
270 DATA 93.75,93.01,93.84
280 DATA 92.99,92.76,93.57
290 DATA 93.52,92.98,94.12
300 DATA 91.89,91.60,93.04
310 DATA 94.02,94.12,95.09
320 END
```

```
ASTRONOMER      AVERAGE DIST.
    1              93.52
    2              93.5333333
    3              93.1066667
    4              93.54
    5              92.1766667
    6              94.41

OVERALL AVERAGE= 93.3811112
```

5.27

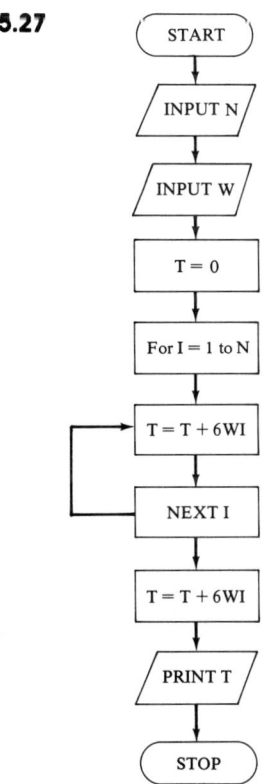

```
100 PRINT "ENTER NO. OF STEPS:"
110 INPUT N
120 PRINT "ENTER WIDTH:"
130 INPUT W
140 LET T=0
150 FOR I=1 TO N
160 LET T=T+I*6*W
170 NEXT I
180 LET T=T+6*N*W
190 PRINT "TOTAL NO. OF BRICKS:"
200 PRINT T

ENTER NO. OF STEPS:
 ? 6
ENTER WIDTH:
 ? 6
TOTAL NO. OF BRICKS:
 972

ENTER NO. OF STEPS:
 ? 10
ENTER WIDTH:
 ? 4
TOTAL NO. OF BRICKS:
 1560
```

5.29
```
100 PRINT "ENTER RESISTANCE"
110 INPUT R1
120 LET R=47000
130 IF R1<0.8*R OR R1>1.2*R THEN 190
140 LET T=5
150 LET R1<0.95*R OR R1>1.05*R THEN LET T=10
160 IF R1<0.9*R OR R1>1.1*R THEN LET T=20
170 PRINT "TOLERANCE= "T
180 END
190 PRINT "REJECT"
200 END
```

(a) ENTER RESISTANCE
   ? 51600
  TOLERANCE=10

(b) ENTER RESISTANCE
   ? 44000
  TOLERANCE=10

(c) ENTER RESISTANCE
   ? 37200
  REJECT

(d) ENTER RESISTANCE
   ? 56100
  TOLERANCE=20

ANSWERS TO SELECTED EXERCISES

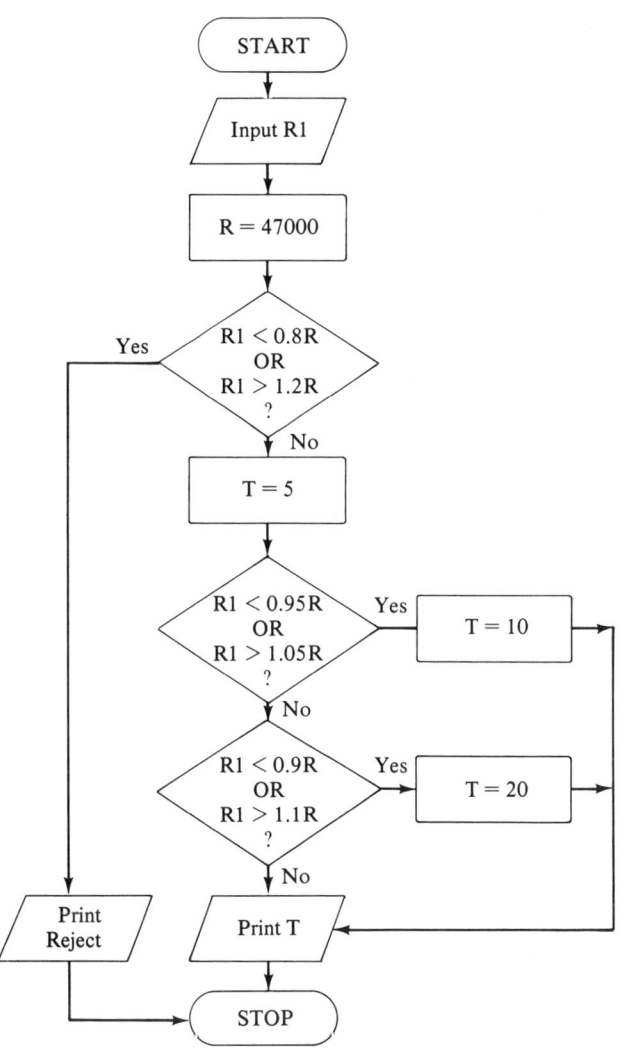

5.31

```
REGION 1        5.51868313
REGION 2        5.24345481
REGION 3        5.5
REGION 4        3.39992441

BUSINESS        4.29998241
INDUSTRIAL      4.05012346
RESIDENTIAL     6.3964409

OVERALL         4.91551559
```

5.33
```
100 LET P=1
110 FOR I=1 TO 10
120 PRINT "COMPONENT"I ":"
130 INPUT "              ";P1
140 LET P=P*P1
150 NEXT I
160 PRINT
170 PRINT "PROB. FAILURE :"1-P
180 END
```

```
COMPONENT  1  :
              ? .99
COMPONENT  2  :
              ? .995
COMPONENT  3  :
              ? .989
COMPONENT  4  :
              ? .999
COMPONENT  5  :
              ? .994
COMPONENT  6  :
              ? .99
COMPONENT  7  :
              ? .987
COMPONENT  8  :
              ? .999
COMPONENT  9  :
              ? .985
COMPONENT 10  :
              ? .995
PROB. FAILURE : .074484735
```

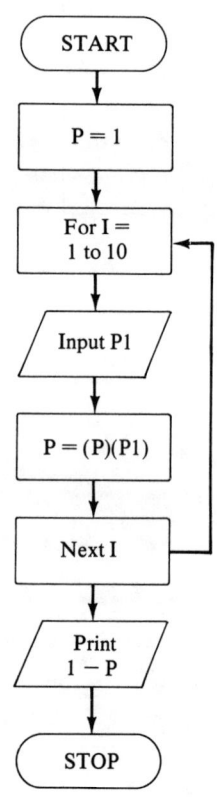

```
5.35  100 INPUT "SPEED DEV.";S
      110 INPUT "BEAR. DEV.";B
      120 INPUT "ALT. DEV.";A
      130 IF A>200 THEN 230
      140 IF A<=50 THEN 210
      150 IF S>50 THEN 230
```

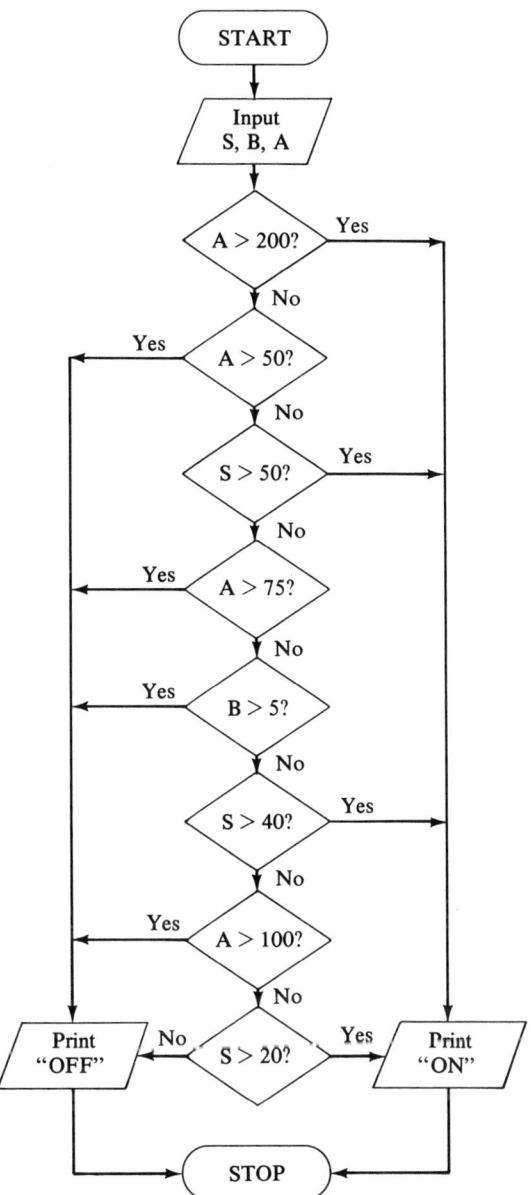

```
160 IF A<=75 THEN 210
170 IF B<=5 THEN 210
180 IF S>40 THEN 230
190 IF A<=100 THEN 210
200 IF S>20 THEN 230
210 PRINT "ALARM OFF"
220 END
230 PRINT "ALARM ON"
240 END

SPEED DEV. ?  33
BEAR. DEV. ?  6.5
ALT. DEV. ?  110
ALARM ON
```

```
SPEED DEV. ?  50
BEAR. DEV. ?  4.2
ALT.  DEV. ?  25
ALARM OFF

SPEED DEV. ?  60
BEAR. DEV. ?  8
ALT.  DEV. ?  60
ALARM ON

SPEED DEV. ?  10
BEAR. DEV. ?  2.3
ALT.  DEV. ?  225
ALARM ON

SPEED DEV. ?  40
BEAR. DEV. ?  7.8
ALT.  DEV. ?  25
ALARM OFF
```

5.37
```
OPERATOR A    50.9842578
OPERATOR B    47.5540486
OPERATOR C    49.3128227
OPERATOR D    50.1930556

MACHINE 1     50.102738
MACHINE 2     47.7870153
MACHINE 3     50.6433852

OVERALL       49.5110462
```

CHAPTER 6

6.1 (a) 26

(b)
```
 3      2     -1
 6      5     -1
11     10     -1
```

(c)
```
R( 1 )= 4
R( 2 )= .5
R( 3 )= 4.5
```

(d)
```
96
97
98
99
100
```

(e)
```
YES
WE
HAVE
NO
BANANAS
```

ANSWERS TO SELECTED EXERCISES

6.3
```
100 LET X=0
110 FOR I=1 TO 10
120 LET X=X+2
130 LET A(I)=X-1
140 LET B(I)=X
150 LET C(I)=A(I)*B(I)
160 PRINT C(I)
170 NEXT I
180 END
```

```
2
12
30
56
90
132
182
240
306
380
```

6.5

1	1	1	1	1
1	2	2	2	2
1	2	3	3	3
1	2	3	4	4
1	2	3	4	5

6.7
```
100 FOR I=1 TO 5
110 READ X(I), Y(I)
120 NEXT I
130 FOR I=1 TO 5
140 LET P=X(I)*Y(I)+P
150 NEXT I
160 PRINT "INNER PRODUCT=" P
170 DATA 2.3, 12.7, 1.7, -14
180 DATA -.9, 6.8, 4.6, 11.9
190 DATA 0,15.5
```

INNER PRODUCT= 54.03

6.9
```
100 FOR I=1 TO 10
110 READ A(I)
120 NEXT I
130 LET M=A(1)
140 FOR I=1 TO 10
150 IF A(I)<M THEN LET M=A(I)
160 NEXT I
170 PRINT "MINIMUM=" M
180 DATA 12, -8, 6.5, -0.4, 17.1
190 DATA -11.9, 2.4, 5.2, 1.8, 0
200 END
```

MINIMUM=-11.9

6.11
```
ENTER ID, SALES, EXPENSES
?  4 , 10050 , 810
ENTER ID, SALES, EXPENSES
?  1 , 6580 , 866
ENTER ID, SALES, EXPENSES
?  2 , 15790 , 1090
ENTER ID, SALES, EXPENSES
?  3 , 1800 , 2000
ENTER ID, SALES, EXPENSES
?  5 , 15010 , 310
SALESMEN 2 AND 5 TIE WITH NET SALES OF $ 14700
```

6.13
```
100 DIM M(12)
110 LET L=12
120 LET Q=500
130 PRINT " X", "MOMENT"
140 FOR I=0 TO 12
150 LET M(I)=I*Q/2*(-I+L)
160 PRINT I, M(I)
170 NEXT I
180 LET A=M(0)
190 FOR I=0 TO 12
200 IF A>M(I) THEN 230
210 LET A=M(I)
220 LET X=I
230 NEXT I
240 PRINT
250 PRINT "MAXIMUM M=" A "AT X=" X
```

```
X         MOMENT
0         0
1         2750
2         5000
3         6750
4         8000
5         8750
6         9000
7         8750
8         8000
9         6750
10        5000
11        2750
12        0

MAXIMUM M= 9000 AT X= 6
```

6.15
```
100 LET E=12
110 LET S=4
120 PRINT "RL","POWER"
130 FOR L=0 TO 10
140 LET P(L)=(E/(S+L))**2*L
150 PRINT L, P(L)
160 NEXT L
```

```
170 PRINT
180 LET M=P(0)
190 FOR I=1 TO 10
200 IF M>P(I) THEN 230
210 LET L=I
220 LET M=P(I)
230 NEXT I
240 PRINT "MAX. POWER=" M "WATTS AT RL=" L "OHMS."
250 END
```

```
RL        POWER
 0         0
 1         5.76
 2         8
 3         8.81632654
 4         9
 5         8.8888889
 6         8.64
 7         8.33057852
 8         8
 9         7.66863906
10         7.34693879
MAX. POWER = 9 WATTS AT RL = 4 OHMS.
```

6.17

AMPLITUDE	REAL PART	IMAG. PART	INSTRUMENT
14.1421356	10	10	OSCILLOSCOPE
44.7213596	40	20	OSCILLOSCOPE
50.0000001	40	-30	OSCILLOSCOPE
58.309519	30	-50	VOM
60.8276254	10	60	DIGITAL VOLTMETER
70.7106781	50	50	SPECTRUM ANALYZER
76.1577311	70	30	VOM
82.4621126	20	-80	OSCILLOSCOPE
90.5538513	10	90	DIGITAL VOLTMETER
100	100	0	DIGITAL VOLTMETER
100	0	-100	SPECTRUM ANALYZER
113.137085	80	80	SPECTRUM ANALYZER

6.19
```
100 DIM M$(5,2), D(5,3)
110 REM : M$ CONTAINS STRING DATA, AND
120 REM : D CONTAINS NUMERIC DATA IN ORDER GIVEN IN
    TABLE.
130 FOR I=1 TO 5
140 READ M$(I,1)
150 FOR J=1 TO 3
160 READ D(I,J)
170 NEXT J
180 READ M$(I,2)
190 NEXT I
200 INPUT "MACHINE TYPE    "; T$
210 PRINT
220 FOR I=1 TO 5
```

```
230 IF T$=M$(I,1) THEN 280
240 NEXT I
250 PRINT "BAD MACHINE TYPE - REDO."
260 PRINT
270 GOTO 200
280 PRINT "MACHINE TYPE   :", "  " T$
290 PRINT "MACHINE AGE    :", D(I,1)
300 PRINT "PRODUCTIVITY   :", D(I,2)/D(I,3)
310 PRINT "SKILL OF OPTR.:", "  " M$(I,2)
320 PRINT
330 GOTO 200
340 DATA A, 7, 1473, 12.4, JOURNEYMAN
350 DATA B, 2, 2060, 11.1, APPRENTICE
360 DATA C, 15, 1389, 10.9, JOURNEYMAN
370 DATA D, 9, 2552, 14.7, MASTER
380 DATA E, 5, 989, 9.6, APPRENTICE

MACHINE TYPE      ? A
MACHINE TYPE   :          A
MACHINE AGE    :          7
PRODUCTIVITY   :          118.790323
SKILL OF OPTR.:           JOURNEYMAN

MACHINE TYPE      ? B
MACHINE TYPE   :          B
MACHINE AGE    :          2
PRODUCTIVITY   :          185.585586
SKILL OF OPTR.:           APPRENTICE

MACHINE TYPE      ? C
MACHINE TYPE   :          C
MACHINE AGE    :          15
PRODUCTIVITY   :          127.431193
SKILL OF OPTR.:           JOURNEYMAN

MACHINE TYPE      ? D
MACHINE TYPE   :          D
MACHINE AGE    :          9
PRODUCTIVITY   :          173.605442
SKILL OF OPTR.:           MASTER

MACHINE TYPE      ? E
MACHINE TYPE   :          E
MACHINE AGE    :          5
PRODUCTIVITY   :          103.020833
SKILL OF OPTR.:           APPRENTICE
```

6.21 PREMIUM BATCHES

```
   5
   7
   2
   9
 AVG. SUB. PER LB.=3.9375
```

```
COMMERCIAL BATCHES

  12
  4
  1
  10
AVG. SUB. PER LB.= 9.5

NONBRAND BATCHES

  8
  3
  6
  11
AVG. SUB. PER LB.= 21.0625
```

6.23
```
100 DIM A(10), W(10)
110 PRINT "TIME","ANG. VEL.","ANG. ACC."
120 PRINT "----","---------","---------"
130 FOR T=0 TO 10
140 LET A(T)=4*T-16
150 LET W(T)=2*T*T-16*T+42
160 PRINT T, W(T), A(T)
170 NEXT T
180 LET M=0
190 FOR T=1 TO 10
200 IF W(T)<W(M) THEN LET M=T
210 NEXT T
220 PRINT
230 PRINT "MINIMUM ANGULAR VELOCITY:"
240 PRINT "TIME               :"; M
250 PRINT "ANGULAR VELOCITY   :"; W(M)
260 PRINT "ANG. ACCELERATION  :"; A(M)
270 END
```

TIME	ANG.VEL.	ANG. ACC.
0	42	-16
1	28	-12
2	18	-8
3	12	-4
4	10	0
5	12	4
6	18	8
7	28	12
8	42	16
9	60	20
10	82	24

```
MINIMUM ANGULAR VELOCITY:
TIME               : 4
ANGULAR VELOCITY   : 10
ANG. ACCELERATION  : 0
```

CHAPTER 7

7.1 (a) 16

 (b) -7.36117E-06

 (c) 1

 (d) 1
 0
 0
 1
 1

 (e) 1

 (f) -45
 -45
 -45
 -45

 (g) 1

 (h) 1
 0
 -1

 (i) 1

 (j) .367879

7.3 ENTER FX AND FY

 ? 10 , 10

 MAGNITUDE: 14.1421356
 ANGLE   : .0137077839

 ENTER FX AND FY

 ? 300 ,-400

 MAGNITUDE: 500
 ANGLE   :- .0161843547

7.5
```
100 PRINT " X" TAB(15) " EXP(X)" TAB(30) " SINH(X)" ;
110 PRINT TAB(45) " COSH(X)" TAB(60) " SUM"
120 PRINT
130 FOR X=0 TO 2 STEP 0.1
140 LET C=(EXP(X)+EXP(-X))/2
150 LET S=(EXP(X)-EXP(-X))/2
160 PRINT X TAB(15) EXP(X) TAB(30) S TAB(45) C TAB(60) C+S
170 NEXT X
180 END
```

X	EXP(X)	SINH(X)	COSH(X)	SUM
0	1	0	1	1
.1	1.10517	.100167	1.005	1.10517
.2	1.2214	.201336	1.02007	1.2214
.3	1.34986	.30452	1.04534	1.34986
.4	1.49182	.410752	1.08107	1.49182
.5	1.64872	.521095	1.12763	1.64872
.6	1.82212	.636654	1.18547	1.82212
.7	2.01375	.758584	1.25517	2.01375
.8	2.22554	.888106	1.33744	2.22554

.9	2.4596	1.02652	1.43309	2.4596
1	2.71828	1.1752	1.54308	2.71828
1.1	3.00417	1.33565	1.66852	3.00417
1.2	3.32012	1.50946	1.81066	3.32012
1.3	3.6693	1.69838	1.97091	3.6693
1.4	4.0552	1.9043	2.1509	4.0552
1.5	4.48169	2.12928	2.35241	4.48169
1.6	4.95303	2.37557	2.57747	4.95303
1.7	5.47395	2.64563	2.82832	5.47395
1.8	6.04965	2.94218	3.10747	6.04965
1.9	6.6859	3.26816	3.41773	6.6859
2	7.38906	3.62686	3.7622	7.38906

7.7
```
10 INPUT "ENTER ANGLE"; T
20 LET L=LOG(TAN(SQR(T)))
30 PRINT "PERCENT ERROR=" ABS(L-.5*LOG(T)-T/3)*100/L
40 END

ENTER ANGLE? 0.5
PERCENT ERROR= - 14.4253

ENTER ANGLE? 0.1
PERCENT ERROR= - .0716298

ENTER ANGLE? 0.01
PERCENT ERROR= - 3.52429E-04
```

7.9 (a) DEF FNF(X)=SQR((X*X+1)**2)

(b) DEF FNF(A)=EXP(-SIN(A))-EXP(-SIN(B))

(c) DEF FNF(K)=ABS(TAN(K-P1/2)**2-ABS(J))

(d) DEF FNF(N)=M-SGN(N)*N

7.11 (a) -2　　　　　　　　　　(c) 0
　　　　-1　　　　　　　　　　　　0
　　　　 0　　　　　　　　　　　　2.71828
　　　　 1
　　　　 2　　　　　　　　　　(d) 1
　　　　　　　　　　　　　　　　　2
　　　(b) 120　　　　　　　　　　3
　　　　　　　　　　　　　　　　　4
　　　　　　　　　　　　　　　　　5
　　　　　　　　　　　　　　　　　6
　　　　　　　　　　　　　　　　　36

7.13 A=INT(B*10+0.5)/10

7.15
```
10 DIM X(21), Y(21)
20 FOR I=1 TO 21
30 LET X(I)=(I-1)/10
```

```
40 LET Y(I)=12*EXP(-2*X(I))
50 NEXT I
60 GOSUB 1000
70 FOR I=1 TO 21
80 LET Y(I)=20-Y(I)
90 NEXT I
100 GOSUB 1000
110 FOR I=1 TO 21
120 LET Y(I)=6-12*EXP(-2*X(I))
130 NEXT I
140 GOSUB 1000
150 STOP
160 REM : "PLOTTER" SUBROUTINE STARTS AT LINE 1000
```

7.17
```
10 DIM X(21), Y(21)
20 PRINT "ENTER A, TAU"
30 INPUT A, T1
40 DEF FNY(I)=A*(1-EXP(-I/T1))
50 FOR I=1 TO 21
60 LET Y(I)=(I-1)/5
70 LET Y(I)=FNY(X(I))
80 NEXT I
90 GOSUB 1000
100 REM : "PLOTTER" SUBROUTINE STARTS ON LINE 1000.
110 STOP

ENTER A, TAU
? 10, 1
ENTER NUMBER OF VALUES TO BE PLOTTED.
? 21
ENTER MAXIMUM COLUMN NO. AVAILABLE FOR PLOT
? 50
```

X	Y
0	0
.2	1.81269
.4	3.2968
.6	4.51188
.8	5.50671
1	6.32121
1.2	6.98806
1.4	7.53403
1.6	7.98104
1.8	8.34701
2	8.64665
2.2	8.89197
2.4	9.09282
2.6	9.25726
2.8	9.3919
3	9.50213
3.2	9.59238
3.4	9.66627

```
              3.6            9.72676
              3.8            9.77629
              4              9.81684
```

7.19
```
100 DIM A(5,5)
110 FOR I=1 TO 5
120 FOR J=1 TO 5
130 READ A(I,J)
140 NEXT J
150 NEXT I
160 PRINT "ROW OR COLUMN?"
170 INPUT T$
180 IF T$="COLUMN" THEN GOSUB 270
190 IF T$="ROW" THEN GOSUB 400
200 FOR I=1 TO 5
210 FOR J=1 TO 5
220 PRINT TAB(6*J) A(I,J) ;
230 NEXT J
240 PRINT
250 NEXT I
260 END
270 REM : TO SORT COLUMNS
280 FOR J=1 TO 5
290 LET E=0
300 FOR I=1 TO 4
310 IF A(I,J)<=A(I+1,J) THEN 360
320 LET S=A(I,J)
330 LET A(I,J)=A(I+1,J)
340 LET A(I+1,J)=S
350 LET E=E+1
360 NEXT I
370 IF E>0 THEN 290
380 NEXT J
390 RETURN
400 REM : TO SORT ROWS
410 FOR I=1 TO 5
420 LET E=0
430 FOR J=1 TO 4
440 IF A(I,J)<=A(I,J+1) THEN 490
450 LET S=A(I,J)
460 LET A(I,J)=A(I,J+1)
470 LET A(I,J+1)=S
480 LET E=E+1
490 NEXT J
500 IF E>0 THEN 420
510 NEXT I
520 RETURN
530 DATA 5,  6,  2,  3,  1
540 DATA 2,  5,  9,  0, -4
550 DATA 1,  0, -9,  2,  8
560 DATA 6,  4,  5, -1,  3
570 DATA-4,  2,  3,  4,  5
```

```
ROW OR COLUMN?
? ROW
         1        2        3        5        6
        -4        0        2        5        9
        -9        0        1        2        8
        -1        3        4        5        6
        -4        2        3        4        5

ROW OR COLUMN?
? COLUMN
        -4        0       -9       -1       -4
         1        2        2        0        1
         2        4        3        2        3
         5        5        5        3        5
         6        6        9        4        8
```

7.21
```
100 DEF FNH(A)=K*I*COS(A)**2+F*COS(A)
110 DEF FNV(A)=K*I*.5*SIN(2*A)+F*SIN(A)
120 LET K=100
130 LET F=1
140 INPUT "ENTER I"; I
150 PRINT " A" TAB(18) "H" TAB(33) "V"
160 FOR A=1 TO 30 STEP .5
170 PRINT A TAB(15) FNH(A/57.295) TAB(30) FNV(A/
57.295)
180 NEXT A
190 END

ENTER I? 1.3
A                H                V
1                130.96           2.28595
1.5              130.911          3.42806
2                130.841          4.56913
2.5              130.752          5.70882
3                130.643          6.84678
3.5              130.514          7.98266
4                130.365          9.11613
4.5              130.197          10.2468
5                130.009          11.3744
5.5              129.801          12.4986
6                129.574          13.619
6.5              129.328          14.7352
7                129.062          15.847
7.5              128.777          16.954
8                128.472          18.0558
8.5              128.149          19.1522
9                127.806          20.2428
9.5              127.445          21.3273
10               127.065          22.4052
10.5             126.666          23.4765
11               126.248          24.5406
11.5             125.813          25.5972
12               125.359          26.6461
12.5             124.886          27.687
13               124.396          28.7194
```

ANSWERS TO SELECTED EXERCISES

13.5	123.888	29.7432
14	123.362	30.758
14.5	122.818	31.7634
15	122.257	32.7592
15.5	121.679	33.7451
16	121.084	34.7208
16.5	120.472	35.686
17	119.844	36.6403
17.5	119.198	37.5836
18	118.537	38.5155
18.5	117.859	39.4357
19	117.166	40.344
19.5	116.457	41.2401
20	115.732	42.1237
20.5	114.992	42.9945
21	114.238	43.8523
21.5	113.468	44.6969
22	112.684	45.5279
22.5	111.885	46.3451
23	111.073	47.1483
23.5	110.246	47.9372
24	109.407	48.7116
24.5	108.553	49.4713
25	107.687	50.216
25.5	106.808	50.9455
26	105.916	51.6596
26.5	105.012	52.358
27	104.096	53.0406
27.5	103.169	53.7071
28	102.23	54.3574
28.5	101.28	54.9912
29	100.319	55.6084
29.5	99.3471	56.2088
30	98.3652	56.7921

ENTER I? 0.62

A	H	V
1	62.981	1.09935
1.5	62.9572	1.64861
2	62.9239	2.19738
2.5	62.8811	2.74548
3	62.8288	3.29276
3.5	62.7671	3.83905
4	62.6959	4.38418
4.5	62.6153	4.92799
5	62.5252	5.47032
5.5	62.4259	6.011
6	62.3171	6.54988
6.5	62.199	7.08678
7	62.0717	7.62155
7.5	61.9352	8.15402
8	61.7894	8.68405
8.5	61.6345	9.21145
9	61.4704	9.73609
9.5	61.2973	10.2578

10	61.1153	10.7764
10.5	60.9242	11.2918
11	60.7243	11.8038
11.5	60.5155	12.3122
12	60.298	12.8169
12.5	60.0718	13.3178
13	59.8369	13.8146
13.5	59.5935	14.3073
14	59.3416	14.7957
14.5	59.0813	15.2797
15	58.8126	15.759
15.5	58.5357	16.2336
16	58.2507	16.7033
16.5	57.9575	17.168
17	57.6564	17.6276
17.5	57.3473	18.0818
18	57.0305	18.5306
18.5	56.7059	18.9738
19	56.3737	19.4113
19.5	56.034	19.843
20	55.6869	20.2687
20.5	55.3325	20.6883
21	54.9709	21.1017
21.5	54.6022	21.5087
22	54.2265	21.9093
22.5	53.844	22.3032
23	53.4547	22.6905
23.5	53.0588	23.071
24	52.6563	23.4445
24.5	52.2475	23.8109
25	51.8324	24.1702
25.5	51.4112	24.5223
26	50.984	24.8669
26.5	50.5509	25.2041
27	50.112	25.5338
27.5	49.6676	25.8557
28	49.2176	26.1699
28.5	48.7623	26.4762
29	48.3018	26.7745
29.5	47.8362	27.0648
30	47.3656	27.347

7.23 (a) LIVING AREA   ? 1654   (b) LIVING AREA   ? 2040
      BATHROOMS   ? 2.5   BATHROOMS   ? 3.5
      UNHEATED AREA? 155   UNHEATED AREA? 240
      FLOORS   ? 1   FLOORS   ? 2

      TOTAL COST = 76227.9   TOTAL COST = 88661.2
      COST/SQ.FT.= 46.087   COST/SQ.FT.= 43.4614

7.25
```
100 DEFINT R,S
110 DEF FNA=SQR((R/(2*L))**2-1/(L*C))
120 DEF FNB=SQR(1/(L*C)-(R/(2*L))**2)
130 INPUT "E"; E
```

```
140 INPUT "R"; R
150 INPUT "L"; L
160 INPUT "C"; C
170 LET S=2*SQR(L/C)
180 PRINT
190 IF R<S THEN 310
200 IF R=S THEN 260
210 PRINT "OVER DAMPED"
220 PRINT
230 PRINT " T", " I(T)"
240 GOSUB 360
250 GOTO 350
260 PRINT "CRITICALLY DAMPED"
270 PRINT
280 PRINT " T", " I(T)"
290 GOSUB 430
300 GOTO 350
310 PRINT "UNDER DAMPED"
320 PRINT
330 PRINT " T", " I(T)"
340 GOSUB 490
350 END
360 REM : OVER DAMPED SUBROUTINE
370 FOR T=0 TO 10*L/R STEP 0.2*L/R
380 LET I=EXP(FNA*T)-EXP(-FNA*T)
390 LET I=I*EXP(-R*T/(2*L))*E/(2*FNA*L)
400 PRINT T,I
410 NEXT T
420 RETURN
430 REM : CRITICALLY DAMPED SUBROUTINE
440 FOR T=0 TO 10*L/R STEP 0.2*L/R
450 LET I=(E/L)*T*EXP(-R*T/(2*L))
460 PRINT T,I
470 NEXT T
480 RETURN
490 REM : UNDER DAMPED SUBROUTINE
500 FOR T=0 TO 10*L/R STEP 0.2*L/R
510 LET I=E/(L*FNB)*EXP(-R*T/(2*L))*SIN(FNB*T)
520 PRINT T,I
530 NEXT T
540 RETURN
```

(a) E? 10
   R? 40
   L? 20E-3
   C? 50E-6

   CRITICALLY DAMPED

T	I(T)
0	0
1E-04	.0452419
2E-04	.0818731
3E-04	.111123
4E-04	.134064
5E-04	.151633

6E-04	.164644
7E-04	.173805
8E-04	.179732
9E-04	.182956
1E-03	.18394
1.1E-03	.183079
1.2E-03	.180717
1.3E-03	.177146
1.4E-03	.172618
1.5E-03	.167348
1.6E-03	.161517
1.7E-03	.155281
1.8E-03	.148769
1.9E-03	.14209
2E-03	.135335
2.1E-03	.128579
2.2E-03	.121883
2.3E-03	.115298
2.4E-03	.108862
2.5E-03	.102606
2.6E-03	.0965556
2.7E-03	.0907274
2.8E-03	.085134
2.9E-03	.0797836
3E-03	.0746805
3.1E-03	.0698262
3.2E-03	.0652195
3.3E-03	.0608572
3.4E-03	.0567345
3.5E-03	.0528454
3.6E-03	.0491826
3.7E-03	.0457385
3.8E-03	.0425044
3.9E-03	.0394717
4E-03	.0366312
4.1E-03	.0339739
4.2E-03	.0314907
4.3E-03	.0291724
4.4E-03	.0270101
4.5E-03	.0249952
4.6E-03	.0231192
4.7E-03	.0213739
4.8E-03	.0197514
4.9E-03	.0182441
5E-03	.0168449

(b) E? 10
    R? 120
    L? 4E-3
    C? 10E-6

OVER DAMPED

T	I(T)
0	0
6.66667E-06	.015103
1.33333E-05	.027453

T	I(T)
2E-05	.0375368
2.66667E-05	.0457548
3.33333E-05	.0524371
4E-05	.0578554
4.66667E-05	.0622331
5.33333E-05	.0657547
6E-05	.0685717
6.66667E-05	.0708091
7.33333E-05	.0725695
8E-05	.0739378
8.66667E-05	.0749836
9.33333E-05	.0757643
1E-04	.0763272
1.06667E-04	.0767112
1.13333E-04	.0769484
1.2E-04	.0770651
1.26667E-04	.077083
1.33333E-04	.0770201
1.4E-04	.076891
1.46667E-04	.0767079
1.53333E-04	.0764807
1.6E-04	.0762177
1.66667E-04	.0759256
1.73333E-04	.07561
1.8E-04	.0752755
1.86667E-04	.0749259
1.93333E-04	.0745641
2E-04	.0741929
2.06667E-04	.0738142
2.13333E-04	.0734299
2.2E-04	.0730413
2.26667E-04	.0726495
2.33333E-04	.0722557
2.4E-04	.0718604
2.46667E-04	.0714645
2.53333E-04	.0710684
2.6E-04	.0706725
2.66667E-04	.0702772
2.73333E-04	.0698828
2.8E-04	.0694895
2.86667E-04	.0690975
2.93333E-04	.068707
3E-04	.0683182
3.06667E-04	.0679309
3.13333E-04	.0675455
3.2E-04	.067162
3.26667E-04	.0667803

(c) E? 10
R? 10
L? 40E-3
C? 100E-6

UNDER DAMPED

T	I(T)
0	0

8E-04	.176477
1.6E-03	.295712
2.4E-03	.351018
3.2E-03	.346071
4E-03	.2925
4.8E-03	.206785
5.6E-03	.107018
6.4E-03	.0100221
7.2E-03	-.0708256
8E-03	-.126883
8.8E-03	-.154623
9.6E-03	-.155209
.0104	-.13348
.0112	-.096589
.012	-.052564
.0128	-8.99792E-03
.0136	.0279586
.0144	.0542153
.0152	.0679546
.016	.0694795
.0168	.060786
.0176	.0449702
.0184	.0255864
.0192	6.05508E-03
.02	-.0108023
.0208	-.0230582
.0216	-.029793
.0224	-.0310438
.0232	-.0276257
.024	-.0208741
.0248	-.0123595
.0256	-3.61969E-03
.0264	4.05376E-03
.0272	9.75621E-03
.028	.0130289
.0288	.013844
.0296	.0125304
.0304	9.66188E-03
.0312	5.93077E-03
.032	2.02735E-03
.0328	-1.4586E-03
.0336	-4.10394E-03
.0344	-5.68253E-03
.0352	-6.16182E-03
.036	-5.67253E-03
.0368	-4.46023E-03
.0376	-2.82945E-03
.0384	-1.08942E-03
.0392	4.91098E-04
.04	1.71484E-03

7.27
```
100 DEF FNL(S)=2*S*COS(FNT(S))+P1*(R1+R2+(R1-R2)*FNT(S)/90)
110 DEF FNX(S)=((R1-R2)/S)
120 DEF FNT(S)=ATN(FNX(S)/SQR(1-FNX(S)*FNX(S)))*P1/180
```

ANSWERS TO SELECTED EXERCISES

```
130 P1=4*ATN(1)
140 INPUT "R1"; R1
150 INPUT "R2"; R2
160 PRINT
170 PRINT " S", " LENGTH"
180 FOR S=48 TO 240 STEP 6
190 PRINT S, FNL(S)
200 NEXT S
210 END
```

(a) R1? 18
   R2? 6

S	LENGTH
48	171.399
54	183.399
60	195.399
66	207.399
72	219.399
78	231.399
84	243.399
90	255.399
96	267.399
102	279.399
108	291.399
114	303.399
120	315.399
126	327.399
132	339.399
138	351.399
144	363.399
150	375.399
156	387.399
162	399.399
168	411.398
174	423.398
180	435.398
186	447.398
192	459.398
198	471.398
204	483.398
210	495.398
216	507.398
222	519.398
228	531.398
234	543.398
240	555.398

(b) R1? 9
   R2? 9

S	LENGTH
48	152.549
54	164.549
60	176.549

66	188.549
72	200.549
78	212.549
84	224.549
90	236.549
96	248.549
102	260.549
108	272.549
114	284.549
120	296.549
126	308.549
132	320.549
138	332.549
144	344.549
150	356.549
156	368.549
162	380.549
168	392.549
174	404.549
180	416.549
186	428.549
192	440.549
198	452.549
204	464.549
210	476.549
216	488.549
222	500.549
228	512.549
234	524.549
240	536.549

7.29
```
     TYPE? A
         P? .08
     PROB. OF ACCEPTANCE= .524931
     PROB. OF REJECTION= .475069

     TYPE? B
         P? .035
     PROB. OF ACCEPTANCE= .965858
     PROB. OF REJECTION= .0341416
```

7.31
```
     100 DEF FNB=B1*B1/(M*M)
     110 DEF FNR=SQR(FNB-4*K/M)*0.5
     120 DEF FNS=B1*0.5/SQR(K*M)
     130 DEF FNW=SQR(K/M)
     140 LET P1=4*ATN(1)
     150 INPUT "F"; F
     160 INPUT "M"; M
     170 INPUT "K"; K
     180 INPUT "B"; B1
     190 LET S=4*K/M
     200 IF FNB<S THEN GOSUB 420
     210 IF FNB=S THEN GOSUB 330
```

ANSWERS TO SELECTED EXERCISES 341

```
220 IF FNB>S THEN GOSUB 240
230 END
240 LET A=B1/M*0.5+FNR
250 LET B2=B1/M*0.5-FNR
260 PRINT " (B/M)**2>4K/M"
270 FOR T=0 TO 10*M/B1 STEP 0.2*M/B1
280 LET X=1-B2*EXP(-A*T)/(B2-A)+A*EXP(-B2*T)/(B2-A)
290 LET X=X*F/K
300 PRINT T,X
310 NEXT T
320 RETURN
330 PRINT " (B/M)**2=4K/M"
340 PRINT " T", " X"
350 LET A=B1/M*0.5
360 FOR T=0 TO 10*M/B1 STEP 0.2*M/B1
370 LET X=1-EXP(-A*T)-A*T*EXP(-A*T)
380 LET X=X*F/(M*A*A)
390 PRINT T, X
400 NEXT T
410 RETURN
420 PRINT " (B/M)**2<4K/M"
430 LET N=SQR(1-FNS*FNS)
440 LET D=-FNS
450 LET A=ATN(ABS(N/D))
460 IF N<0 AND D<0 THEN A=P1-A
470 IF N>0 AND D<0 THEN A=A-P1
480 IF N<0 AND D>0 THEN A=-A
490 PRINT " T", "X"
500 FOR T=0 TO 10*M/B1 STEP 0.2*M/B1
510 LET X=SIN(FNW*N*T-A)
520 PRINT T,X
530 NEXT T
540 RETURN
```

(a) F? 200
M? 2
K? 5000
B? 200
  (B/M)**2=4K/M

T	X
0	0
2E-03	1.87153E-04
4E-03	7.00923E-04
6E-03	1.47745E-03
8E-03	2.46208E-03
.01	3.60816E-03
.012	4.87606E-03
.014	6.2322E-03
.016	7.64832E-03
.018	9.10071E-03
.02	.0105696
.022	.0120388
.024	.0134949
.026	.0149271

.028	.0163267
.03	.017687
.032	.0190028
.034	.0202702
.036	.0214865
.038	.02265
.04	.0237598
.042	.0248154
.044	.0258172
.046	.0267658
.048	.0276624
.05	.0285081
.052	.0293046
.054	.0300536
.056	.0307569
.058	.0314164
.06	.0320341
.062	.0326119
.064	.033152
.066	.0336561
.068	.0341263
.07	.0345645
.072	.0349724
.074	.035352
.076	.0357048
.078	.0360326
.08	.0363369
.082	.0366192
.084	.0368809
.086	.0371235
.088	.0373481
.09	.037556
.092	.0377484
.0939999	.0379263
.0959999	.0380907
.0979999	.0382426
.1	.0383829

(b) F? 20
   M? 2
   K? 5000
   B? 100
     (B/M)**2<4K/M

T	X
0	.866025
4E-03	.766897
8E-03	.64482
.012	.503446
.016	.347006
.02	.180182
.024	7.96662E-03
.028	-.164488
.032	-.332019
.036	-.489615

.04	-.63256
.044	-.756574
.048	-.857949
.052	-.933649
.056	-.98141
.06	-.999802
.064	-.988275
.068	-.947173
.072	-.877728
.076	-.782016
.08	-.662902
.084	-.523951
.088	-.369321
.092	-.203639
.096	-.0318625
.1	.140867
.104	.309381
.108	.468637
.112	.613869
.116	.740731
.12	.845426
.124	.924822
.128	.976543
.132	.99904
.136	.991642
.14	.954568
.144	.888929
.148	.796689
.152	.680608
.156	.54416
.16	.391428
.164	.226987
.168	.0557442
.172	-.117163
.176	-.286562
.18	-.447387
.184	-.594824
.188	-.72446
.192	-.832418
.196	-.915464
.2	-.971116

(c) F? 200
   M? 2
   K? 5000
   B? 500
   (B/M)**2>4K/M

0	0
8E-04	2.99645E-05
1.6E-03	1.12453E-04
2.4E-03	2.37827E-04
3.2E-03	3.98126E-04
4E-03	5.86786E-04
4.8E-03	7.98397E-04
5.6E-03	1.02849E-03

6.4E-03	1.27339E-03
7.2E-03	1.53004E-03
8E-03	1.79595E-03
8.8E-03	2.06905E-03
9.6E-03	2.34765E-03
.0104	2.63034E-03
.0112	2.91596E-03
.012	3.20359E-03
.0128	3.49243E-03
.0136	3.78185E-03
.0144	4.07132E-03
.0152	4.36041E-03
.016	4.64878E-03
.0168	4.93613E-03
.0176	5.22223E-03
.0184	5.5069E-03
.0192	5.78999E-03
.02	6.07136E-03
.0208	6.35092E-03
.0216	6.62859E-03
.0224	6.90432E-03
.0232	7.17806E-03
.024	7.44977E-03
.0248	7.71942E-03
.0256	7.987E-03
.0264	8.25249E-03
.0272	8.51588E-03
.028	8.77719E-03
.0288	9.03639E-03
.0296	9.29351E-03
.0304	9.54855E-03
.0312	9.80151E-03
.032	.0100524
.0328	.0103012
.0336	.010548
.0344	.0107928
.0352	.0110355
.036	.0112763
.0368	.011515
.0376	.0117518
.0384	.0119866
.0392	.0122194
.04	.0124504

CHAPTER 8

8.1 (a) Dependent
(b) $x = y = -11$
(c) $a = -14$, $b = -2$, $c = -21$
(d) Nonlinear
(e) Two equations with three unknowns
(f) Dependent

ANSWERS TO SELECTED EXERCISES

8.3 (a) = −4
(b) = 300.16
(c) = 0

8.5
```
10 PRINT "ENTER D1 & S1"
20 INPUT D1, S1
30 PRINT "ENTER D2 & S2"
40 INPUT D2, S2
50 PRINT
60 LET M=D1-D2
70 PRINT "COST= " (S2*D1-S1*D2)/M
80 PRINT "NUMBER=" (S2-S1)/M
90 END

ENTER D1 & S1
? 20, 500
ENTER D2 & S2
? 30, 250

COST  = 1000
NUMBER= 25
```

8.7 (a)
$$\det \begin{bmatrix} a & b & c \\ d & e & f \\ g & h & i \end{bmatrix} = c \det \begin{bmatrix} d & e \\ g & h \end{bmatrix} - f \det \begin{bmatrix} a & b \\ g & h \end{bmatrix} + i \det \begin{bmatrix} a & b \\ d & e \end{bmatrix}$$

$$= c(dh - ge) - f(ah - bg) + i(ae - bd)$$

(b)
$$\det \begin{bmatrix} a & b & c \\ d & e & f \\ g & h & i \end{bmatrix} = a \det \begin{bmatrix} e & f \\ h & i \end{bmatrix} - b \det \begin{bmatrix} d & f \\ g & i \end{bmatrix} + c \det \begin{bmatrix} d & e \\ g & h \end{bmatrix}$$

$$= a(ei - hf) - b(di - fg) + c(dh - eg)$$

8.9
```
10 PRINT "ENTER A1, B1, C1, & K1"
20 INPUT A, B, C, K1
30 PRINT "ENTER A2, B2, C2, & K2"
40 INPUT D, E, F, K2
50 PRINT "ENTER A3, B3, C3, & K3"
60 INPUT G, H, I, K3
70 GOSUB 1020
80 PRINT
90 PRINT " A="; -LOG(C1)
100 PRINT " B=" EXP(C2)
110 PRINT " C=" ATN(C3)
120 END
1020 LET D1=A*(E*I-F*H)-D*(B*I-C*H)+G*(B*F-C*E)
1030 LET D2=K1*(E*I-F*H)-K2*(B*I-C*H)+K3*(B*F-C*E)
1040 LET C1=D2/D1
1050 LET D3=A*(K2*I-F*K3)-D*(K1*I-C*K3)+G*(K1*F-C*K2)
1060 LET C2=D3/D1
1070 LET D4=A*(E*K3-K2*H)-D*(B*K3-K1*H)+G*(B*K2-K1*E)
```

```
1080 LET C3=D4/D1
1090 RETURN

ENTER A1, B1, C1, & K1
? 5, 6, 3, 8
ENTER A2, B2, C2, & K2
? 2, -9, -2, 0
ENTER A3, B3, C3, & K3
? -1, 4, 6, 5

A= 0
B= 1
C= .785398
```

8.11 (a) 4×4 identity matrix, because $I \times I = I$.
(b) No; improper dimensions.
(c) The dimension of AB is (3×3). BA is defined and has a dimension of (2×2).
(d) The dimension of AB is (1×1). BA is defined and has a dimension of $(n \times n)$.

8.13 (a) $\begin{bmatrix} X \\ Y \\ Z \end{bmatrix} = \begin{bmatrix} 1 & -9 & -1 \\ -2 & -3 & 4 \\ -1 & 1 & -2 \end{bmatrix}^{-1} \begin{bmatrix} 0 \\ -17 \\ 10 \end{bmatrix}$

(b) $\begin{bmatrix} a_1 \\ a_2 \end{bmatrix} = \begin{bmatrix} 3 & 3 \\ -12 & 4 \end{bmatrix}^{-1} \begin{bmatrix} 5 \\ -6 \end{bmatrix}$

(c) $\begin{bmatrix} R_1 \\ R_2 \\ R_3 \\ R_4 \end{bmatrix} = \begin{bmatrix} 0 & 0 & 1.5 & -2.5 \\ 1 & 0 & -2 & -1 \\ 0 & 9 & -1 & 2 \\ 1 & 1 & -1 & -1 \end{bmatrix}^{-1} \begin{bmatrix} 0 \\ -7.6 \\ 6 \\ 0 \end{bmatrix}$

8.15 (a)
```
100 DIM A(4,4), E(4,4), R(4,4), S(4,4), I(4,4)
110 MAT READ S, I
120 MAT A=S
130 MAT E=INV(S)
140 PRINT "MATRIX A IS :"
150 MAT PRINT A
160 PRINT
170 PRINT
180 PRINT
190 PRINT "INVERSE OF A IS :"
200 MAT PRINT E
210 PRINT
220 PRINT
230 PRINT I
240 MAT R=I*A
250 PRINT " I*A="
260 MAT PRINT R
270 PRINT
280 PRINT
290 PRINT
300 MAT R=A*I
```

```
310 PRINT " A*I="
320 MAT PRINT R
330 PRINT
340 PRINT
350 PRINT
360 MAT R=A*E
370 PRINT " A*INV(A)="
380 MAT PRINT R
390 PRINT
400 PRINT
410 PRINT
420 MAT R=E*A
430 PRINT " INV(A)*A="
440 MAT PRINT R
450 MAT R=A*E
460 MAT S=R
470 MAT R=S*A
480 PRINT
490 PRINT
500 PRINT
510 PRINT " A*INV(A)*A="
520 MAT PRINT R
530 DATA .20, .35, .40, .30
540 DATA .15, .30, .50, .10
550 DATA .50, .35, .10, .25
560 DATA .40, .20, .30, .20
570 DATA 1, 0, 0, 0
580 DATA 0, 1, 0, 0
590 DATA 0, 0, 1, 0
600 DATA 0, 0, 0, 1
610 END
```

```
MATRIX A IS :
 .200000          .350000          .400000          .300000
 .150000          .300000          .500000          .100000
 .500000          .350000          .100000          .250000
 .400000          .200000          .300000          .200000

INVERSE OF A IS :
-2.53076          .246046          .773286         2.70650
-.386643         3.09315          4.00703         -5.97540
-.140598         1.12478         -2.17926          2.37258
 5.65905        -5.27241         -2.28471          2.00351

I*A=
 .200000          .350000          .400000          .300000
 .150000          .300000          .500000          .100000
 .500000          .350000          .100000          .250000
 .400000          .200000          .300000          .200000

A*I=
 .200000          .350000          .400000          .300000
 .150000          .300000          .500000          .100000
 .500000          .350000          .100000          .250000
 .400000          .200000          .300000          .200000
```

```
A*INV(A)=
 1                 -2.22045E-16       -4.16334E-17        1.80411E-16
 2.22045E-16        1                 -3.19189E-16        1.38778E-16
 2.22045E-16       -4.44089E-16        1                  6.93889E-17
 2.22045E-16       -2.22045E-16        5.55112E-17        1

INV(A)*A=
 1                 -9.71445E-17       -1.11022E-16       -5.55112E-17
 0                  1.00000            6.66134E-16        2.22045E-16
 2.77556E-17        2.77556E-17        1                  5.55112E-17
 6.93889E-17       -6.93889E-17       -1.38778E-16        1.00000

A*INV(A)*A=
 .200000            .350000            .400000            .300000
 .150000            .300000            .500000            .100000
 .500000            .350000            .100000            .250000
 .400000            .200000            .300000            .200000
```

(b)
```
100 DIM A(3,3), B(3,3)
110 MAT READ A
120 PRINT "ORIGINAL MATRIX :"
130 MAT PRINT A
140 MAT B=INV(A)
150 PRINT
160 PRINT
170 PRINT
180 PRINT "THE INVERSE IS :"
190 MAT PRINT B
200 PRINT
210 DATA 2, 8, -4, 6, 0, 3, 4, 16, -8
220 END

ORIGINAL MATRIX :

 2                  8                 -4
 6                  0                  3
 4                 16                 -8

           140 SINGULAR MATRIX
```

The determinant of this matrix is zero. It is singular and has no inverse, so the INV function causes an error and program termination in statement 140.

8.17 SOLUTIONS
```
 2.93069E-02        .280297            .228911
```
CHECK
```
-3.66000            .480000            .110000
```

8.19
```
100 DIM A(3,3), C(3,1), D(1,3)
110 DIM B(3,3)
120 DIM R(1,1)
130 DIM K(3,1)
140 MAT READ C, A, D
150 MAT B=INV(A)
```

```
       160 MAT K=B*C
       170 PRINT
       180 PRINT
       190 PRINT
       200 FOR I=1 TO 3
       210 PRINT "K" I "=" K(I,1)
       220 NEXT I
       230 MAT R=D*K
       240 PRINT
       250 PRINT "COST=" R(1,1)
       260 DATA 283800, 248800, 594200
       270 DATA 10500, 470, 55000
       280 DATA 8400, 520, 60000
       290 DATA 25670, 880, 110200
       300 DATA 15000, 670, 85000
       310 END

       K 1= 23.0349
       K 2= 1314.64
       K 3=-10.4718

       COST= 336232.
```

8.21
```
       10 DIM A(2,2), V(2), K(2), E(2,2)
       20 PRINT "ENTER I1, I2, R1, R2, & R3"
       30 INPUT I1, I2, R1, R2, R3
       40 LET A(1,1)=1/R1+1/R2
       50 LET A(1,2)=-1/R2
       60 LET A(2,1)=-1/R2
       70 LET A(2,2)=-1/R2-1/R3
       80 LET K(1)=I1
       90 LET K(2)=I2
       100 MAT E=INV(A)
       110 MAT V=E*K
       120 PRINT "THE CURRENT IN R1=" V(1)/R1 " AMPS."
       130 PRINT "THE CURRENT IN R2="(V(1)-V(2))/R2 " AMPS."
       140 PRINT "THE CURRENT IN R3="V(2)/R3 " AMPS."
       150 END

       ENTER I1, I2, R1, R2, & R3
       ?25E-3, 40E-3, 3300, 1500, 1000
       THE CURRENT IN R1= 2.20588E-03 AMPS.
       THE CURRENT IN R2= 2.27941E-02 AMPS.
       THE CURRENT IN R3=-2.69118E-02 AMPS.
```

8.23
```
       10 DIM A(3,3), E(3,3)
       15 DIM C(1), K(3), V(3)
       20 MAT READ A, K
       30 MAT E=INV(A)
       40 MAT V=E*K
       50 MAT READ K
       60 MAT C=K*V
       70 PRINT "TOTAL COST ($)=" C(1)+5780
```

```
80 DATA 8.2, 10.1, -1, 12, -11.4
90 DATA 0, -78, 3, 5
100 DATA -220.45, 390.9, 658.5
105 DATA 9.75, 18.2, 0.15
110 END

TOTAL COST ($)= 7421.80
```

8.25
```
10 DIM A(2,2), E(2,2), V(2), K(2)
20 PRINT "INPUT L, L1, L2"
30 INPUT L, L1, L2
40 PRINT "DO YOU WANT F2 OR F3 TO BE AN UNKNOWN?"
50 INPUT T$
60 IF T$="F2" THEN 111
65 PRINT "INPUT F2 AND F4"
66 INPUT F2, F4
70 LET A(1,2)=-1
80 LET A(2,2)=L-L2
90 LET K(1)=F4-F2
100 LET K(2)=F2*L2
110 GOTO 160
111 PRINT "INPUT F3 AND F4"
112 INPUT F3, F4
120 LET A(1,2)=1
130 LET A(2,2)=L3
140 LET K(1)=F3+F4
150 LET K(2)=-F3*(L-L2)
160 LET A(1,1)=1
170 LET A(2,1)=L-L1-L2
180 MAT E=INV(A)
190 MAT V=E*K
200 IF T$="F2" THEN 230
210 PRINT "F3=" V(1)
220 GOTO 240
230 PRINT "F2=" V(1)
240 PRINT "F1=" V(2)
250 END
```

(a) INPUT L, L1, L2
?18, 6, 6
DO YOU WANT F2 OR F3 TO BE AN UNKNOWN?
?F3
INPUT F2 AND F4
?10, 20
F3=10.0000
F1=-2.22045E-16

(b) INPUT L, L1, L2
?24, 8, 6
DO YOU WANT F2 OR F3 TO BE AN UNKNOWN?
?F2
INPUT F3 AND F4
?10, 30
F2=-18.0000
F1=58.0000

CHAPTER 9

9.1
```
10 PRINT "INPUT A, M, N, X"
20 INPUT A, M, N, X
30 PRINT "DERIVATIVE=" A*M/N*X**(M/N-1)
40 PRINT
50 END
```

(a) INPUT A, M, N, X
    ? -5, 1, 3, 8
    DERIVATIVE=-.416667

(b) INPUT A, M, N, X
    ? 3.75, -3, 4, 17.63
    DERIVATIVE=-.0185418

9.3
```
100 PRINT "ENTER DEGREE OF POLYNOMIAL"
110 INPUT D
120 PRINT "ENTER UNIT OF LENGTH"
130 INPUT L$
140 PRINT "ENTER UNIT OF TIME"
150 INPUT T$
160 GOSUB 270
170 PRINT "ENTER T"
180 INPUT X
190 LET M=1
200 GOSUB 360
210 PRINT "VALUE OF 1ST DERIVATIVE=" S; L$ "/" T$
220 LET M=2
230 GOSUB 430
240 GOSUB 360
250 PRINT "VALUE OF 2ND DERIVATIVE=" S; L$ "/(" T$ "*" T$ ")"
260 END
270 REM : INPUT POLYNOMIAL
280 DIM A(10), B(10), N(10)
290 FOR I=D TO 0 STEP -1
300 PRINT "ENTER COEFFICIENT OF A" I
310 INPUT A(I)
320 LET N(I)=I
330 LET B(I)=I
340 NEXT I
350 RETURN
360 REM : EVALUATE DERIVATIVE
370 LET S=0
380 FOR L=D TO 0 STEP -1
390 LET S=S+A(L)*B(L)*X**(L-M)
400 NEXT L
410 REM : ANSWER IS S
420 RETURN
430 FOR K=0 TO D
440 LET B(K)=B(K)*(N(K)-1)
450 LET N(K)=N(K)-1
460 NEXT K
470 RETURN
```

(a) ENTER DEGREE OF POLYNOMIAL
    ? 3
    ENTER UNIT OF LENGTH
    ? FT
    ENTER UNIT OF TIME
    ? MIN
    ENTER COEFFICIENT OF A 3
    ? 6.2
    ENTER COEFFICIENT OF A 2
    ? -4
    ENTER COEFFICIENT OF A 1
    ? 1
    ENTER COEFFICIENT OF A 0
    ? 10
    ENTER T
    ? 1.4
    VALUE OF 1ST DERIVATIVE=26.256 FT/MIN
    VALUE OF 2ND DERIVATIVE=44.08 FT/(MIN*MIN)

(b) ENTER DEGREE OF POLYNOMIAL
    ? 1
    ENTER UNIT OF LENGTH
    ? MILE
    ENTER UNIT OF TIME
    ? HOUR
    ENTER COEFFICIENT OF A 1
    ? 3
    ENTER COEFFICIENT OF A 0
    ? -7
    ENTER T
    ? 0.25
    VALUE OF 1ST DERIVATIVE=3 MILE/HOUR
    VALUE OF 2ND DERIVATIVE=0 MILE/(HOUR*HOUR)

(c) ENTER DEGREE OF POLYNOMIAL
    ? 4
    ENTER UNIT OF LENGTH
    ? KM
    ENTER UNIT OF TIME
    ? SEC
    ENTER COEFFICIENT OF A 4
    ? 0.004
    ENTER COEFFICIENT OF A 3
    ? 0
    ENTER COEFFICIENT OF A 2
    ? -1
    ENTER COEFFICIENT OF A 1
    ? 3
    ENTER COEFFICIENT OF A 0
    ? 0
    ENTER T
    ? 162
    VALUE OF 1ST DERIVATIVE=67703.5 KM/SEC
    VALUE OF 2ND DERIVATIVE=1257.71 KM/(SEC*SEC)

9.5
```
10 DEF FNK(X)=(8*X**3+3*X*X+4*X+7)
20 DEF FNF(X)=FNK(X)**0.8
30 DEF FND(X)=0.8*FNK(X)*(24*X*X+6*X+4)
40 PRINT " X", " F(X)", " F'(X)"
50 FOR X=0 TO 10 STEP 0.25
60 PRINT X, FNF(X), FND(X)
70 NEXT X
80 END
```

X	F(X)	F'(X)
0	4.74328	22.4
.25	5.44234	46.55
.5	6.68539	111.8
.75	8.75625	265.1
1	11.856	598.4
1.25	16.1249	1266.65
1.5	21.6683	2505.8
1.75	28.5736	4650.8
2	36.9176	8153.6
2.25	46.771	13601.2
2.5	58.1997	21733.4
2.75	71.2658	33461.3
3	86.0283	49884.8
3.25	102.543	72310.9
3.5	120.865	102271
3.75	141.044	141541
4	163.13	192157
4.25	187.172	256433
4.5	213.216	336981
4.75	241.306	436728
5	271.485	558935
5.25	303.795	707211
5.5	338.278	885538
5.75	374.973	1.09828E+06
6	413.919	1.35021E+06
6.25	455.154	1.64653E+06
6.5	498.713	1.99287E+06
6.75	544.635	2.39532E+06
7	592.952	2.86046E+06
7.25	643.701	3.39535E+06
7.5	696.915	4.00758E+06
7.75	752.627	4.70525E+06
8	810.867	5.49702E+06
8.25	871.671	6.39213E+06
8.5	935.067	7.40038E+06
8.75	1001.09	8.53219E+06
9	1069.76	9.79859E+06
9.25	1141.12	1.12113E+07
9.5	1215.19	1.27825E+07
9.75	1292.01	1.45254E+07
10	1371.59	1.64536E+07

9.7
```
10 DEF FNK(X)=5*X*X-3*X
20 DEF FNF(X)=FNK(X)**3*(X*X-2)
```

```
30 DEF FND(X)=3*FNK(X)**2*(10*X-3)*(X*X-2)+FNK(X)**3*2*X
40 LET X1=1
50 PRINT " X2", "DELTA Y/DELTA X"
60 FOR X2=1.1 TO 1.01 STEP -0.005
70 PRINT X2, (FNF(X2)-FNF(X1))/(X2-X1)
80 NEXT X2
90 PRINT
100 PRINT "ACTUAL F'(1)=" FND(X1)
110 END
```

X2	DELTA Y/DELTA X
1.1	-84.2953
1.095	-83.6172
1.09	-82.9168
1.085	-82.1963
1.08	-81.4566
1.075	-80.6992
1.07	-79.9253
1.065	-79.1363
1.06	-78.3331
1.055	-77.5175
1.05	-76.6899
1.045	-75.8521
1.04	-75.0046
1.035	-74.1487
1.03	-73.2853
1.025	-72.4157
1.02	-71.54
1.015	-70.6603
1.01	-69.7757

ACTUAL F'(1)=-68

9.9
```
10 DEF FNF(T)=A1*SIN(W*T)+A2*COS(W*T)
20 DEF FNM(T)=ATN(A1/A2)/W
30 DEF FNS(T)=-W*W*(A1*SIN(W*T)+A2*COS(W*T))
31 PRINT "INPUT A1, A2, & W"
32 INPUT A1, A2, W
40 LET T=FNM(T)
50 IF FNS(T)<0 THEN M$="MAXIMUM"
60 IF FNS(T)>0 THEN LET M$="MINIMUM"
70 PRINT "THE FIRST " M$ " OCCURS AT T=" T
80 PRINT "THE FIRST " M$ " VALUE OF F(T)=" FNF(T)
```

(a) INPUT A1, A2, & W
    ? 1, 2, 4
    THE FIRST MAXIMUM OCCURS AT T=.115912
    THE FIRST MAXIMUM VALUE OF F(T)=2.23607

(b) INPUT A1, A2, & W
    ? 1.57, -2.03, 165
    THE FIRST MINIMUM OCCURS AT T=-3.98975E-03
    THE FIRST MINIMUM VALUE OF F(T)=-2.56628

9.11
```
100 DEF FNI(T)=M*EXP(-A*T)*(-A*SIN(W*T)-W*COS(W*T))/(A*A+W*W)
110 LET P1=4*ATN(1)
120 PRINT "INPUT M, A, W"
130 INPUT M, A, W
140 LET A1=FNI(P1/W)-FNI(0)
150 LET A2=FNI(2*P1/W)-FNI(P1/W)
160 PRINT "AREA A1=" A1
170 PRINT "AREA A2=" A2
180 PRINT "NET AREA=" A1+A2
190 END

INPUT M, A, W
? 10, 2, 12
AREA A1= 1.29112
AREA A2=-.764842
NET AREA=.526281
```

9.13
```
100 PRINT "Y=X EXP(KX) , ENTER K"
110 INPUT K
120 PRINT "ENTER LOWER & UPPER LIMITS"
130 INPUT A1, B1
140 PRINT
150 PRINT " N", "  VALUE", " % ERROR"
160 DEF FNY(X)=X*EXP(K*X)
170 DEF FNZ(X)=EXP(K*X)*(X-1/K)/K
180 FOR N=2 TO 10 STEP 2
190 LET D=(B1-A1)/N
200 LET S=FNY(A1)/2+FNY(B1)/2
210 FOR I=1 TO N-1
220 LET X=A1+I*D
230 LET S=S+FNY(X)
240 NEXT I
250 LET E=FNZ(B1)-FNZ(A1)
260 PRINT N, S*D, (S*D-E)*100/E
270 NEXT N
280 END
```

(a) Y=X EXP(KX) , ENTER K
? 1
ENTER LOWER & UPPER LIMITS
? 1, 4

N	VALUE	% ERROR
2	211.518	29.136
4	176.176	7.55898
6	169.337	3.38355
8	166.92	1.90804
10	165.797	1.22258

(b) Y=X EXP(KX) , ENTER K
? -1
ENTER LOWER & UPPER LIMITS
? 1, 4

N	VALUE	% ERROR
2	.638675	-.854643
4	.641928	-.349681
6	.643101	-.167651
8	.643557	-.0967656
10	.643777	-.0626691

9.15
```
100 DEF FNY(X)=1/X
110 DIM Y(101)
120 PRINT "ENTER B"
130 INPUT B1
140 LET A1=1
150 REM : TRY TRAPEZOID RULE FOR DIFFERENT
160 REM : NUMBERS OF INTERVALS, AND STOP
170 REM : WHEN ACCURACY IS WITHIN 0.1%.
180 FOR N=1 TO 100
190 GOSUB 360
200 GOSUB 430
210 IF ABS(T-LOG(B1))<=0.001 THEN 230
220 NEXT N
230 PRINT "TRAPEZOID SUM=" T
240 PRINT "NO. OF INTERVALS=" N
250 REM : TRY SIMPSON'S RULE FOR DIFFERENT
260 REM : NUMBERS OF INTERVALS, AND STOP
270 REM : WHEN ACCURACY IS WITHIN 0.1%.
280 FOR N=1 TO 100
290 GOSUB 360
300 GOSUB 500
310 IF ABS(S-LOG(B1))<=0.001 THEN 330
320 NEXT N
330 PRINT "SIMPSON SUM=" S
340 PRINT "NO. OF INTERVALS=" N
350 END
360 REM : EVALUATE Y1, Y2, Y3, . . .
370 LET D=(B1-A1)/N
380 FOR I=1 TO N+1
390 LET X=A1+(I-1)*D
400 LET Y(I)=FNY(X)
410 NEXT I
420 RETURN
430 REM : TRAPEZOID RULE SUMMATION :
440 LET T=(Y(1)+Y(N+1))/2
450 FOR J=2 TO N
460 LET T=T+Y(J)
470 NEXT J
480 LET T=T*D
490 RETURN
500 REM : SIMPSON'S RULE SUMMATION :
510 LET S=Y(1)+4*Y(N)+Y(N+1)
520 FOR J=2 TO N/2
530 LET S=S+4*Y(2*J-2)
540 LET S=S+2*Y(2*J-1)
550 NEXT J
```

```
560 LET S=S*D/3
570 RETURN
```

(a) ENTER B
  ? 1.5
  TRAPEZOID SUM   = .406187
  NO. OF INTERVALS= 4
  SIMPSON SUM   = .405471
  NO. OF INTERVALS= 4

(b) ENTER B
  ? 10
  TRAPEZOID SUM   = 2.30358
  NO. OF INTERVALS= 82
  SIMPSON SUM   = 2.30357
  NO. OF INTERVALS= 20

9.17 Preliminary math:
$$y = b(1 - ax^2)^{1/2}$$
$$y' = -abx(1 - ax^2)^{-1/2}$$
$$y'' = -ab(1 - ax^2)^{-1/2} - a^2bx(1 - ax^2)^{-3/2}$$
$$= \frac{-ab}{(1 - ax^2)^{3/2}}$$

```
10 DEF FNK(X)=1-A*X*X
20 DEF FND(X)=-A*B*X/SQR(FNK(X))
30 DEF FNS(X)=-A*B/FNK(X)**(3/2)
40 DEF FNR(X)=SQR(1+FND(X)**2)/FNS(X)
50 PRINT "INPUT A & B"
60 INPUT A, B
70 PRINT " X", " RADIUS"
80 FOR X=0 TO 1/SQR(A) STEP 0.1/SQR(A)
90 PRINT X, FNR(X)
100 NEXT X
110 END

INPUT A & B
? 1E-4, 200

X            RADIUS
0            -50
10           -50.237
20           -50.7984
30           -51.2759
40           -51.0952
50           -49.6079
60           -46.1511
70           -40.0764
80           -30.7584
90           -17.5943
100          -1.19209E-05
```

9.19
```
TRAPEZOID RESULT= 1.09143E+06
SIMPSON'S RESULT= 1.45524E+06
PERCENT DIFF.   = 25
```

9.21 Preliminary math:
$$Q(t) = te^{-at}$$
$$Q'(t) = e^{-at}(1-at)$$
$$Q''(t) = ae^{-at}(at-2)$$
$$Q'''(t) = a^2 e^{-at}(3-at)$$

if $Q'''(t) = 0$ then $3 - at = 0$ or $t = \dfrac{3}{a}$

```
100 DEF FNQ(T)=T*EXP(-A*T)
110 DEF FNI(T)=(1-A*T)*EXP(-A*T)
120 DEF FND(T)=A*(A*T-2)*EXP(-A*T)
130 REM : MAXIMUM DI/DT AT T=3/A
140 PRINT "ENTER A"
150 INPUT A
160 PRINT "MAXIMUM DI/DT=" FND(3/A)
170 PRINT "CHARGE       =" FNQ(3/A)
180 PRINT "CURRENT      =" FNI(3/A)
190 PRINT "TIME         =" 3/A
200 END

ENTER A
? 0.2
MAXIMUM DI/DT=9.95742E-03
CHARGE      =.746806
CURRENT     =-.0995742
TIME        =15
```

9.23 (a)
```
ENTER A1, A2, B1, B2, W, THETA, & PHI
? 0, 10, 0, 5, 10, 0, 0
TRAPEZOID RESULT  = 5
SIMPSON RESULT    = 5
PERCENT DIFFERENCE= 0
```

(b)
```
ENTER A1, A2, B1, B2, W, THETA, & PHI
? 10, 5, 2, 6, 377, 30, -45
TRAPEZOID RESULT  = 7.49794
SIMPSON RESULT    = 7.51039
PERCENT DIFFERENCE= 1.65729E-03
```

9.25
```
10 DEF FNP(R)=A*R*EXP(-B*R*R-1)
20 DEF FNR=1/SQR(2*B)
30 PRINT "INPUT A AND B"
40 INPUT A, B
50 PRINT "MAXIMUM PROFIT=" FNP(FNR)
60 PRINT "WHEN RATE IS  =" FNR
70 END

INPUT A AND B
```

```
? 20, 0.02
MAXIMUM PROFIT= 22.313
WHEN RATE IS  = 5
```

9.27 (a)
```
ENTER MU, SIGMA, D1, & D2 :
? 1.00, 1, 0.9, 1.1
TRAPEZOID VALUE=.0795151
SIMPSON'S VALUE=.0795233
PERCENT DIFF.  =.0103622 %
```

(b)
```
ENTER MU, SIGMA, D1, & D2 :
? 1.00, 0.1, 0.9, 1.1
TRAPEZOID VALUE=.592816
SIMPSON'S VALUE=.595904
PERCENT DIFF.  =.518354 %
```

(c)
```
ENTER MU, SIGMA, D1, & D2 :
? 1.00, 10, 0.9, 1.1
TRAPEZOID VALUE=7.97857E-03
SIMPSON'S VALUE=7.97858E-03
PERCENT DIFF.  =9.33823E-05 %
```

9.29
```
100 REM : VX'(T)=0 AT T=2/A
110 REM = VY'(T)=0 AT T=2/B
120 DEF FNX(T)=A1*EXP(-A*T)*(1-A*T)
130 DEF FNY(T)=B1*EXP(-B*T)*(1-B*T)
140 LET P1=4*ATN(1)
150 PRINT "INPUT CAP-A, CAP-B, A, & B"
160 INPUT A1, B1, A, B
170 LET T=2/A
180 GOSUB 290
190 PRINT "MAXIMUM X-VELOCITY=" V1"AT T=" T
200 PRINT "MAGNITUDE         =" V
210 PRINT "ANGLE (DEGREES)   =" P
220 LET T=2/B
230 GOSUB 290
240 PRINT
250 PRINT "MAXIMUM Y-VELOCITY=" V2 " AT T=" T
260 PRINT "MAGNITUDE         =" V
270 PRINT "ANGLE (DEGREES)   =" P
280 END
290 LET V1=FNX(T)
300 LET V2=FNY(T)
310 LET V=SQR(V1*V1+V2*V2)
320 LET P=ATN(V2/V1)
330 IF V1<0 THEN P=P+P1
340 IF V2<0 AND V1>0 THEN P=P+2*P1
350 LET P=P*180/P1
360 RETURN

INPUT CAP-A, CAP-B, A, & B
? 12, 9, 0.2, 0.1
MAXIMUM X-VELOCITY=-1.62402 AT T=10
```

```
        MAGNITUDE          = 1.62402
        ANGLE (DEGREES)    = 180

        MAXIMUM Y-VELOCITY= -1.21802 AT T=20
        MAGNITUDE          = 1.38504
        ANGLE (DEGREES)    = 241.571
```

9.31
```
    ENTER A, B, T0, V(0)
    ? 120, 16, 1, 0
    TRAPEZOID RESULT  = 58.1544
    SIMPSON'S RESULT  = 58.1635
    PERCENT DIFFERENCE= .015616
```

INDEX

ABS function, 153
Absolute value, 277
   function (ABS), 153
Acceleration, 42, 237, 251
   angular, 76, 152, 273
Acceptance sampling, 200
Account number, 23, 277
Acoustic coupler, 3, 277
ACS function, 164
Address, 278
Algorithm, 8, 278
Alloys, of iron, 151
Alphanumeric, 22, 278
Amplitude, of an ac voltage, 148
AND, 85, 278
Angle
   of a complex number, 173
   radian measure of, 155
Annual cost of inventory, 150
ANSI (American National Standards Institute), 12, 278
Answers, to selected exercises, 301

Antiderivative, 250
APL, 278
Arc functions, 159, 278
Architecture, 278
Arctangent computation, correction for, 160
Area, under a curve, by integration, 252
Argument, of a function, 153, 278
Array, 127, 278
ASCII (American Standard Code for Information Interchange), 278
ASN function, 164
Assembler, 279
Assembly language, 279
Assignment statement, 16
   multiple, 17
Asynchronous, 279
ATN function, 160
Auxiliary memory, 279

Base, of a number system, 279
Batch programming, 7, 279

Baud, 279
BCD, 279
Best straight line, through experimental data points, 142
Binary number system, 5, 279
Bit, 5, 279
Body, of a loop, 107
Boolean algebra, 279
Bootstrap, 280
Branching, 79, 280
Break key, 22, 80
Breakpoint, 280
Bubble sort, 139
Bug, 280
Byte, 5, 280

CAD (computer-aided design), 180, 281
Calculator mode, 38, 280
Call, 172, 280
Capacitance, voltage across, 196, 272
Card, 280
Card punch, 280
Card reader, 280
Cassette, 4, 280
Central Processing Unit, 4, 282
Character, 281
   alphanumeric, 22, 278
   control, 282
   data, 5
Chip, 281
Clock, 281
Code, 281
Color code, for resistors, 149
Color graphics, 281
Command, 281
   system, 8, 23, 24, 51
Comment, 39, 281
Compiler, 5, 281
Complement, 281
Complex number, 148
   polar form of, 173, 194
   rectangular form of, 194
Components, of a force vector, 155, 165
Compound logical expression, 85, 101
Concatenate, 281
Conditional branch, 81, 282
Conditions, for conditional branch statements, 82
Connector, 282
   symbol on a flowchart, 97, 99, 282
Console, 282
Control character, 23, 282
Core, 282
Correction program for arctan computations, 160
COS function, 154
Cosh function, 165
COT function, 156
Counter variable, 88, 107, 282
Cpi, 282

CPM, 282
CP/M, 282
CPU, 4, 282
Cramer's rule, 207, 210
Critically damped electrical circuit, 197
Cross assembler, 282
CRT (cathode ray tube), 280, 282
CSC function, 164
Current amplification factor, β, 148
Cursor, 60, 282
Cycle, 282

Daisy-wheel printer, 282
Damped natural frequency, of a spring-mass-damper system, 50
Damped sine wave, 157
Data, 7, 283
   alphanumeric, 22
   character, 281
Data base, 283
Data processing, 8, 283
DATA statement, 64
Debounce, 283
Debug, 24, 283
Decision block, in a flowchart, 91
Deck, 283
Decrement, 89, 283
Dedicated, 5, 283
DEF FN statement, 66
Define function statement, 166
Definite integral, 251
Deflection, of a beam, 195
Delimiter, 14, 39
Delta, determinant, 208
Density
   of a body, 26
   double, of a disk, 284
Derivative, definition, 245
Determinant, 207, 210
   delta, 208
   test, for solvability of linear equations, 212
Development system, 283
Deviation, of a point, from a line, 142
Differentiation
   of polynomials, 229
   of transcendental functions, 246
Digit, 283
Digital, 283
Dimension, 129, 283
DIM statement, 129
Direct mode, 38, 283
Disassembler, 284
Discrete, 284
Disk, 4, 284
Disk operating system, 284
Diskette, 5
Distance, between points on a straight line, 43
Division, mathematical, 33

Documentation, 103, 284
DOS, 284
Dot matrix printer, 284
Down, 284
Dummy variable, 166, 284
Dump, 284

e, 157
E format, for number representation, 12, 284
Economic lot size, 48
Editor, 284
Effective value, of a periodic voltage, 273
Efficiency, 30, 31, 75, 152, 213
Emulator, 284
END statement, 20
ENTER key, 14, 23, 24
Equality, of matrices, 215
Equilibrium condition, 67
Equivalent resistance, of a parallel network, 48, 122
Error message, 6, 285
Exclusive OR, 285
Execute, 285
EXP function, 157
Exponent, 285
Exponential probability density function, 274
Exponentiation, 34, 285
Expressions, 35

"Fall through," 108
Field, 285
File, 7, 285
Firmware, 285
FIX function, 164
Fixed point, 5, 285
Floating point, 5, 12, 285
Floppy disk, 4, 285
Flowcharts, 91, 286
FOR-TO, NEXT statement, 107
   representation on a flowchart, 111
Force
   average, 276
   on a beam, 224
   between charges, 47
   on a connecting rod, 227
   developed by a dashpot, 275
   in a free body diagram, 67
   horizontal and vertical components of, 155, 165
   on a lever, 67
   resultant, 70, 165
Formatted disk, 286
FORTRAN, 286
Friction loss, in a circular water pipe, 203
Friendly, 286
FSK, 286
Full duplex, 286

Functions, 286
   intrinsic, 154
   library, 153
   multiline, 168
   table of, 164
   user-defined, 166

G (giga), 286
Gas law, 77
Gate, 286
Gear
   train, inertia reflected by, 125
   velocity, 31
General-purpose computer, 286
GIGO, 66, 286
Glossary, 277
GOSUB statement, 171
GOTO statement, 79
Graphics, 180, 287
Gravitational constant g, 71
Greater than ($>$), 82

Half duplex, 287
Hard copy, 287
Hard-sectored disk, 287
Hardware, 4, 287
Hard-wired, 287
Heat, required to overcome infiltration losses, 72
Hexadecimal, 287
Hierarchy of mathematical operators, 36, 287
High-level language, 5, 287
Hollerith code, 280, 287
Hooke's law, 30
Hyperbolic functions, 165

Identities
   for arc trigonometric functions, 160
   trigonometric, 156
Identity matrix, 215, 287
IF-THEN statement, 81
IF-THEN-ELSE statement, 84
Imaginary part, of a complex number, 148
Immediate mode, 38
Impedance, 198
Increment, 89, 287
Indefinite integral, 251
Independent equations, 206
Index, 128, 287
Inductance, voltage across, 271
Inertia, 76, 125
Infinite loop, 80, 287
Initial velocity and displacement, 42, 252
Initialize, a variable, 88, 288
Inner product, 145
INPUT statement, 19, 62

Input/Output, 4, 51, 288, 289
  representation of, on a flowchart, 92
Instruction, 288
INT function, 164, 182
Integer, 288
Integrated circuit, 288
Integration, 250
  numerical, 261
  by parts, 258
Interactive, 6, 19, 288
Intercept, of a straight line, 27
Interface, 288
Interpreter, 5, 288
Intrinsic functions, 154
Inverse, 289
  of a trigonometric function, 278
  of a matrix, 216
Inverse trigonometric functions, 159
Invert, 289
I/O. See Input/Output
Iron alloys, 151
Iteration, 289

JCL (job control language), 289

k, 289
K, 5, 289
Key, 289
Keyboard, 2, 21, 289
Keypunch, 289
Keyword, 15, 289
Kinetic energy, 45, 193
Kirchhoff's voltage law, 73

Label, 290
Language, computer, 5, 281
LCD, 290
LED, 290
Less than (<), 82
LET statement, 15
Library functions, 153, 290
  table of, 164
Light pen, 290
Line printer, 290
Linear equations, 205
Linearity principle
  for differentiation, 229
  for integration, 251
LIST command, 52
Listing, of a program, 39, 290
Load, 290
LOAD command, 52
Local maximum and minimum, 237
LOG function, 164
Log on, log off, 24, 290
Logic circuitry, 290
Logical operator, 290

Loops, 87, 290
  FOR-TO, NEXT, 107
  infinite, 80
  nested, 113
  representation of, on a flowchart, 92, 111
Low-level language, 5, 290

M (mega), 291
Machine dependent, 11, 52, 291
Machine language, 5, 291
Magnetic disk, 4, 291
Magnetic tape, 4, 291
Magnitude, 291
  of a complex number, 173, 194
Mainframe, 2, 291
Map, memory, 291
Mass storage, 4, 291
MAT operations, in BASIC, 220, 291
Mathematical operators, 33, 291
Matrix, 130, 291
  algebra, 214
  determinant of, 207
  inverse of, 216
  multiplication, 214
  operations in BASIC, 220
  singular and nonsingular, 216
  square, 214
  transpose of, 145
MAX function, 164
Maximum, local, 237
Maximum power transfer theorem, 147
  ac, 198
Memory, 4, 291
  auxiliary, 279
  capacity, 5
  external, 285
  nonvolatile, 292
  random access (RAM), 295
  read-only (ROM), 295
  speed, 4
Mesh analysis, 225
Micro, 291
Microcomputer, 2, 291
Microprocessor, 4, 291
MIN function, 164
Minicomputer, 291
Minimal BASIC, 12, 292
Minimum, local, 237
Mnemonic, 292
Modem, 3, 292
Modulus of elasticity, 30, 72
Moment, of a force, 29
  bending, on a beam, 147, 270
  summing, 227
Moment of inertia
  of a circular shaft, 49
  of a triangle, 270
Monitor, 292
Multiline function, 168

Multiple assignment statement, 17
Multiple statements, 14
Multiplex, 292
Multiplication
   matrix, 214
   operator, 25

n (nano), 292
Name
   array, 127
   file, 7
   string variable, 63
   variable, 15
Nested
   loops, 113, 292
   subroutines, 175, 292
Newton's law, 25
Nibble, 292
Nodal analysis, 225
Nonexecutable statement, 20
Nonsingular matrix, 216
NOR, 292
Normal probability function, 250, 269
NOT, 85, 292
Not equal to (<>), 82
Number representation, in BASIC, 12
Numeric, 293
Numerical analysis, 293

Object program, 6, 293
Octal, 293
OEM, 293
Off line, 293
Ohm's law, 30, 74
ON-GOTO statement, 118
One-pass assembler, 293
On line, 293
Operating system, 284, 293
Operators, mathematical, 33
   hierarchy of, 36
   relational, 82
OR, 85, 293
Order, of a derivative, 233
Ordering, of data, 138
Output, 4, 293
Overdamped electrical circuit, 197
Overflow, 293

p (pico), 294
Paper tape, 293
Parabola, straight-line segment
   approximation of, 119
Parallel, 293
Parentheses, in BASIC expressions, 37
Pascal, 294
Percolation test, 71
Peripheral, 4

PLOTTER subroutine, 185
Point-slope form of a line, 230
Polar form of a complex number, 173
Polar moment of inertia, of a hollow circular
   shaft, 49
Port, 294
Power, 294
   of a computer or computer language, 11
   electrical, 30, 48, 74, 147, 198
   mathematical, 34
Power rule, for differentiation, 229
   for differentiation of functions, 240
Powerful, 11
Precision, 294
PRINT statement, 18, 52
Priority, of mathematical operators, 36
Process block, in a flowchart, 91
Product, of matrices, 214
Product rule, for differentiation, 242
Program, 5, 20, 294
PROM, 294
Prompt, 14
   message for, 63
Proportional spacing, 294
Pseudo-random number, 164, 294

Quadratic formula, 41, 82, 119, 158
Queue, 295
Quotient rule, for differentiation, 242

Radians, 155
Radius of curvature, 270
RAM, 295
Random number, 295
Rate of change, of a function, 230
RC network, 196
Read, 295
READ statement, 64
Real part, of a complex number, 148
Real-time solution, 295
Record, 8, 295
Recursion, 295
Relational operator, 82, 295
REM statement, 39, 296
Remarks, in a program, 39
Resident (in), 296
RESTORE statement, 65
Return key, 14, 23, 24
RETURN statement, 172, 296
Reverse display, 296
Reverse Polish, 296
RLC network, 197
RMS value, 273
RND function, 164
Rollover, 296
Round off, 296
   error due to, 296

Round off (continued)
  of numbers displayed, 13
  using the INT function for, 182
Routine, 296
RS 232, 296
Run-time error, 24, 296

Sag correction, in surveyor's tape, 47
SAVE command, 52
Scientific notation, 12, 296
Scientific programming, 8, 296
Scroll, 297
Search, 143
Second derivative, 232
Second derivative test, for a minimum or maximum, 238
Sector, 297
Serial, 297
Settlement, of a foundation, 46
SGN function, 158
Shift key, 22
SI system of units, 25, 201
Silicon Valley, 297
Simplex, 297
Simpson's rule, 266
Simulator, 297
Simultaneous equations, 205
SIN function, 154
Singular matrix, 216
Sinh function, 165
Slope, of a line, 27, 42
Smart terminals, 297
Software, 4, 297
Solid state, 297
Sorting, 138
Source program, 7, 297
Spacing, in a PRINT statement, 52, 60
Special-purpose computer, 297
Spring, force developed by, 31
Spring-damper system, 201
Spring-mass-damper system, 50, 202
SQR function, 154, 164
Stadia surveying, 194
Standard deviation, 75, 110, 128
Start block, on a flowchart, 93
Statement number, 14
Statements, 14. (For specific statements, refer to their keywords.)
Steel alloys, 151
STEP, in a FOR-TO, NEXT loop, 109
Stop block, on a flowchart, 93
STOP statement, 173
Store, 297
Stored program computer, 7, 297
Straight line, equation of, 27
  best, in fitting to data, 142
Straight-line segment approximation, of a parabola, 119
Strain, 30, 125

Stress, 30, 125, 199
  on a beam, 72
  on a rod, 46
String, 18, 297
  in an array, 128
String variable, 63, 298
Structured programming, 298
Subroutines, 171, 298
Subscript, 128, 298
Supervisor, 298
Support, 298
Synchronous data transmission, 298
Syntax, 11, 298
System commands, 8, 23, 24, 51
  table of, 61

TAB, 60
TAN function, 156
Tan^{-1} function, 159
  correction program for, 160
Tape drive, 298
Teletypewriter (TTY), 298
Temperature, conversion between Fahrenheit and Celsius, 29
Terminal, 3, 298
Text editor, 298
Time-constant, 167
Time-shared computer, 2, 23, 299
Tolerance assignment problem, 99, 122
Torque, 49, 76
Track, 299
Transcendental functions
  differentiation of, 246
  integration of, 256
Transparent, 299
Transpose, of a matrix, 145
Trapezoidal rule, 261
Trigonometric identities, 156
Trucks, 18-wheel, high acccident rate of, 131
Truncate, 182, 299
Two-dimensional array, 130
Two-pass assembler, 299

Unconditional branch, 81, 299
Underdamped electrical circuit, 197
Underflow, 299
Unfriendly, 299
Upper control limit, 48
User-defined functions, 166

Variable, 14, 299
  array, 127
  dummy, 166, 284
  name, 15
  string, 63
  subscripted, 128

Vectors, 130, 299
  inner product of, 144
Velocity
  angular, of a gear, 31
  as the derivative of displacement, 237, 251, 273, 275
  effect of, on mass, 43
  resultant, 275
Versions, of BASIC, 11
Video display, 2
Volatile memory, 299

Weight flow rate, of a fluid, 76
Word, 5, 299
Word processor, 299
Workspace, 51, 300
Write, 300

Yield strength, 125
Young's modulus, 72